Design and Applications of Adaptive Impedance Converters

Lei Wang · Man-Chung Wong · Ziyi Bai

Design and Applications of Adaptive Impedance Converters

Springer

Lei Wang
State Key Laboratory of High-Efficiency and High-Quality Conversion for Electric Power
Hunan University
Changsha, China

Man-Chung Wong
State Key Laboratory of Internet of Things for Smart City and Department of Electrical and Computer Engineering
University of Macau
Macau, China

Ziyi Bai
State Key Laboratory of Internet of Things for Smart City and Department of Electrical and Computer Engineering
University of Macau
Macau, China

ISBN 978-981-97-5318-5 ISBN 978-981-97-5319-2 (eBook)
https://doi.org/10.1007/978-981-97-5319-2

This Springer imprint is published by the registered company Springer Nature Singapore Pte Ltd.
The registered company address is: 152 Beach Road, #21-01/04 Gateway East, Singapore 189721, Singapore

Preface

With the increasing development of power electronics devices, adaptive impedance converter is widely implemented using passive and active electronic components to achieve impedance transformation, optimize system performance, and minimize power losses. Both passive and active adaptive impedance converters can be applied to different applications, including distribution systems, renewable energy integration, AC/DC microgrids, traction supply systems, and power amplifiers.

In this book, the content starts by overviewing different coupling impedance converters, and the potential advantages of adaptive impedance converters have been explored in different applications. According to the aspects of topology structure, characteristic analysis, parameter design, control system, and experimental verifications, the different applications of adaptive impedance have been developed. In addition, the analysis method of system characteristics, parameter design, and control techniques presented in this book can be extendable to other converters.

This book was compiled by Prof. Lei Wang. Chapter 1 was written by Lei Wang, Man-Chung Wong, and Ziyi Bai. Chapter 2 was written by Lei Wang, Man-Chung Wong, Ziyi Bai, and Chi-Seng Lam, while Chap. 3 was written by Lei Wang and Man-Chung Wong. Chapters 4 and 5 were written by Lei Wang and Ying Pang. Chapter 6 was written by Lei Wang, Wei Tian, and Jiawen Yu. Finally, Chap. 7 was written by Lei Wang, Man-Chung Wong, and Ziyi Bai.

The book is tailored for encompassing researchers, Ph.D. candidates, postgraduate students, and electrical power engineers specializing in power electronics. It offers them a comprehensive understanding of the intricacies of designing and controlling adaptive impedance converters, providing specialized knowledge tailored to their advanced expertise. Additionally, it serves as a valuable resource for bachelor students and postgraduate learners in electrical engineering, imparting foundational knowledge on power electronics converters, parameter design principles, and the essentials of pulse width modulation (PWM), among other fundamental concepts.

Changsha, China — Lei Wang
Macao, China — Man-Chung Wong
Macao, China — Ziyi Bai

Contents

Chapter 1
Introduction

1.1 Develop of Adaptive Impedance Converters

AC-DC converters, also known as rectifiers and inverters, are widely applied in various applications across different industries and sectors [1]. These converters play a crucial role in converting the voltage and frequency of the input alternating current (AC) power to a desired level of direct current (DC) voltage, which is essential for powering a wide range of electronic devices and systems [2, 3]. From power quality compensators to renewable energy systems, AC–DC converters are ubiquitous and indispensable components that enable the efficient and reliable operation of countless devices and systems. Generally, the AC–DC converter is composed of coupling impedance and inverter. The coupling impedance is responsible for the initial connection between the AC input source and the converter [4]. The coupling impedance is designed to match the impedance of the AC power source and facilitate the transfer of power to the converter. It helps to maintain proper voltage and current levels throughout the conversion process, minimizing power losses and maximizing efficiency [5, 6]. Nowadays, the design of coupling impedance [7–10] and the control of converters [11–13] have become focal points of extensive research among scholars and researchers. Recognizing the critical role these components play in AC–DC converters, numerous studies have been conducted to advance the understanding and improve the performance of coupling impedance design and converter control techniques. In coupling impedance design, researchers have explored various approaches to optimize its characteristics and ensure efficient power transfer.

Conventionally, the coupling impedance for AC–DC converters is inductive, including the L [14, 15] from 1970 and LCL [16–18] from 2003. The inductive L connects the grid side and active converter and can reduce the current ripple. Since the inductive impedance of L is small and its withstand voltage is also small, the active inverter part requires a high dc voltage and voltage rating. Therefore, the L-coupling AC–DC converter requires high active part capacity, which leads to low reliability and large operational power loss, thus limiting their widespread applications. Meanwhile,

L. Wang et al., *Design and Applications of Adaptive Impedance Converters*,
https://doi.org/10.1007/978-981-97-5319-2_1

a large inductance usually needs to be adopted to reduce the current ripple, which is costly, bulky, and inefficient. To further reduce the current ripple and the special high-frequency harmonic current and decrease the inductance value, the higher order LCL coupling filter has been widely investigated to connect the grid and active inverter due to their better harmonic attenuation ability with lower total inductance. The LCL coupling impedance is very sensitive to filter parameter variation [19, 20], which is preferred in practical applications because of their robustness against filter parameter variation [21]. However, despite its reduction in the total inductance compared to L filters, two individual magnetic cores are required. In this case, filter inductors still account for the large weight and volume of the overall power conversion system [22]. Moreover, the LCL coupling converter still requires a high DC voltage and voltage rating of the active inverter part, resulting in high cost, low reliability, and large operational power loss. To reduce the power rating (cost) and DC voltage (power loss) of the active inverter part, a capacitive LC coupling impedance network was proposed to connect the grid side and active converter in 2003 [23, 24]. The LC coupling network generally has a large impedance, which can withstand a certain voltage and provide a certain reactive power, thereby reducing the inverter part power rating and DC voltage [25]. The LC coupling impedance can reduce the active inverter power rating but only suits certain inductive reactive loads. It limits the compensating range of the inverter for dynamically changing loads and capacitive loads [26]. In 2020, a capacitive LCLC coupling impedance network was proposed to connect the grid side and inverter for reactive power and current harmonics compensation [27]. The LCLC coupling impedance network combines the harmonic suppression ability of inductive LCL and the reactive power support of capacitive LC. Compared with the existing coupling impedance network, the LCLC coupling impedance provides better high-frequency harmonic suppression capability, lower inverter power rating, lower switching loss, and smaller inductor value requirement. However, the LCLC coupling impedance network also suits a certain inductive reactive load and limits the compensating range of the inverter for dynamic loads and capacitive loads. In 2012, Luo et al. [28] proposed a novel coupling impedance network to connect the converter, which consists of passive power filters (PPFs) in series with fundamental series resonant circuit (FSRC), called PPF + FSRC coupling impedance network. In the PPF + FSRC, the fixed capacitors (FCs) are usually connected in series with inductors tuned to be passive power filters (PPFs) that can suppress limited characteristic frequency harmonic current. The PPFs can compensate for a fixed capacity of capacitive reactive power, corresponding to FCs. Because the FSRC is resonant at the fundamental frequency, the PPFs mainly sustain fundamental voltage at the point of common connection. In contrast, the active converter only supports harmonic voltage, which greatly minimizes the voltage level of power electronic devices of the inverter. Due to the existence of FCs, the PPF + FSRC coupling impedance network can only play to its advantage in applications with certain inductive reactive and harmonic loads. From 2014, the adaptive coupling impedance network in [29–31] is proposed in series/parallel with the active converter, which is composed of an inductor, a parallel capacitor, and a pair of anti-parallel thyristors connected in series with another inductor. The adaptive coupling impedance network can be changed to inductive and capacitive by controlling the thyristor firing angles, and it can be considered a controllable inductor and capacitor. The adaptive coupling impedance network has the characteristics of a wide reactive power compensation range and a low voltage rating of the active converter. It cannot only be applied to dynamic inductive and capacitive load scenarios but also can lead to a decrease in the total cost and power loss of the active inverter. The comparisons among different coupling impedance networks in terms of compensation range and converter voltage rating are summarized in Fig. 1.1. The characteristic comparisons among different coupling impedance networks for AC–DC converters are shown in Table 1.1.

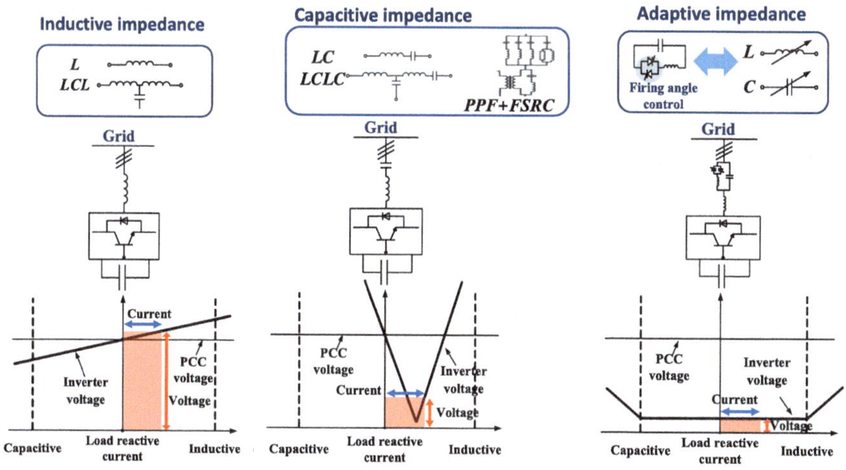

Fig. 1.1 Structures of conventional APF, HAPF and SVC-HAPF

Table 1.1 Comparisons among different coupling impedance networks for AC–DC converters

Coupling impedance type		Year	Compensation range	Inverter power rating	Inverter voltage rating	Power loss
Inductive type	L	1976	****	****	****	****
	LCL	2003	****	***	***	***
Capacitive type	LC	2003	*	**	**	**
	LCLC	2020	*	*	*	*
	PPF + FSRC	2012	*	*	*	*
Adaptive type	TCLC	2014	****	**	**	**

Notes The shade area means the undesirable characteristic

1.2 Active Adaptive Impedance

Besides of the above-mentioned passive adaptive impedance converter, the active adaptive impedance has also been widely studied. Active adaptive impedance refers to a technique used in electronic circuits or systems to actively control and adapt the impedance presented to a signal source or load based on specific requirements [32, 33]. Unlike passive impedance matching techniques that rely on passive components like resistors, capacitors, or inductors, active adaptive impedance matching actively modifies the impedance to achieve optimal performance [34]. The concept of active adaptive impedance arises from the fact that the impedance of a circuit can vary with frequency, load conditions, or other factors. By utilizing active components, the impedance can be actively controlled and adjusted to match the impedance of the source or load, providing efficient power transfer and minimizing power loss.

Active adaptive impedance matching is commonly employed in various applications, including radio frequency systems, audio systems, and communication networks [35]. In radio frequency systems, active impedance matching circuits are used to match the impedance between the radio frequency amplifier and antenna, ensuring maximum power transfer and minimizing signal loss [36]. In audio systems, active adaptive impedance matching can be utilized to match the output impedance of an audio amplifier with the input impedance of a speaker or headphone, optimizing the audio signal transfer and minimizing distortion. The advantage of active adaptive impedance matching is its ability to adapt to changing conditions dynamically. By continuously monitoring the impedance and adjusting the active control outputs accordingly, optimal impedance matching can be achieved even in the presence of varying load conditions or frequency changes [37]. This technique offers flexibility and improved performance compared to passive impedance matching, particularly when impedance characteristics are nonlinear. Overall, active adaptive impedance matching is a technique that utilizes active components to dynamically adjust the impedance of a circuit, enabling efficient power transfer and improved system performance in various applications.

The active converters have the potential advantages of fast response, wider matching range, high precision, and smaller size [38]. Therefore, the active converters have a high potential to be developed as active adaptive impedance matching topologies, which are barely discussed in the literature. Referring to different connection methods of active converters, the active adaptive impedance matching topologies can be developed as series and parallel types. While the conventional converters operate at the fundamental frequency 50/60 Hz, the active adaptive impedance matching network can operate at a wide frequency range to realize broadband impedance matching. the previous research for active converters can hardly extend to a wide frequency range, new controllers also need to be explored to control the active adaptive impedance.

1.3 Potential Advantages of Adaptive Impedance Converters in Different Applications

Adaptive impedance converters are designed to transform the impedance of an electronic system to optimize its performance. The adaptive impedance converter is typically implemented using passive and active electronic components to achieve impedance transformation. The main advantage of adaptive impedance converters is their ability to dynamically adjust the impedance to optimize system performance. This can help improve signal quality, reduce interference, and minimize power losses. Both passive and active adaptive impedance converters can be applied to different applications, including distribution systems, renewable energy integration, AC/DC microgrids, traction supply systems, and power amplifiers. The passive-based adaptive impedance converters can effectively reduce the converter rating and power loss, maintaining a large operational range. The active adaptive impedance can change its impedance, varying with frequency, load conditions, or other factors to match the impedance of the source or load, providing efficient power transfer and minimizing signal reflections.

Traditional large-capacity converters have many switching devices, resulting in high system costs, high losses, and difficult maintenance. Low-cost, high-performance, and large-capacity converters are an urgent need in the international industry. The concept of adaptive impedance coupling converter is proposed, and corresponding topologies and related products for distribution network systems, railway power supply systems, renewable energy grid-connected systems, and AC/DC microgrids are developed, which strongly supports low-cost, low-loss, easy-to-maintain large-capacity conversion application of the device.

The adaptive hybrid static synchronous compensator and adaptive hybrid active filter are applied in distribution systems. The adaptive passive coupling impedance is used to provide dynamic reactive power effectively and reduce the converter's DC side voltage, thereby significantly reducing the number of switches and breaking through the bottleneck of low efficiency and low reliability of large-capacity converters.

The adaptive hybrid grid-connected inverter (HGCI) is used in renewable energy integration. The DC side voltage of the inverter with a traditional multi-photovoltaic power supply series structure is as high as 1000 V, resulting in high operating costs and low reliability. The applicant proposed to use HGCI to reduce voltage and developed a low DC side voltage inverter (200 V) with a photovoltaic power supply in parallel, which reduced the cost of system switching devices and improved reliability.

In a hybrid AC/DC microgrid, the interlinking converter is the tie to connect the AC and DC microgrid. To reduce the power switch capacity, control complexity and cost, an adaptive hybrid impedance coupling interlinking converter is proposed for the hybrid AC/low voltage DC microgrid. The key feature of the proposed system is that the coupling impedance is considered as one type of static var compensator with a controllable impedance. This part enables the voltage drop between the AC microgrid and the low voltage DC (LVDC) microgrid. It can effectively regulate active/reactive powers and improve power quality with higher efficiency.

In view of the complex and changeable characteristics of railway loads, the adaptive hybrid coupled impedance in series with a large-capacity converter is proposed, which greatly broadens the scope of application. It effectively solves the contradiction between converter capacity and wide-scale adjustment under large-scale fluctuations.

The active adaptive impedance converter refers to the concept of parallel and series converters that operate at a wide frequency range to realize broadband impedance matching. Compared with the above passive impedance matching methods, the active converters have the potential advantages of fast response, wider matching range, high precision, and smaller size. The active adaptive impedance converter is controllable and can be adjusted flexibly to provide different modes and realize the broadband impedance matching instead of the fixed 50/60 Hz system.

1.4 Book Organization

In this book, this chapter presents an overview of impedance matching and gives the potential advantages of impedance matching converters. Then, the applications of adaptive impedance converters in distribution systems for power quality compensation are given in Chap. 2. After that, the adaptive impedance grid connected inverters for renewable energy and power quality integration are fully provided in Chap. 3. Then, an adaptive impedance interlink converter for the AC/DC microgrid is provided in Chap. 4, and the adaptive impedance converter in the co-phase railway traction supply system is given in Chap. 5. The active adaptive impedance matching for power amplifiers is explored in Chap. 6. Finally, the conclusion and the perspective for further works are given in Chap. 7.

References

1. H. Wu, L. Zhu, F. Yang, Three-port-converter-based single-phase bidirectional AC–DC converter with reduced power processing stages and improved overall efficiency. IEEE Trans. Power Electron. **33**(12), 10021–10026 (2018)
2. R. Oruganti, M. Palaniapan, Inductor voltage control of buck-type single-phase AC-DC converter. IEEE Trans. Power Electron. **15**(2), 411–416 (2000)
3. B. Singh, B.N. Singh, A. Chandra, K. Al-Haddad, A. Pandey, D.P. Kothari, A review of three-phase improved power quality AC–DC converters. IEEE Trans. Ind. Electron. **51**(3), 641–660 (2004)
4. B. Singh, S. Gairola, B.N. Singh, A. Chandra, K. Al-Haddad, Multipulse AC–DC converters for improving power quality: a review. IEEE Trans. Power Electron. **23**(1), 260–281 (2008)
5. T.C. Beh, M. Kato, T. Imura, S. Oh, Y. Hori, Automated impedance matching system for robust wireless power transfer via magnetic resonance coupling. IEEE Trans. Ind. Electron. **60**(9), 3689–3698 (2013)

6. K.E. Koh, T.C. Beh, T. Imura, Y. Hori, Impedance matching and power division using impedance inverter for wireless power transfer via magnetic resonant coupling. IEEE Trans. Ind. Appl. **50**(3), 2061–2070 (2014)
7. K.-W. Lao, M.-C. Wong, N. Dai, C.S. Lam, C.-K. Wong, L. Wang, Analysis in the effect of co-phase traction railway HPQC coupled impedance on its compensation capability and impedance-mapping design technique based on required compensation capability for reduction in operation voltage. IEEE Trans. Power Electron. **32**(4), 2631–2646 (2017)
8. N. Inagaki, Theory of image impedance matching for inductively coupled power transfer systems. IEEE Trans. Microw. Theory Tech. **62**(4), 901–908 (2014). https://doi.org/10.1109/TMTT.2014.2300033
9. J. Kim, H.-C. Son, D.-H. Kim, Y.-J. Park, Impedance matching considering cross coupling for wireless power transfer to multiple receivers, in *2013 IEEE Wireless Power Transfer (WPT), Perugia, Italy* (2013), pp. 226–229
10. Z.N. Low, R.A. Chinga, R. Tseng, J. Lin, Design and test of a high-power high-efficiency loosely coupled planar wireless power transfer system. IEEE Trans. Ind. Electron. **56**(5), 1801–1812 (2009)
11. K. Zeb, W. Uddin, M.A. Khan, Z. Ali, M.U. Ali, N. Christofides, H.J. Kim, A comprehensive review on inverter topologies and control strategies for grid connected photovoltaic system. Renew. Sustain. Energy Rev. **94**, 1120–1141 (2018)
12. J. Rodriguez et al., Predictive current control of a voltage source inverter. IEEE Trans. Ind. Electron. **54**(1), 495–503 (2007)
13. D.B. Rathnayake et al., Grid forming inverter modeling, control, and applications. IEEE Access **9**, 114781–114807 (2021)
14. Y. Terriche, M.U. Mutarraf, M. Mehrzadi, A. Lashab, J.M. Guerrero, J.C. Vasquez, D. Kerdoun, Adaptive CDSC-based open-loop synchronization technique for dynamic response enhancement of active power filters. IEEE Access **7**, 96743–96752 (2019)
15. A. Luo, C. Tang, Z.K. Shuai et al., A novel three-phase hybrid active power filter with a series resonance circuit tuned at the fundamental frequency. IEEE Trans. Ind. Electron. **56**(7), 2431–2440 (2009)
16. L. Yang, J. Yang, A robust dual-loop current control method with a delay-compensation control link for LCL-type shunt active power filters. IEEE Trans. Power Electron. **234**(7), 6183–6199 (2018)
17. W. Yeetum, V. Kinnares, Parallel active power filter based on source current detection for antiparallel resonance with robustness to parameter variations in power systems. IEEE Trans. Ind. Electron. **66**(2), 876–886 (2019)
18. T. Liu, J. Liu, Z. Liu, et al., A study of virtual resistor-based active damping alternatives for LCL resonance in grid-connected voltage source inverters. IEEE Trans. Power Electron. **35**(1), 247–262 (2020)
19. M. Liserre, F. Blaabjerg, R. Teodorescu, Grid impedance estimation via excitation of $ LCL $-filter resonance. IEEE Trans. Ind. Appl. **43**(5), 1401–1407 (2007)
20. W. Wu, J. Liu, Y. Li et al., Individual channel design-based precise analysis and design for three-phase grid-tied inverter with LCL-filter under unbalanced grid impedance. IEEE Trans. Power Electron. **35**(5), 5381–5396 (2019)
21. S. Jayalath, M. Hanif, An LCL-filter design with optimum total inductance and capacitance. IEEE Trans. Power Electron. **33**(8), 6687–6698 (2017)
22. X. Li, J. Fang, P. Lin, Y. Tang, Active magnetic decoupling for improving the performance of integrated LCL-filters in grid-connected converters. IEEE Trans. Ind. Electron. **65**(2), 1367–1376 (2018)
23. C.-S. Lam, W.-H. Choi, M.-C. Wong, Y.-D. Han, Adaptive DC-link voltage-controlled hybrid active power filters for reactive power compensation. IEEE Trans. Power Electron. **27**(4), 1758–1772 (2012)
24. W.U.K. Tareen, S. Mekhielf, Three-phase transformerless shunt active power filter with reduced switch count for harmonic compensation in grid-connected applications. IEEE Trans. Power Electron. **33**(6), 4868–4881 (2018)

25. C.-S. Lam, M.-C. Wong, W.-H. Choi, X.-X. Cui, H.-M. Mei, J.-Z. Liu, Design and performance of an adaptive low-DC-voltage-controlled LC-hybrid active power filter with a neutral inductor in three-phase four-wire power systems. IEEE Trans. Ind. Electron. **61**(6), 2635–2647 (2014)
26. C. Gong, W.-K. Sou, C.-S. Lam, Second-order sliding-mode current controller for LC-coupling hybrid active power filter. IEEE Trans. Ind. Electron. **68**(3), 1883–1894 (2021)
27. Z. Xiang, Y. Pang, L. Wang et al., Design, control and comparative analysis of an LCLC coupling hybrid active power filter. IET Power Electron. **13**(6), 1207–1217 (2020)
28. A. Luo, S. Peng, C. Wu, J. Wu, Z. Shuai, Power electronic hybrid system for load balancing compensation and frequency-selective harmonic suppression. IEEE Trans. Ind. Electron. **59**(2), 723–732 (2012)
29. L. Wang, C.-S. Lam, M.-C. Wong, Design of a thyristor controlled LC compensator for dynamic reactive power compensation in smart grid. IEEE Trans. Smart Grid **8**(1), 409–417 (2017)
30. L. Wang, Y. Pang, W. Tian, Q. Xu, S. Liao, J. Yang, A new adaptive broadband thyristor-controlled impedance matching network. IEEE Trans. Ind. Electron. **69**(3), 2521–2532 (2022)
31. L. Wang, C.-S. Lam, M.-C. Wong, Multifunctional hybrid structure of SVC and capacitive grid-connected inverter (SVC//CGCI) for active power injection and nonactive power compensation. IEEE Trans. Ind. Electron. **66**(3), 1660–1670 (2019)
32. J. Yu, L. Wang, W. Tian, K. Zhang, S. Liao, Q. Xu, An adaptive broadband active impedance matching network for underwater electroacoustic transduction system. IEEE Trans. Ind. Electron. **70**(6), 5433–5443 (2023)
33. M.A. Jensen, B.K. Lau, Uncoupled matching for active and passive impedances of coupled arrays in MIMO systems. IEEE Trans. Antennas Propag. **58**(10), 3336–3343 (2010)
34. Y. Li, N. Duan, Z. Liu, J. Hu, Z. He, Impedance-matching-based maximum power tracking for magnetic field energy harvesters using active rectifiers. IEEE Trans. Ind. Electron. **70**(10), 10730–10739 (2023)
35. A. Al Bastami, A. Jurkov, D. Otten, D.T. Nguyen, A. Radomski, D.J. Perreault, A 1.5 kW radio-frequency tunable matching network based on phase-switched impedance modulation. IEEE Open J. Power Electron. **1**(1), 124–138 (2020)
36. H. Lee, W. Lee, T. Kim, M. Helaoui, F.M. Ghannouchi, Y. Yang, 6–18 GHz gaas pHEMT broadband power amplifier based on dual-frequency selective impedance matching technique. IEEE Access **7**, 66275–66280 (2019)
37. S.C. Dutta Roy, Characteristics of single- and multiple-frequency impedance matching networks. IEEE Trans. Circ. Syst. II Exp. Briefs **62**(3), 222–225 (2015)
38. W. Tian, S. Liao, L. Wang, X. Zhou, Q. Xu, New adaptive broadband full-bridge converter impedance matching network for giant magnetostrictive transducer. IEEE Trans. Power Electron. **37**(7), 7652–7662 (2022)

Chapter 2
Adaptive Impedance Converters for Power Quality Compensation

Abstract This chapter presents the applications of adaptive impedance converters in distribution systems for power quality compensation, including the related adaptive hybrid static synchronous compensator and adaptive hybrid active filter from the aspects of topology structure, characteristic analysis, parameter design, control system, and experimental verifications. This chapter provides different compensation topologies composed of adaptive coupling impedance-connected converters in different connected methods. The adaptive passive coupling impedance is used to provide dynamic reactive power effectively and reduce the converter's DC side voltage, thereby significantly reducing the number of switches/voltage resistance and breaking through the bottleneck of low efficiency and low reliability of large-capacity converters.

2.1 Introduction of Different Converters in the Distribution Systems

Electric power distribution is delivering electricity from the generation source to the end-users [1]. This process involves converting the high-voltage electricity generated by power plants into low-voltage electricity suitable for residential, commercial, and industrial consumers. The common industrial, commercial, residential, and transportation power conversion systems generally comprise passive devices (inductor, capacitor, and resistor), active power electronic devices (inverter, rectifier, and other converters), transformers, lines, etc. The distribution loads usually are not pure resistive and linear, which would bring numerous reactive power, harmonic components, voltage drops, etc., resulting in current and voltage quality problems [2, 3].

Low power factor and current harmonic distortion are considered two of the major current quality problems in distribution systems [2, 3]. This chapter introduces and compares different converters in terms of topology, structure, and characteristics for power quality problems in distribution systems.

Static var compensators (SVCs) are traditionally used to compensate reactive power dynamically [4, 5]. However, SVCs have many inherent problems, such as

L. Wang et al., *Design and Applications of Adaptive Impedance Converters*,
https://doi.org/10.1007/978-981-97-5319-2_2

resonance problems, poor harmonic compensation ability, and slow response [6]. To overcome these inherent problems, the active power filters (APFs) and distribution static compensators (DSTATCOMs) based on L coupling impedance for power quality compensation have drawn much attention since 1976 [7, 8]. However, during the operation of APFs and DSTATCOMs, the active inverter part requires a high voltage rating. Therefore, the APFs and DSTATCOMs require costly multilevel structures for medium voltage level applications, thus limiting their widespread applications. To reduce the voltage rating of the active inverter part, an *LC* coupling hybrid active power filter (HAPF) and *LC* coupling distribution static compensator (C-DSTATCOM) were proposed in 2003 [9, 10]. Even though the large *LC* impedance can reduce the inverter part voltage rating, it also limits the compensating range. In 2014, the hybrid structures of SVC in series with APF (SVC-HAPFs) and DSTATCOMs (SVC-DSTATCOMs) were proposed in [11–13]. These SVCs can be seen as adaptive impedances, and the SVC-HAPFs and SVC-DSTATCOMs can be seen as adaptive impedance converters. The SVC-HAPF and SVC-DSTATCOM have the characteristics of a wide reactive power compensation range and a low voltage rating of the active inverter part. The above are the series adaptive impedance converters in distribution systems for power quality compensation.

For the centralized DSTATCOMs integrated with a transformer in the distribution system, two possible connections are located on the primary or secondary side of the distribution transformer. When the DSTATCOM is connected to the low-voltage secondary side, its current level is higher due to the low coupling voltage, which gives rise to higher power loss and overheating issues. The current compensation level and power loss can be largely reduced when the DSTATCOM is connected to the primary side with a high coupling voltage. However, the high coupling voltage requires a high operating voltage of DSTATCOM, which may not be cost-effective in applications. Extensive efforts have been made to address the high coupling voltage at the primary side of a distribution transformer. The DSTATCOM connects a step-down transformer (DSTATCOM-T) on the primary side, which increases the potential investment cost and floor space [14]. As a common topology, it has been applied to microgrids and distribution systems for reactive power compensation, voltage support, and reduction of renewable energy fluctuations. For the above issues, a cost-effective method is integrating DSTATCOM and the distribution transformer [15, 16]. Among them, the DSTATCOM connects the customized winding of a three-winding transformer by coupling L or LC filters [17, 18], which is named controllably inductive power filters (CIPFs) or hybrid inductive and active filters (HIAFs). To adjust the coupling voltage of CIPFs and HIAFs, the power transformer needs to be additionally customized to contain the high-cost third winding. Compared with CIPF, the HIAP can decrease the voltage source inverter (VSI) DC voltage and operational loss. The CIPF was further applied to non-linear load scenarios and photovoltaic power plants, while the HIAF was applied to the shipboard and DC grid. Motivated by the structure of auto-transformer [19–21], a STATCOM integrated with the primary tapped transformer (PTT) in Fig. 2.1e is proposed, which is also called L- or LC-DSTATCOM PTTs based on the coupling L or LC filters [22, 23]. The L-DSTATCOM PTT has been reported in harmonic and reactive compensation, while the LC-DSTATCOM PTT has been applied in reactive and unbalanced compensation of the distribution system. Instead of the auxiliary step-down transformer or third winding, the STATCOM connects with the transformer via primary winding taps. Its coupling voltage can be reasonably reduced by adjusting the tap ratio to accommodate the inverter's rated voltage and current. Compared with L-DSTATCOM PTT, the inverter DC voltage and power rating of the LC-DSTATCOM PTT can be reduced, which leads to lower initial cost and power loss. However, it is only favorable to the fixed inductive load, and its operational range is narrow, which limits its compensation capability for varying loads. This chapter proposes adaptive impedance converters integrated with a tapped transformer for centralized reactive power compensation, which is called thyristor-controlled LC-coupling distribution static synchronous compensator integrated with the primary tapped transformer (TCLC-DSTATCOM PTT). It can deal with low DC voltage and a wide operational range for both inductive and capacitive loads, as introduced in Sect. 2.2. According to the varying load, the thyristor firing angle and impedance of TCLC can be adaptively adjusted, thereby changing the provided voltage drop and reactive power to maintain a low inverter DC voltage.

Though the SVC-HAPFs can reduce the voltage rating of the active inverter part, the star-connected SVC-HAPFs still suffer from the high current rating requirement of the active inverter part. This Chapter proposes a delta-connected adaptive impedance-based hybrid active power filter (AI-HAPF), including its structure and

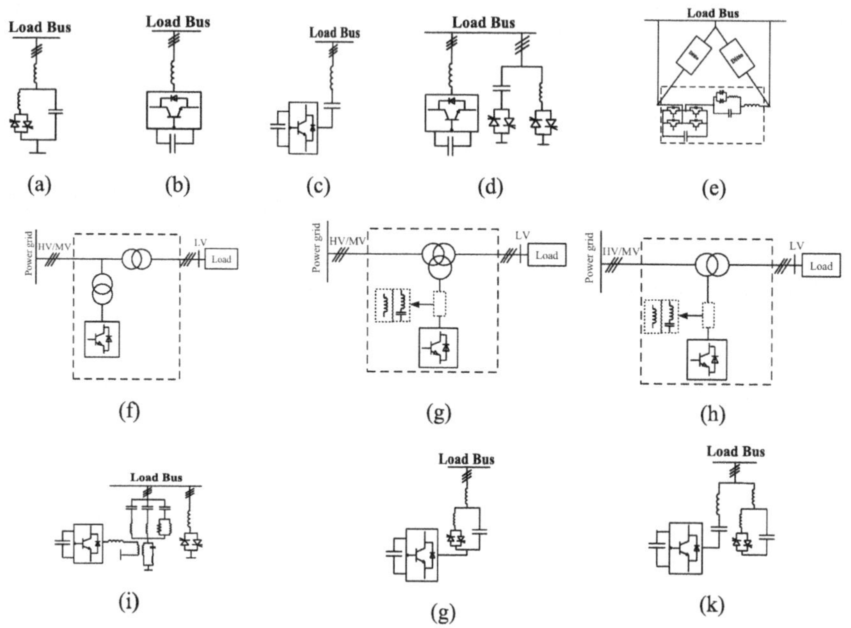

Fig. 2.1 The simplified circuit configurations of the **a** SVC, **b** APF, **c** HAPF, **d** SVC//APF, **e** Delta-connected AI-HAPF, **f** DSTATCOM-T, **g** CIPF and HIPF, **h** L-DSTATCOM PTT and LC-DSTATCOM PTT, **i** TCR//RITHAPF, **g** SVC-HAPF, and **k** SVC//HAPF

control method, which will be introduced in Sect. 2.3 [24]. The delta-connected AI-HAPF aims to reduce the compensation current in each phase under unbalanced loads. For example, a delta-connected inverter is being proposed to compensate for the negative sequence current caused by the unbalanced load [25–27]. In 2012, a modular multilevel cascade converter was proposed to accomplish negative-sequence reactive-power control [26]. In 2015, a harmonic reference current balancing method was proposed for a delta-connected converter [27]. However, the compensation capability is not flexible enough in these studies. Compared with the traditional star-connected APFs and SVC-HAPFs, the proposed delta-connected AI-HAPF provides extra flexibility in unbalanced current compensation by controlling the circulating currents flowing inside the structure, and the current rating can be reduced to $1/\sqrt{3}$.

For the heavy loads compensation, the current rating of the series SVC-HAPF can be high. In 2003, Dixon et al. [28] proposed the combined system of SVC in parallel with APF (SVC//APF). The SVC part of this structure is used to compensate for most of the reactive power. Thus, the current rating of the APF can be significantly reduced. However, the voltage rating of the APF is still high, so the SVC//APF still requires a costly multilevel structure for medium voltage level applications. In 2012, Luo et al. [29] proposed a novel hybrid system that consists of a thyristor-controlled reactor (TCR) and a resonant impedance-type hybrid active power filter (RITHAPF) (TCR//RITHAPF). The RITHAPF part is an APF crossing over a matching transformer connected in parallel with a fundamental series resonant circuit. As most of the loads are inductive and the TCR part is also inductive, the required capacitive compensating current generated by the RITHAPF part should be larger in order to count in the TCR part. Due to this reason, the current rating of the RITHAPF part can be high for heavy load compensation. In addition, the high current rating matching transformer in TCR//RITHAPF can also significantly drive up the system cost. This Chapter proposes a hybrid structure of a static var compensator in parallel with a hybrid active power filter (SVC//HAPF) for medium voltage heavy loads compensation, which is introduced in Sect. 2.4 [30]. The proposed SVC//HAPF has a wide reactive power compensation range and obtains both low voltage and current inverter ratings. As the cost of the SVC part is much lower than that of the active inverter part, thus the reduction of the active inverter rating (voltage and current) can lead to a decrease in the total cost of the SVC//HAPF system. Therefore, the total cost of the hybrid structure SVC//HAPF can be significantly reduced for medium voltage level heavy loads compensation.

The simplified circuit configurations of the above-mentioned power quality compensators are provided in Fig. 2.1. The comparisons among different power quality compensators in terms of reactive power compensation range, inverter voltage, and current ratings are summarized in Table 2.1.

The chapter introduction is provided in Sect. 2.1, which presents the development of different converters in the distribution system. Then, the applications of adaptive impedance-coupling DSTATCOM integrated with a primary tapped transformer are given in Sect. 2.2. After that, the delta-connected adaptive impedance-based hybrid active power filter and its control method are provided in Sect. 2.3. Then, a hybrid structure of static var compensator and hybrid active power filter is provided in Sect. 2.4. Finally, the Chapter summary is drawn in Sect. 2.5.

Table 2.1 Comparisons between series and parallel adaptive impedance converters

Adaptive impedance converter		Year	Reactive power Compensation range	Inverter current rating	Inverter voltage rating
Series type	SVC	1960s	****	–	–
	APF	1976	****	***	****
	HAPF	2003	*	***	**
	SVC-HAPF	2014	****	***	**
	Delta-connected AI-HAPF	2017	****	**	**
Integrated with transformer type	DSTATCOM-T	1981	****	*	****
	CIPF	2012	****	**	****
	HIPF	2015	*	**	**
	L-DSTATCOM PTT	2016	****	**	****
	LC-DSTATCOM PTT	2022	*	**	**
	TCLC-DSTATCOM PTT	2023	****	**	**
Parallel type	SVC//APF	2003	*****	*	****
	TCR//RITHAPF	2012	***	**	**
	SVC//HAPF	2017	****	*	**

2.2 Adaptive Impedance-Coupling DSTATCOM Integrated with Primary Tapped Transformer

This chapter proposes a thyristor-controlled LC (adaptive impedance) coupling DSTATCOM integrated with the primary tapped transformer, called TCLC-DSTATCOM PTT, to achieve low DC voltage operation under a wide load range. According to the varying load, the thyristor firing angle and impedance of TCLC can be adaptively adjusted, thereby changing the provided voltage drop and reactive power to maintain a low inverter DC voltage. Residual reactive and harmonic power would be compensated via the inverter part. To develop such TCLC-DSTATCOM PTT topology, some potential technical issues should be considered, including:

(1) Quantifying the dc voltage and the operational range compared with existing L-DSTATCOM PTT in 2016 and LC-DSTATCOM PTT in 2022 are addressed;
(2) The influence of component parameters, including the transformer winding tap ratio, thyristor firing angle, and coupling impedance to the compensation performance is revealed;
(3) Power loss comparisons among L-DSTATCOM PTT, LC-DSTATCOM PTT, and the proposed one are given.

Based on the above issues, the contributions of this section are summarized as follows.

(1) A centralized TCLC-DSTATCOM PTT is proposed, which integrates the DSTATCOM with transformer taps by TCLC filters. Compared with the L- and LC-DSTATCOM PTTs, it can achieve adaptive compensation for different inductive and capacitive loads under low DC voltage.
(2) The V-Q characteristic is formulated to reveal the DC voltage reduction mechanism and quantify the compensation range of TCLC-DSTATCOM PTT. It shows the DC voltage of TCLC-DSTATCOM PTT can be reduced by 87.5% compared with L-DSTATCOM PTT. Its compensation range can reach 88.3 times that of the LC-DSTATCOM PTT at 1000 V DC voltage.
(3) A thyristor firing angle design method with respect to the required DC voltage is given to ensure the adaptive control of TCLC-DSTATCOM PTT, thereby achieving the compensation operation for varying loads.

2.2.1 Circuit and Modeling of TCLC-STATCOM PTT

Figure 2.2 shows the circuit configuration of the proposed TCLC-DSTATCOM PTT, which consists of adaptive TCLC and VSI parts. The TCLC is composed of an L, a parallel capacitor C_{PF}, and a pair of anti-parallel thyristors (T_{x1} and T_{x2}) connected in series with L_{PF}. The TCLC impedance can be changed by controlling the thyristor firing angles α_x. The VSI is connected to the adjustable taps (labeled as u_0, v_0, and w_0) on the primary winding (labeled as a_0–b_0, b_0–c_0, and c_0–a_0) of the Dyn11 connected distribution transformer through a coupling TCLC.

The subscript 'x' in Fig. 2.2 means phase $x = a, b, c$. v_{Sx}, v_{Lx}, v_{cx}, i_{Sx}, i_{Lx}, and i_{cx} are the source voltage, the load voltage, the coupling voltage, the source current, the load current and the compensation current. C_{dc} and V_{dc} represent the DC capacitor and DC voltage, respectively. The primary and secondary winding turns of the transformer are W_P and W_S. K (=W_P/W_S) represents the transformer turn ratio. The winding turns on both sides of the taps are W_{P1x} and W_{P2x}. k (= W_{P1x}/W_P) is the winding tap ratio. The tap ratios of the three phases are the same.

A. *Current Modeling*

Referring to Fig. 2.2, the relationship among the source, load, and compensation currents is shown by applying ampere-turns balance theory as,

$$\begin{bmatrix} i_{Sa} \\ i_{Sb} \\ i_{Sc} \end{bmatrix} = \frac{1}{K}\left[\begin{bmatrix} 1 & 0 & -1 \\ -1 & 1 & 0 \\ 0 & -1 & 1 \end{bmatrix} \cdot \begin{bmatrix} i_{La} \\ i_{Lb} \\ i_{Lc} \end{bmatrix} + \begin{bmatrix} 1-k & 0 & k \\ k & 1-k & 0 \\ 0 & k & 1-k \end{bmatrix} \cdot \begin{bmatrix} i_{ca} \\ i_{cb} \\ i_{cc} \end{bmatrix}\right] \tag{2.1}$$

Then, the i_{Sx} only contains the active component when the i_{cx} is expressed as,

$$\begin{bmatrix} i_{ca} \\ i_{cb} \\ i_{cc} \end{bmatrix} = N \cdot \begin{bmatrix} i_{La} - i_{Lap} \\ i_{Lb} - i_{Lbp} \\ i_{Lc} - i_{Lcp} \end{bmatrix} \tag{2.2}$$

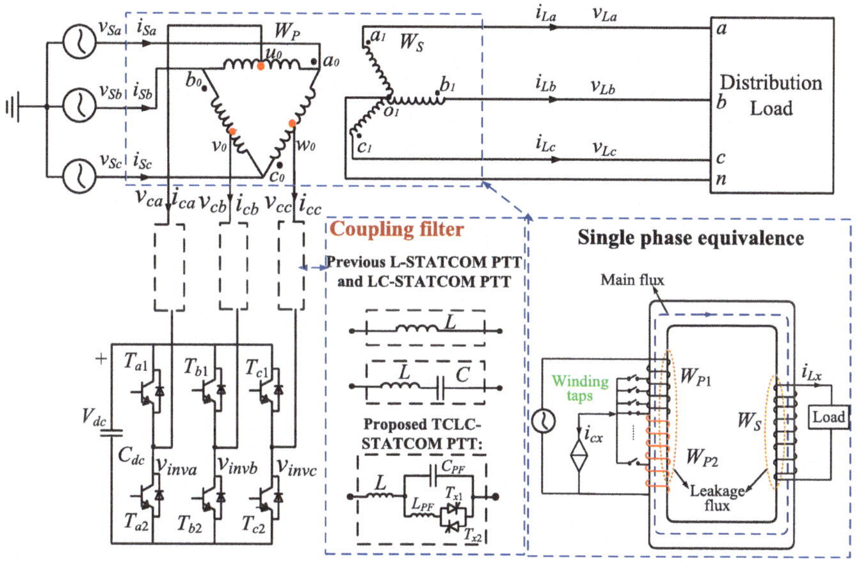

Fig. 2.2 Structures of the L-DSTATCOM PTT, LC-DSTATCOM PTT, and proposed TCLC-DSTATCOM PTT

where the i_{Lxp} and i_{Lxq} (=$i_{Lx} - i_{Lxp}$) are the active and reactive components of load current, and the N can be obtained as,

$$N = \frac{1}{K\left(3k^2 - 3k + 1\right)} \begin{bmatrix} 2k-1 & -k & 1-k \\ 1-k & 2k-1 & -k \\ -k & 1-k & 2k-1 \end{bmatrix} \tag{2.3}$$

Based on (2.1)–(2.3), the compensation current of TCLC-DSTATCOM PTT can be calculated. The phase relationship among the source, load, and compensation currents is shown in Fig. 2.3a at the numerical examples $K = 1$ and $k = 0.5$.

B. *Impedance Modeling*

For coupling impedance of TCLC-DSTATCOM PTT, the TCLC part can be divided into thyristor-controlled reactors (TCR), capacitor C_{PF}, and inductor L. According to different thyristor firing angle $\alpha_x = (90° \le \alpha_x \le 180°)$, the impedance X_{TCLC} (α_x) of the TCLC part is given as,

$$X_{TCLCx}(\alpha_x) = \frac{\pi\omega L_{PF}}{(2\pi - 2\alpha_x + \sin 2\alpha_x) - \pi\omega^2 L_{PF} C_{PF}} + \omega L \tag{2.4}$$

where $\omega(= 2\pi f)$ is the angular frequency. When the firing angles $\alpha_x = 90°$ and $\alpha_x = 180°$, the values of X_{TCLCx} (α_x) achieve the minimum inductive (positive)

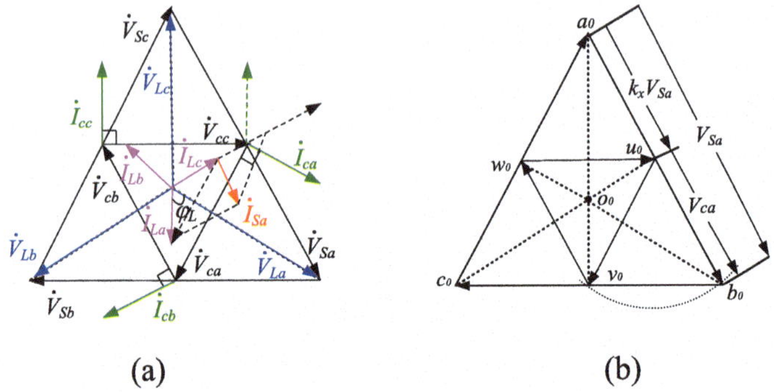

Fig. 2.3 Vector diagram about **a** current relationship; **b** voltage relationship

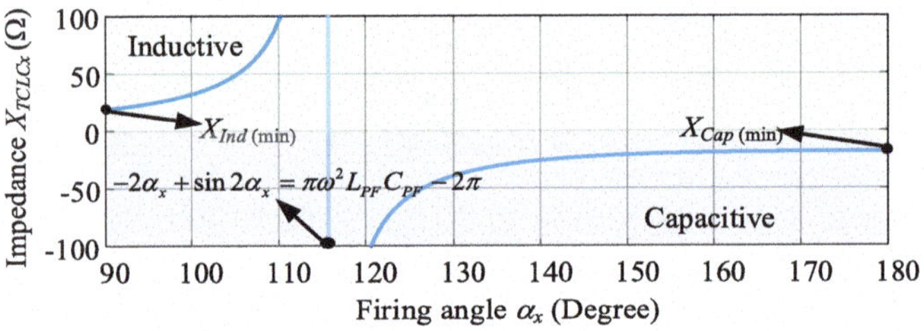

Fig. 2.4 TCLC impedance in terms of firing angle

and capacitive impedances (negative), which can be expressed as (2.5) and (2.6). Figure 2.4 clearly show the relationship between $X_{TCLCx}\ (\alpha_x)$ and α_x.

$$X_{Ind\,(\min)}(\alpha_x = 90^\circ) = \frac{X_{L_{PF}} X_{C_{PF}}}{X_{C_{PF}} - X_{L_{PF}}} + X_L \tag{2.5}$$

$$X_{Cap(\min)}(\alpha_x = 180^\circ) = -\left(X_{C_{PF}} - X_L\right) \tag{2.6}$$

C. *Voltage Modeling*

Figure 2.3b shows the primary side voltage vector diagram. It indicates the ratio of V_{cx} and V_{Sx} can be expressed based on the *law of cosines* as,

$$\frac{V_{cx}}{V_{Sx}} = \frac{u_0 v_0}{a_0 b_0} = \sqrt{3k^2 - 3k + 1} \tag{2.7}$$

Thus, the required inverter voltage V_{invx} of the TCLC-DSTATCOM PTT can be computed as,

$$V_{invx} = V_{cx} - X_{TCLCx}(\alpha_x) \cdot I_{cx} \tag{2.8}$$

The load reactive power Q_{loadx} can be expressed as,

$$Q_{loadx} = V_{Lx} \cdot I_{Lxq} = V_{cx} \cdot (-I_{cx}) \tag{2.9}$$

where the direction of I_{Lxq} is opposite to I_{cx}, and the Q_{loadx} is equal to reactive power provided by TCLC-DSTATCOM PTT under full compensation. Based on (2.8)–(2.9), the V_{invx} can be further derived as,

$$V_{invx} = V_{cx} \cdot \left(1 + \frac{Q_{loadx}}{Q_{TCLCx}(\alpha_x)}\right) \tag{2.10}$$

where the base reactive power $Q_{TCLCx}(\alpha_x)$ of TCLC-DSTATCOM PTT can be calculated as,

$$Q_{TCLCx}(\alpha_x) = \frac{V_{cx}^2}{X_{TCLCx}(\alpha_x)} \tag{2.11}$$

In (2.10), when the Q_{loadx} is inductive ($Q_{loadx} > 0$), the $Q_{TCLCx}(\alpha_x)$ is controlled as capacitive ($Q_{TCLCx} < 0$) and vice versa. Assuming the modulation index m is 1, based on (2.7)–(2.10), the required dc voltage V_{dcTCLC} of TCLC-DSTATCOM PTT can be expressed as,

$$V_{dcTCLC} = \sqrt{6} \cdot \sqrt{3k^2 - 3k + 1} \cdot V_{Sx} \cdot \left|1 + \frac{Q_{loadx}}{Q_{TCLCx}(\alpha_x)}\right| \tag{2.12}$$

where the gain value of $\sqrt{6}$ can be expressed as $\sqrt{6} = \sqrt{2} \cdot \sqrt{3}$; $\sqrt{3}$ means the gain to transfer V_{invx} to its line-to-line voltage, and $\sqrt{2}$ means the gain to transfer the line-to-line voltage to its peak value. The range of $Q_{TCLCx}(\alpha_x)$ is within the maximum inductive reactive power $Q_{TCLCx(maxInd)}$ and the maximum capacitive reactive power $Q_{TCLCx(maxCap)}$. The $Q_{TCLCx(maxInd)}$ and $Q_{TCLCx(maxCap)}$ can be obtained as (2.13)–(2.14) when the thyristor firing angle is selected as 90° and 180°. If the Q_{loadx} is within the range of $Q_{TCLCx(maxInd)}$ to $Q_{TCLCx(maxCap)}$, the $Q_{TCLCx}(\alpha_x)$ can be equal to $-Q_{loadx}$, and thus the required V_{dcTCLC} can be close to zero in ideal conditions based on (2.12). If the Q_{loadx} is outside the range, the $Q_{TCLCx}(\alpha_x)$ is controlled to be $Q_{TCLCx(maxCap)}$ or $Q_{TCLCx(maxInd)}$ when the load is inductive or capacitive, respectively.

$$Q_{TCLCx(maxInd)} = \frac{V_{cx}^2}{X_{Ind(\min)(\alpha_x=90^\circ)}} = \frac{\left(3k^2 - 3k + 1\right) \cdot V_{Sx}^2}{\frac{X_{LPF} X_{CPF}}{X_{CPF} - X_{LPF}} + X_L} \tag{2.13}$$

$$Q_{TCLCx(\max\ Cap)} = \frac{V_{cx}^2}{X_{Cap(\min)(\alpha_x=180^\circ)}} = \frac{\left(3k^2 - 3k + 1\right) \cdot V_{Sx}^2}{X_L - X_{CPF}} \tag{2.14}$$

2.2.2 V-Q Characteristics Comparison

In the above section, the required dc voltage of TCLC-DSTATCOM PTT is given. This section compares the V-Q characteristics of the three DSTATCOM PTTs.

A *L-DSTATCOM PTT*

The coupling filter of L-DSTATCOM PTT is the L filter, and its impedance is X_L. Similar to (2.12), the required dc voltage V_{dcL} of L-DSTATCOM PTT can be expressed as,

$$V_{dcL} = \sqrt{6} \cdot \sqrt{3k^2 - 3k + 1} \cdot V_{Sx} \cdot \left| 1 + \frac{Q_{loadx}}{Q_{Lx}} \right| \quad (2.15)$$

where the base reactive power Q_L of L-DSTATCOM PTT can be expressed as,

$$Q_{Lx} = \frac{V_{cx}^2}{X_L} = \frac{\left(3k^2 - 3k + 1\right) \cdot V_{Sx}^2}{\omega L} \quad (2.16)$$

where the $Q_{Lx} > 0$. Based on the (2.15)–(2.16), the V_{dcL} of L-DSTATCOM PTT in terms of tap ratio k and Q_{Loadx} can be plotted in Fig. 2.5a.

Figure 2.5a shows the required dc voltage of L-DSTATCOM PTT increases as the load changes from capacitive to inductive. The dc voltage is close to $\sqrt{6}$ times the coupling voltage V_{cx} under a larger load range since the slope of V_{dcL} with respect to Q_{Loadx} is small, caused by the small X_L impedance. Figure 2.5a also shows the required dc voltage of L-DSTATCOM PTT can reach its minimum at k equal to 0.5.

B. *LC-DSTATCOM PTT*

The coupling filter and its impedance of LC-STATCOM PTT are LC filter and X_{LC}. The required dc voltage V_{dcLC} of LC-STATCOM PTT can be expressed as,

$$V_{dcLC} = \sqrt{6} \cdot \sqrt{3k^2 - 3k + 1} \cdot V_{Sx} \cdot \left| 1 + \frac{Q_{Loadx}}{Q_{LCx}} \right| \quad (2.17)$$

where the Q_{LC} is the base reactive power of LC-STATCOM PTT. It can be calculated as,

$$Q_{LCx} = \frac{V_{cx}^2}{X_{LC}^2} = \frac{\left(3k^2 - 3k + 1\right) \cdot V_{Sx}^2}{\frac{1}{\omega C} - \omega L} \quad (2.18)$$

where the $Q_{LC} < 0$. Due to the $X_{LC} >> X_L$, the $|Q_L| >> |Q_{LC}|$. Based on the (2.17)–(2.18), the V_{dcLC}–Q_{Loadx} characteristic of LC-STATCOM PTT is shown in Fig. 2.5b.

From Fig. 2.5b, the required DC voltage of LC-STATCOM PTT can be decreased under a narrow inductive load range. The DC voltage can be close to zero when the capacitive LC filter provides the reactive power required by the inductive load. Once

Fig. 2.5 V–Q characteristics of **a** L-DSTATCOM PTT; **b** LC-DSTATCOM PTT; **c** TCLC-DSTATCOM PTT

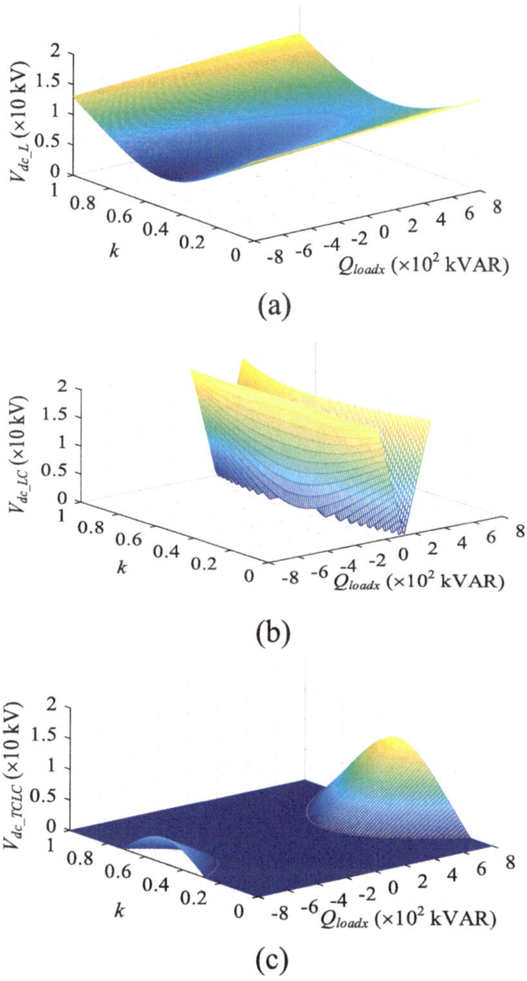

the load exceeds the narrow range, the V_{dcLC} increases sharply since the slope of V_{dcLC} with respect to Q_{Loadx} is large, caused by large X_{LC}. There is a different tap ratio, k, to achieve the minimum V_{dcLC} according to load conditions.

C. *TCLC-DSTATCOM PTT*

The $V_{dcTCLC} - Q_{Loadx}$ characteristic of TCLC-DSTATCOM PTT under different tap ratio k is shown in Fig. 2.5c. It shows the V_{dcTCLC} can be closed to zero theoretically in wide inductive or capacitive load ranges. Once the load exceeds this range, the DC voltage is increased slightly to satisfy the compensation requirement. For the TCLC-DSTATCOM PTT, its compensation range changes with k.

In summary, the L-DSTATCOM PTT is favorable to the capacitive load, and the overall DC voltage level is relatively high under different loads. The LC-DSTATCOM PTT is only favorable to a fixed inductive load. Compared with L-

and LC-DSTATCOM PTTs, the proposed TCLC-DSTATCOM PTT can provide a large range to compensate for inductive and capacitive reactive loads at a low DC voltage, thereby reducing the power loss. The following quantitatively compares the compensation range and power loss among the L-, LC-, and TCLC-DSTATCOM PTTs.

D. *Compensation Range Comparison*

The compensation range can be defined as the load reactive power range where the required dc voltage is lower than the given dc voltage V_{dc}. Based on (2.12), (2.15), (2.17), the compensation ranges of L-, LC-, and TCLC-DSTATCOM PTTs can be respectively calculated as,

$$Q_{Range_L} \in \left(-\frac{V_{cx}^2}{X_L} - \frac{V_{dc} \cdot V_{cx}}{\sqrt{6}X_L}, -\frac{V_{cx}^2}{X_L} + \frac{V_{dc} \cdot V_{cx}}{\sqrt{6}X_L} \right) \tag{2.19}$$

$$Q_{Range_{LC}} \in \left(-\frac{V_{cx}^2}{X_{LC}} + \frac{V_{dc} \cdot V_{cx}}{\sqrt{6}X_{LC}}, -\frac{V_{cx}^2}{X_{LC}} - \frac{V_{dc} \cdot V_{cx}}{\sqrt{6}X_{LC}} \right) \tag{2.20}$$

$$Q_{Range_{TCLC}} \in \left(-\frac{V_{cx}^2}{X_{Ind(\min)}} - \frac{(V_{dc} - V_{\text{margin}}) \cdot V_{cx}}{\sqrt{6}X_{Ind(\min)}}, -\frac{V_{cx}^2}{X_{Cap(\min)}} - \frac{\left(V_{dc} - V_{\text{margin}}\right) \cdot V_{cx}}{\sqrt{6}X_{Cap(\min)}} \right) \tag{2.21}$$

where V_{margin} is the margin voltage of TCLC-DSTATCOM PTT considering the potential firing angle mistuning and calculation deviation. The k of traditional L-DSTATCOM PTT is designed to be 0.5 to achieve its minimum DC voltage. Thus, taking $k = 0.5$ as an example, it gives the compensation ranges of L-, LC-, TCLC-DSTATCOM PTTs under different DC voltages and $V_{margin} = 1000$ V. Figure 2.6a shows the compensation range of L-DSTATCOM PTT is large. However, it mainly compensates for the capacitive loads at low DC voltage, while high DC voltage is required for the inductive loads. Figure 2.6b shows the LC-DSTATCOM PTT can compensate for the inductive loads at low dc voltage, and the compensation range is narrow. Figure 2.6c indicates TCLC-DSTATCOM PTT can simultaneously compensate for a relatively wide range of inductive and capacitive loads at low DC voltage.

Table 2.2 gives the numerical example of the compensation range according to different dc voltage and tap ratios. From the above comparisons of compensation range among the L-, LC-, and TCLC-DSTATCOM PTTs, the following summarizes several important points:

- Compared with L- and LC-DSTATCOM PTTs, the TCLC-DSTATCOM PTT requires the lowest dc voltage to compensate for inductive and capacitive loads simultaneously.
- With low dc voltage (1000 V), the compensation range of TCLC-DSTATCOM PTT can reach 88.3 times that of LC-DSTATCOM PTT and about 1.1 ~ 2.1

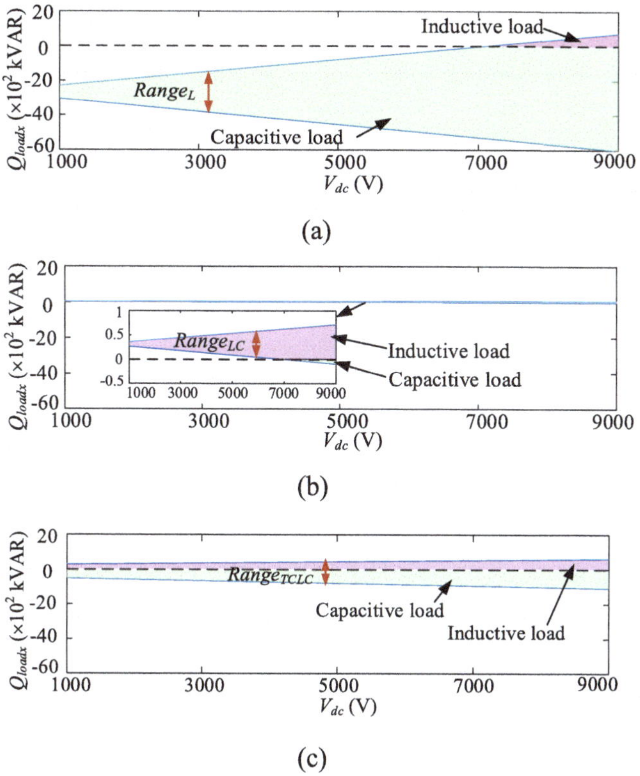

Fig. 2.6 Compensation range comparison of **a** L-DSTATCOM PTT; **b** LC-DSTATCOM PTT; **c** TCLC-DSTATCOM PTT at $k = 0.5$

Table 2.2 Compensation Range Comparison under Different V_{dc} and k_x

V_{dc} (V)	k	Q_{RangeL} (10^2 kVAR)	$Q_{RangeLC}$ (10^2 kVAR)	$Q_{RangeTCLC}$ (10^2 kVAR)
1000	0.5	[−30. 01, −22.82]	[0.27, 0.36]	[−5.05, 2.90]
	0.3/0.7	[−43.82, −34.71]	[0.42, 0.53]	[−7.47, 4.30]
	1	[−113.74, −98.72]	[1.18, 1.36]	[−20.19, 11.61]
5000	0.5	[−45.34, −7.82]	[0.09, 0.54]	[−7.90, 4.55]
	0.3/0.7	[−62.17, −16.53]	[0.19, 0.74]	[−10.94, 6.29]
	1	[−143.73, −68.64]	[0.82, 1.72]	[−25.89, 14.90]
8000	0.5	[−56.62, 3.51]	[−0.04, 0.68]	[−10.04, 5.78]
	0.3/0.7	[−75.81, −2.85]	[0.03, 0.91]	[−13.55, 7.79]
	1	[−166.21, −46.17]	[0.55, 1.99]	[−30.18, 17.36]

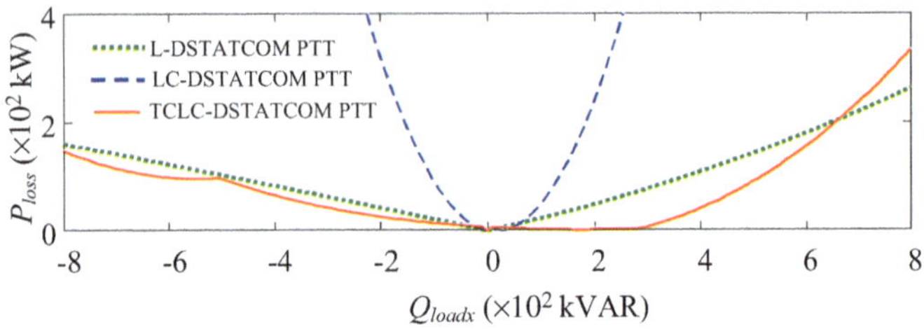

Fig. 2.7 Power loss comparison of L-DSTATCOM PTT, LC-DSTATCOM PTT, and TCLC-DSTATCOM PTT at k = 0.5

times of L-DSTATCOM PTT. Besides, the L-DSTATCOM PTT cannot compensate inductive loads, and the LC-DSTATCOM PTT cannot compensate capacitive loads at $V_{dc} = 1000$ V. (Fig. 2.7)

E. *Power Loss Comparison*

Based on the research on power loss, switching loss and component conduction loss mainly account for the total power loss for the three DSTATCOM PTTs.

(a) *Switching Loss*

While the switching losses of L- and LC-DSTATCOM PTTs are brought by the IGBT components of the VSI, the switching losses of TCLC-DSTATCOM PTT are from IGBTs and thyristors of the TCLC part. The switching loss $P_{sw\,(\mathrm{IGBT})}$ caused by IGBTs can be expressed as,

$$P_{sw(\mathrm{IGBT})} = 6 \cdot V_{dc} I_{loss} = V_{dc} I_{CM} f_{sw} \left(\frac{1}{8} t_{on} \frac{I_{CM}}{I_{CN}} + t_{off} \left(\frac{1}{3\pi} + \frac{1}{24} \frac{I_{CM}}{I_{CN}} \right) \right) \tag{2.22}$$

where I_{CM}, I_{CN}, t_{on}, t_{off}, and f_{sw} are the maximum collector current, rated collector current, rated rise time, rated fall time, and switching frequency, respectively. It indicates that less dc-link voltage leads to less $P_{sw\,(\mathrm{IGBT})}$. The switching loss $P_{sw\,(\mathrm{thyristor})}$ of thyristors is expressed as,

$$P_{sw(\mathrm{thyristor})} = 6 \cdot \left(U_{TO} \times I_{TRMSx} + r_T \times I_{TRMSx}^2 \right) \tag{2.23}$$

where I_{TRMSx} is the RMS value of the current through thyristor; U_{TO} is the thyristor threshold voltage; r_T is the thyristor slope resistance. Based on the datasheet, $r_T = 0.44$ mΩ and $U_{TO} = 0.25$ V.

(b) *Component Conduction Loss*

The component conduction losses of L-, LC-, and TCLC-DSTATCOM PTTs can be calculated as (2.24)–(2.26).

$$P_{com(L)} = 3 \cdot ESR_L \cdot I_{cx}^2 \tag{2.24}$$

$$P_{com(LC)} = 3 \cdot \left(ESR_L \cdot I_{cx}^2 + ESR_C \cdot I_{cx}^2\right) \quad (2.25)$$

$$P_{com(TCLC)} = 3 \cdot \left(ESR_L \cdot I_{cx}^2 + ESR_{C_{PF}} \cdot I_{C_{PFx}}^2 + ESR_{L_{PF}} \cdot I_{TRMSx}^2\right) \quad (2.26)$$

where ESR_L (=0.18Ω), ESR_C (=0.73Ω), ESR_{CPF} (=0.08Ω), and ESR_{LPF} (=0.41Ω) are the equivalent series resistances of L, C, C_{PF}, and C_{PF}. I_{CPFx} is the current passing through C_{PF}. According to (2.24)–(2.26), the total power losses P_{loss} (=P_{sw} + P_{com}) of L-, LC-, and TCLC-DSTATCOM PTTs under different load reactive power Q_{loadx} when $k = 0.5$ is given in Fig. 2.7. It indicates the total power loss of TCLC-DSTATCOM PTT is lower than that of L- and LC-DSTATCOM PTTs under a large inductive and capacitive load range. Although the thyristor and component losses of TCLC-DSTATCOM PTT are larger than those of L- and LC-DSTATCOM PTTs, its IGBT loss is much smaller, and the IGBT loss mainly accounts for the total losses.

2.2.3 Parameter Design

A. *Design of C_{PF}, L_{PF}, and L*

According to (2.13) and (2.14), the TCLC can provide the maximum inductive and capacitive reactive powers when its inductive and capacitive impedances are the minimum. Thus, the design of C_{PF} and L_{PF} can be given as,

$$C_{PF} = \frac{Q_{Loadx(\max Ind)}}{\omega^2 Q_{Loadx(\max Ind)} L + \omega\left(3k^2 - 3k + 1\right) V_{Sx}^2} \quad (2.27)$$

$$L_{PF} = \frac{\left(3k^2 - 3k + 1\right) V_{Sx}^2 + \omega L Q_{Loadx(\max Cap)}}{-\omega Q_{Loadx(\max Cap)} + \omega^3 C_{PF} L Q_{Loadx(\max Cap)} + \omega^2 C_{PF}\left(3k^2 - 3k + 1\right) V_{Sx}^2} \quad (2.28)$$

The design of coupling L is to smooth the current ripple brought by the switches, which can be designed as,

$$L \geq \frac{5\%}{\omega^2 \cdot C_{PF}} \quad (2.29)$$

B. *Selection of thyristor firing angle*

The coupling TCLC needs to fully deal with the reactive power of load, thus minimizing the dc voltage .($Q_{TCLCx}(\alpha_x) = -Q_{loadx}$) Based on (2.4) and (2.11), the impedance $X_{TCLCx}(\alpha_x)$ and corresponding firing angles α_x can be calculated. Since (2.4) is a transcendental equation, a look-up table is used to get the thyristor firing angles. The selection of α_x and corresponding dc voltage under different loads at $k = 0.5$ is shown in Fig. 2.8.

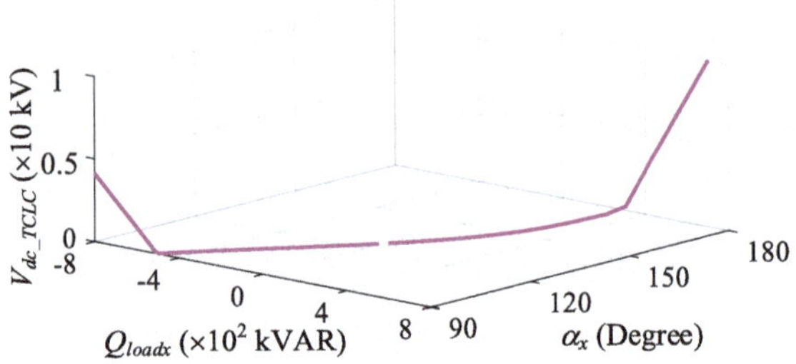

Fig. 2.8 Firing angle and corresponding dc voltage under different loads

2.2.4 *Control System*

This part gives the control strategy of TCLC-DSTATCOM PTT, including the control of TCLC and VSI parts. The overall control block is given in Fig. 2.9.

C. *TCLC Part Control*

The required $X_{TCLCx}(\alpha_x)$ in real-time can be obtained based on q_{Loadx}. The real-time q_{Load} can be calculated by (2.30). The real-time load voltage v_{Lx} and current i_{Lx} can

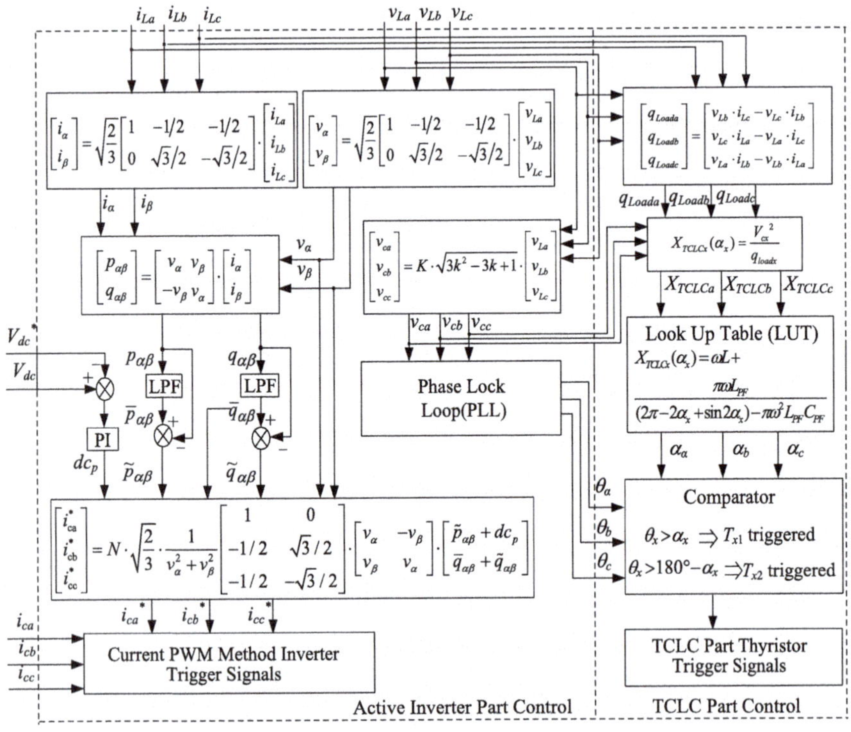

Fig. 2.9 Control block diagram of the TCLC-DSTATCOM PTT

be detected by sensors. After the calculation of $X_{TCLCx}(\alpha_x)$, the corresponding α_x can be found based on the look-up table.

$$\begin{bmatrix} q_{Loada} \\ q_{Loadb} \\ q_{Loadc} \end{bmatrix} = \begin{bmatrix} v_{Lb} \cdot i_{Lc} - v_{Lc} \cdot i_{Lb} \\ v_{Lc} \cdot i_{La} - v_{La} \cdot i_{Lc} \\ v_{La} \cdot i_{Lb} - v_{Lb} \cdot i_{La} \end{bmatrix} \tag{2.30}$$

By comparing the α_x and phase angle θ_x of the v_{cx}, trigger signals can be generated to control the thyristors. The θ_x can be detected through the phase lock loop.

D. *VSI Part Control*

The VSI part limits the compensation current i_{cx} to track its reference $i_{cx}{}^*$. The $i_{cx}{}^*$can be obtained as,

$$\begin{bmatrix} i^*_{\mathrm{ca}} \\ i^*_{\mathrm{cb}} \\ i^*_{\mathrm{cc}} \end{bmatrix} = N \cdot \sqrt{\frac{2}{3}} \cdot \frac{1}{v_\alpha^2 + v_\beta^2} \begin{bmatrix} 1 & 0 \\ -1/2 & \sqrt{3}/2 \\ -1/2 & -\sqrt{3}/2 \end{bmatrix} \cdot \begin{bmatrix} v_\alpha & -v_\beta \\ v_\beta & v_\alpha \end{bmatrix} \cdot \begin{bmatrix} \tilde{p}_{\alpha\beta} + dc_p \\ \overline{q}_{\alpha\beta} + \tilde{q}_{\alpha\beta} \end{bmatrix} \tag{2.31}$$

where the load voltages (v_α and v_β) in the α–β plane can be expressed as,

$$\begin{bmatrix} v_\alpha \\ v_\beta \end{bmatrix} = \sqrt{\frac{2}{3}} \begin{bmatrix} 1 & -1/2 & -1/2 \\ 0 & \sqrt{3}/2 & -\sqrt{3}/2 \end{bmatrix} \cdot \begin{bmatrix} v_{La} \\ v_{Lb} \\ v_{Lc} \end{bmatrix} \tag{2.32}$$

In (2.31), the $p_{\alpha\beta}$ and $q_{\alpha\beta}$ are the instantaneous active and reactive powers of the load, which contain dc components $\overline{p}_{\alpha\beta}, \overline{q}_{\alpha\beta}$ and ac components $\tilde{p}_{\alpha\beta}, \tilde{q}_{\alpha\beta}$. dc_p is used to control the V_{dc} to its reference value. $p_{\alpha\beta}$ and $q_{\alpha\beta}$ can be calculated as,

$$\begin{bmatrix} p_{\alpha\beta} \\ q_{\alpha\beta} \end{bmatrix} = \begin{bmatrix} v_\alpha & v_\beta \\ -v_\beta & v_\alpha \end{bmatrix} \cdot \begin{bmatrix} i_\alpha \\ i_\beta \end{bmatrix} \tag{2.33}$$

where the i_α and i_β can be obtained by

$$\begin{bmatrix} i_\alpha \\ i_\beta \end{bmatrix} = \sqrt{\frac{2}{3}} \begin{bmatrix} 1 & -1/2 & -1/2 \\ 0 & \sqrt{3}/2 & -\sqrt{3}/2 \end{bmatrix} \cdot \begin{bmatrix} i_{La} \\ i_{Lb} \\ i_{Lc} \end{bmatrix} \tag{2.34}$$

Finally, by comparing i_{cx} with $i_{cx}{}^*$ through the pulse-width modulation (PWM) strategy, trigger signals can be generated to control the IGBTs of the VSI.

2.2.5 Simulation Results

In this section, the simulation results are given to verify the proposed TCLC-DSTATCOM PTT can compensate different inductive and capacitive reactive loads at a low dc voltage in comparison with the L-DSTATCOM PTT and LC-DSTATCOM PTT. Three different loads are tested, which are (1) light inductive load; (2) heavy inductive load; and (3) capacitive load. Table 2.3 summarizes the simulation parameters.

The compensation performances for different loads and detailed simulation results are given in Fig. 2.10 and Table 2.4. It can be found that L-DSTATCOM PTT with 8000 V dc voltage can provide acceptable compensating results for the three different loads. After compensation, the source-side PF_{Isx} is 1, and the worst total harmonic distortion (THD_{Isx}) is 13.9%, which satisfies the international standard. For LC-DSTATCOM PTT with 1000 V dc voltage, the compensation performance for the first inductive load is better with $THD_{Isx} = 4.9\%$ and $PF_{Isx} = 1$ since the coupling LC is designed specifically for this load. Once the load changes, the dc voltage 1000 V is not enough for compensation, so $THD_{Isx} = 22.0\%$, $PF_{Isx} = 0.84$ for the second load, and $THD_{Isx} = 21.5\%$, $PF_{Isx} = 0.72$ for the third load. Compared with L-DSTATCOM PTT, LC-DSTATCOM PTT can reduce the DC voltage. However, its compensation range is narrow. For TCLC-DSTATCOM PTT, it all provides good compensation performances with 1000 V dc voltage, when the α_x are selected as 128.5°, 131.0°, 122.9°, and the $X_{TCLCx}(\alpha_x)$ are designed as −276.0 Ω, −138.75 Ω, 201.3 Ω respectively for three different loads. After compensation, the PF_{Isx} becomes 1 from 0.71, 0.64, 0.84, and the THD_{Isx} are all lower than 5%. It can be concluded that the TCLC-DSTATCOM PTT with the firing angle selection method can perform with lower dc voltage and wider compensation range. More specifically, the required dc voltage of TCLC-DSTATCOM PTT is only 12.5% of L-DSTATCOM PTT for various inductive and capacitive loads. On the other hand, since the first load is specifically designed for LC-DSTATCOM PTT, it has the smallest power loss among the three DSTATCOM PTTs due to its low dc voltage. For the three different loads, the power loss of TCLC-DSTATCOM PTT is always lower than that of L-DSTATCOM PTT. All in all, TCLC-DSTATCOM PTT shows superiority in dc voltage, power loss,

Table 2.3 Simulation and experimental system parameters

	Parameters	Physical values
System parameters	V_{sx}, K, k	10 kV/√3, 10 kV/380 V, 0.5
L-STATCOM PTT	L	10 mF
LC-STATCOM PTT	L, C	10 mF, 12 μF
TCLC-STATCOM PTT	L, L_{PF}, C_{PF}	10 mF, 30 mF, 100 μF
1st load	P_{Loadx}, Q_{Loadx}	30 kW, 30 kVAR
2nd load	P_{Loadx}, Q_{Loadx}	50 kW, 60 kVAR
3rd load	P_{Loadx}, Q_{Loadx}	60 kW, −40 kVAR

Fig. 2.10 Simulation performance under three different loads by applying **a** L-DSTATCOM PTT; **b** LC-DSTATCOM PTT; **c** TCLC-DSTATCOM PTT

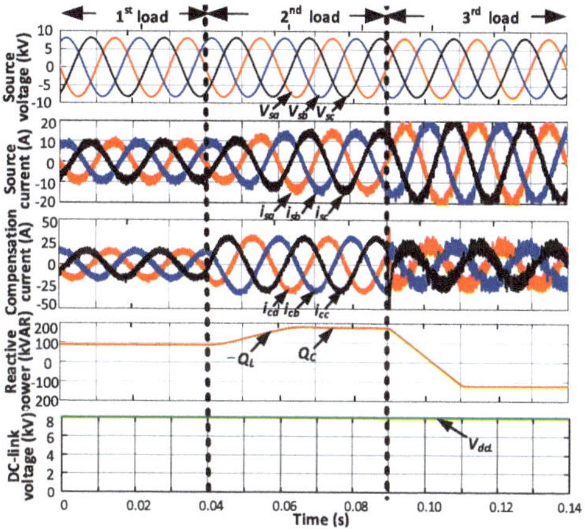

(a)

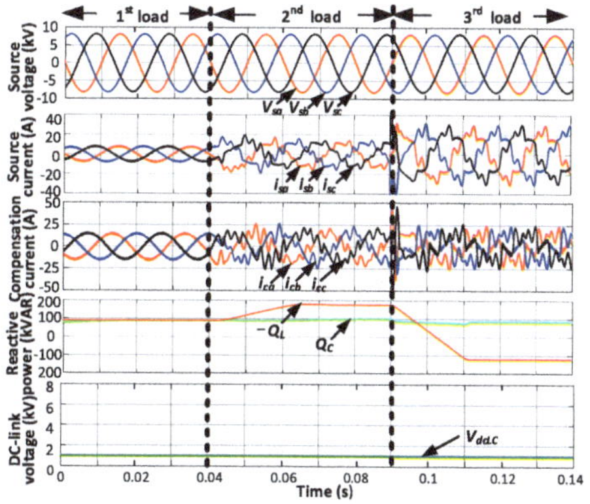

(b)

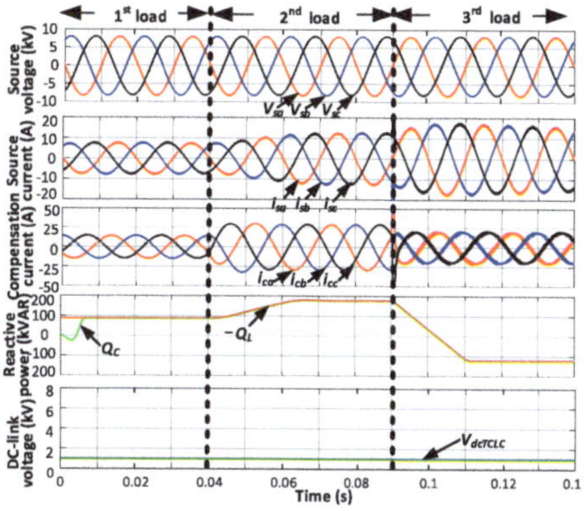

(c)

Table 2.4 Simulation results for three loads

Load type	Situation	I_{Sx} (A)	PF_{ISx}	THD_{ISx} (%)	V_{dc} (V)	P_{loss} (kW)
1st load (inductive)	Before comp	10.46	0.71	0	–	–
	L-DSTATCOM PTT	8.87	1.00	13.9	8000	8.93
	LC-DSTATCOM PTT	7.67	1.00	4.9	1000	1.35
	TCLC-DSTATCOM PTT	7.65	1.00	1.9	1000	6.79
2nd load (inductive)	Before comp	19.42	0.64	0	–	–
	L-DSTATCOM PTT	11.70	1.00	11.1	8000	14.97
	LC-DSTATCOM PTT	12.16	0.84	22.0	1000	–
	TCLC-DSTATCOM PTT	11.54	1.00	3.8, Table 2.4	1000	6.85
3rd load (capacitive)	Before comp	18.09	0.84	0	–	–
	L-DSTATCOM PTT	18.05	1.00	11.3	8000	10.94
	LC-DSTATCOM PTT	24.07	0.72	21.5	1000	–
	TCLC-DSTATCOM PTT	17.03	1.00	2.6	1000	10.05

and compensation performance under the wider load range compared with L- and LC-DSTATCOM PTTs.

2.2.6 *Experiment Results*

A scale-down small rating 220 V–5 kVA TCLC-DSTATCOM PTT experimental prototype was built in the laboratory, as shown in Fig. 2.11. The experimental parameters are shown in Table 2.5. The DSPs TMS320F2812s are used for VSI and TCLC parts control. The Mitsubishi IGBT intelligent power modules PM300DSA60 part and SanRexPK110FG160 thyristors are employed for VSI and TCLC parts, respectively. The dynamic experimental waveforms of v_{Sx}, and i_{sx} before and after TCLC-DSTATCOM PTT for inductive and capacitive loads are shown in Fig. 2.12. The source-side power before and after TCLC-DSTATCOM PTT for inductive and capacitive loads are shown in Fig. 2.13.

Figures 2.12a and 2.13a, b show the TCLC-DSTATCOM PTT can well compensate for inductive load at a 60 V (much lower than $\sqrt{6}$*220 V) dc voltage with $THD_{Isx} = 9.2\%$ and $PF_{Isx} = 0.99$ from 0.5~0.56. Figures 2.12b and 2.13c, d show the TCLC-DSTATCOM PTT can well compensate for capacitive load at a 60 V dc voltage with $THD_{Isx} = 7.8\%$ and $PF_{Isx} = 0.99$~1 from 0.89~0.94. Thus, this article evaluates the lower dc voltage and wider compensation range, including the inductive and capacitive loads of the proposed TCLC-DSTATCOM PTT.

Fig. 2.11 Experimental prototype of a scale-down adaptive impedance coupling converter in the laboratory

Table 2.5 Experimental system parameters

	Parameters	Physical values
System parameters	V_{sx}, K, k, V_{dc}	220 V, 2, 0.5, 60 V
TCLC-DSTATCOM PTT	L, L_{PF}, C_{PF}	5 mH, 30 mH, 160 μF
Sampling time	T_s	100 μs
Inductive load	P_{Loadx}, Q_{Loadx}	730 W, 600 VAR
Capacitive load	P_{Loadx}, Q_{Loadx}	450 W, −190 VAR

2.2.7 Section Summary

In this section, a centralized TCLC-DSTATCOM PTT is proposed in a smart substation for inductive and capacitive reactive power compensation, and the mathematical modelings are formulated to reveal the low dc voltage and wide compensation range. The remarkable improvement of the proposed topology is the TCLC for flexible impedance and reactive power under different loads. It can be concluded from the simulation and experimental results that (1) compared with L-DSTATCOM PTT is favorable for capacitive load and LC-DSTATCOM PTT is favorable for inductive load at a narrow range, TCLC-DSTATCOM PTT can both compensate inductive and capacitive loads at low dc volt-age; (2) the required dc voltage of TCLC-DSTATCOM PTT can be reduced by 87.5% compared with L-DSTATCOM PTT in experiment results; (3) the compensation range of TCLC-DSTATCOM PTT is much larger than that of LC-DSTATCOM PTT.Please confirm if the section headings identified are correct.Sorry, it's our mistake. The original section 2.2.7 should be section 2.2.6, and the original section 2.2.8 should be section 2.2.7.

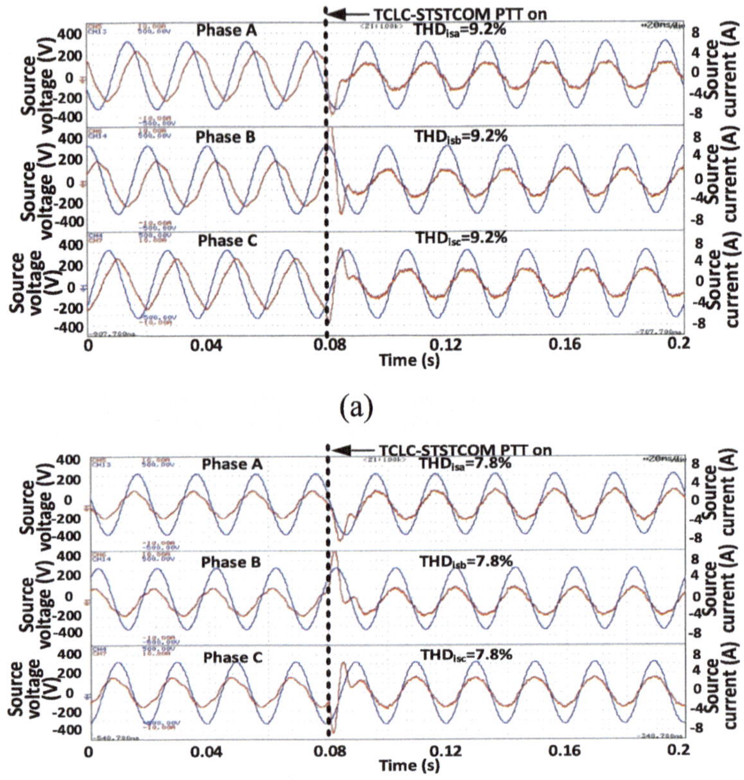

Fig. 2.12 Dynamic experimental waveforms of vSx, and isx before and after TCLC-DSTATCOM PTT for **a** inductive, and **b** capacitive loads

2.3 Delta-Connected Adaptive Impedance-Based Hybrid Active Power Filter and Its Control Method

In order to reduce the voltage rating of the active inverter part, the star-connected adaptive impedance-based hybrid active power filter has been proposed for reactive current compensation. However, the star-connected topology still suffers from the high current rating requirement of the active inverter part. This Section proposes the structure and control method of a delta-connected adaptive impedance-based hybrid active power filter called AI-HAPF. The delta-connected AI-HAPF aims to reduce the compensation current in each phase under an unbalanced load. For example, an inverter with a delta-connected inverter is being proposed to compensate for the negative sequence current caused by the unbalanced load. Compared with the traditional star-connected APFs and SVC-HAPFs, the proposed delta-connected AI-HAPF provides extra flexibility in unbalanced current compensation by controlling

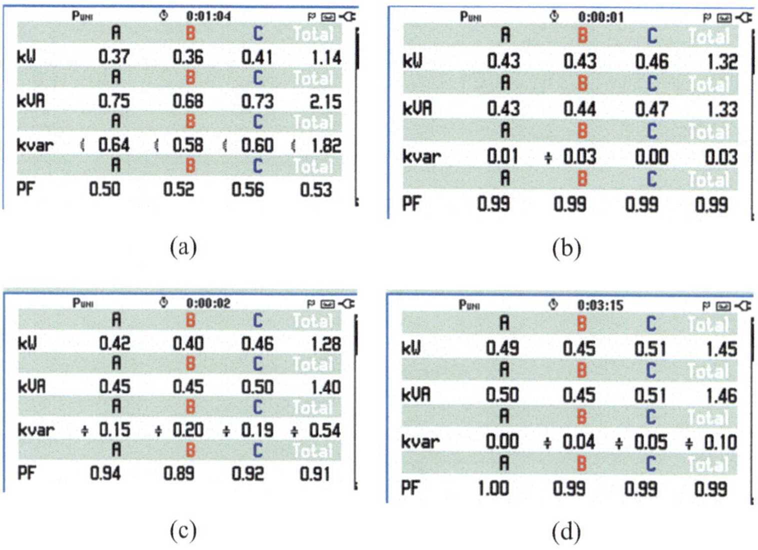

Fig. 2.13 Source-side power **a** without TCLC-DSTATCOM PTT for inductive load, **b** with TCLC-DSTATCOM PTT for inductive load, **c** without TCLC-DSTATCOM PTT for capacitive load, and **d** with TCLC-DSTATCOM PTT for capacitive load

the circulating currents flowing inside the structure, and the current rating can be reduced to $1/\sqrt{3}$.

2.3.1 Circuit Structure of Delta-Connected AI-HAPF

Figure 2.14 shows the circuit configuration of the delta-connected AI-HAPF, in which the subscript "x" stands for phases a, b, and c in the following analysis. v_{sx} and v_x are the source and load voltages; i_{sx}, i_{Lx}, and i_{cx} are the source, load, and compensating currents, respectively. L_s is the transmission line impedance. The delta-connected AI-HAPF consists of an AI part and an active inverter part.

The AI part is a SVC, including a coupling inductor Lc, a parallel capacitor C_{PF}, and a thyristor-controlled reactor with L_{PF}. The AI part provides a wide and continuous inductive and capacitive reactive power compensation range that is controlled by controlling the firing angles of the thyristors. The active inverter part is composed of a voltage source inverter with a DC-link capacitor C_{DC}, and the small rating active inverter part is used to improve the compensation performance of the AI part and compensate the harmonics currents.

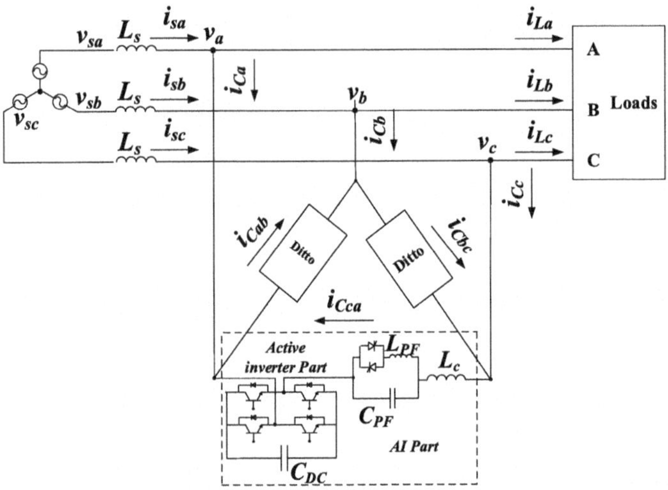

Fig. 2.14 The proposed delta-connected AI-HAPF

2.3.2 Control System

In this section, a control strategy for the delta-connected AI-HAPF is proposed by coordinating the control of the AI part and the active inverter part so that the two parts can complement each other's disadvantages and the overall performance of the delta-connected AI-HAPF can be improved. In the following, the control strategy of the delta-connected AI-HAPFs is discussed. The discussion is separated into two parts: *A*. AI control and *B*. active inverter control. The overall control block diagram of the AI-HAPF is shown in Fig. 2.15.

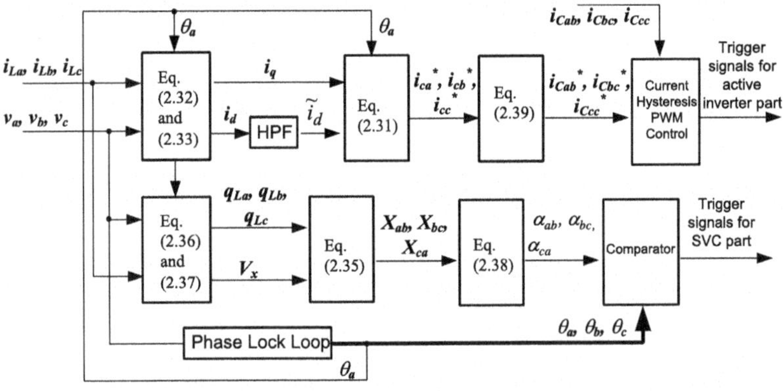

Fig. 2.15 The overall control block of the delta-connected AI-HAPF

A. *AI part control*

The purpose of the AI part is to compensate the fundamental reactive power and unbalanced power by providing different firing angles for each AI part. In order to obtain firing angles of AI part, the AI part impedance X_{xy} (x and y stand for phase a, b or c) in real-time is calculated first. The expression of X_{xy} can be rewritten as:

$$\begin{bmatrix} X_{ab} \\ X_{bc} \\ X_{ca} \end{bmatrix} = \begin{bmatrix} \frac{3 \cdot V_x^2}{Q_{ab}^c} \\ \frac{3 \cdot V_x^2}{Q_{bc}^c} \\ \frac{3 \cdot V_x^2}{Q_{ca}^c} \end{bmatrix} = \begin{bmatrix} \frac{3 \cdot V_x^2}{Q_{Lc} - Q_{La} - Q_{Lb}} \\ \frac{3 \cdot V_x^2}{Q_{La} - Q_{Lb} - Q_{Lc}} \\ \frac{3 \cdot V_x^2}{Q_{Lb} - Q_{Lc} - Q_{La}} \end{bmatrix} \tag{2.35}$$

where V_x is the rms value of the load voltage, and Q_{Lx} is the phase load reactive power (x stands for phase a, b or, c). The V_x and $Q_{Lx} \approx -\overline{q}_{Lx}/2$ can be calculated in real-time as:

$$\begin{bmatrix} q_{La} \\ q_{Lb} \\ q_{Lc} \end{bmatrix} = \begin{bmatrix} v_a \cdot i_{La}^D - v_a^D \cdot i_{La} \\ v_b \cdot i_{Lb}^D - v_b^D \cdot i_{Lb} \\ v_c \cdot i_{Lc}^D - v_c^D \cdot i_{Lc} \end{bmatrix} \tag{2.36}$$

$$V_x = \|v\|/\sqrt{3} = \sqrt{v_a^2 + v_b^2 + v_c^2}/\sqrt{3} \tag{2.37}$$

where ${v_x}^D$ and ${i_{Lx}}^D$ can be obtained by delaying v_x and i_{Lx} by a phase angle of *90°*. q_{Lx} is the phase instantaneous reactive power. Then $Q_{Lx} \approx -\overline{q}_{Lx}/2$ can be obtained by passing the q_{Lx} in (2.36) though low pass filters (LPFs). $\|v\|$ is the norm of the three-phase instantaneous load voltage. After obtaining the AI impedance X_{xy} through (2.35)–(2.37), the firing angle α_{xy} can be determined by solving the equation below.

$$\begin{bmatrix} X_{ab}(\alpha_{ab}) \\ X_{bc}(\alpha_{bc}) \\ X_{ca}(\alpha_{ca}) \end{bmatrix} = \begin{bmatrix} \frac{\pi X_{LPF} X_{CPF}}{X_{CPF}[2\pi - 2\alpha_{ab} + \sin(2\alpha_{ab})] - \pi X_{LPF}} + X_{L_c} \\ \frac{\pi X_{LPF} X_{CPF}}{X_{CPF}[2\pi - 2\alpha_{bc} + \sin(2\alpha_{bc})] - \pi X_{LPF}} + X_{L_c} \\ \frac{\pi X_{LPF} X_{CPF}}{X_{CPF}[2\pi - 2\alpha_{ca} + \sin(2\alpha_{ca})] - \pi X_{LPF}} + X_{L_c} \end{bmatrix} \tag{2.38}$$

where α_{xy} is the firing angle for AI part, X_{Lc}, X_{CPF}, and X_{LPF} are the impedances of the coupling inductor, the AI capacitor, and inductor, respectively. With the calculated X_{xy} from (2.35)–(2.37), the firing angle α_{xy} can be determined by solving (2.38). Since (2.38) has a term of $-2\alpha_{xy} + \sin(2\alpha_{xy})$ it does not have a closed-form solution. Therefore, a look-up table is installed inside the controller. The trigger signals can be generated by comparing the firing angle α_{xy} with the phase angle of the load voltage θ_x to control the AI part. And, θ_x can be obtained by using a phase lock loop.

B. *Active inverter part control*

The active inverter part is used to improve the AI part's characteristics and compensate the current harmonics by limiting the compensating current i_{Cx} to its reference value $i_{Cx}{}^{*}$. For the delta-connected structure, the reference phase compensating current $i_{Cx}{}^{*}$ is required to transfer to the line-to-line reference current $i_{Cxy}{}^{*}$ in order to get the trigger signals for the active inverter part. The transform function can be expressed as:

$$\begin{bmatrix} i^*_{\text{Cab}} \\ i^*_{\text{Cbc}} \\ i^*_{\text{Cca}} \end{bmatrix} = \frac{1}{3} \cdot \begin{bmatrix} 1 & -1 & 0 \\ 1 & 2 & 0 \\ -2 & -1 & 0 \end{bmatrix} \cdot \begin{bmatrix} i^*_{\text{Ca}} \\ i^*_{\text{Cb}} \\ i^*_{\text{Cc}} \end{bmatrix} \tag{2.39}$$

The $i_{Cx}{}^{*}$ can be obtained by (2.31)–(2.34), and the reference current $i_{Cxy}{}^{*}$ can then be obtained. By comparing the $i_{Cxy}{}^{*}$ with the i_{Cxy} (measured by transducers), the trigger signal for the active inverter part can be generated by using the current hysteresis PWM method. Based on the discussion above, the overall control block diagram of the delta-connected AI-HAPF is shown in Fig. 2.15.

2.3.3 Simulation Results

The system configuration for the proposed delta-connected AI-HAPF simulation is shown in Fig. 2.16, and the AIHAPF with low dc-link voltage (V_{DC} = 90 V) is connected to a 380 V grid. The simulation results are obtained using PSCAD/EMTDC.

Figure 2.16 shows the simulated waveforms of the load voltages, dc-link voltage, load, and source currents for balanced harmonic inductive and capacitive loads by using the delta-connected AI-HAPF compensation. At 40 ms, the capacitive load is connected to the system. And, Fig. 2.17 shows the corresponding harmonic spectrums of the load and source currents. Figures 2.18, 2.19, 2.20 show the simulated waveforms, source current harmonic spectrums, and phasor diagrams of the source voltages and currents under unbalanced loads compensation by using the delta-connected AI-HAPF. The simulation results are summarized in Table 2.1. At 40 ms, capacitive loads are connected and cause disturbance on the load current.

The purpose of Fig. 2.16 is to verify the dynamic performance of AI-HAPF due to the loads changing situation. The loads are changing from capacitive to inductive at 40 ms. From Fig. 2.16 and Table 2.6, after the delta-connected AI-HAPF compensation, the source power factors (PFs) are compensated to unity from the original 0.87 and 0.82 for inductive and capacitive loads by using the delta-connected AI-HAPF compensation. From Figs. 2.16, 2.17, and Table 2.6, it can be seen that the source current THD_{isx} are improved from 10.2 and 13.7% to 2.9 and 4.9%. Based

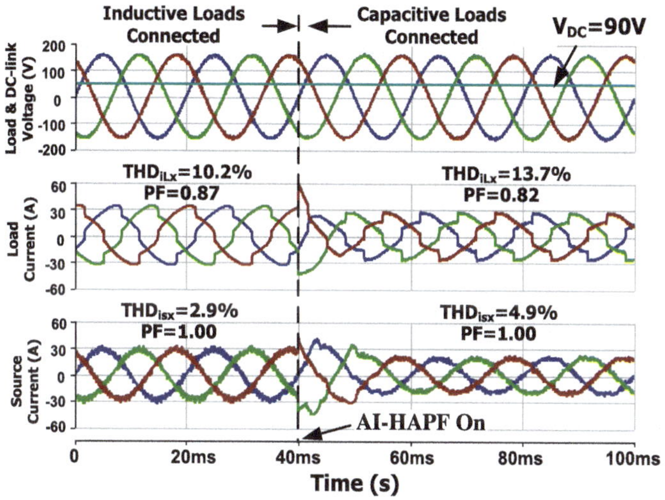

Fig. 2.16 Simulated waveforms of the load voltages, dc-link voltage, load current, and source current for inductive and capacitive loads by using the delta-connected AI-HAPF compensation under balanced harmonic loads

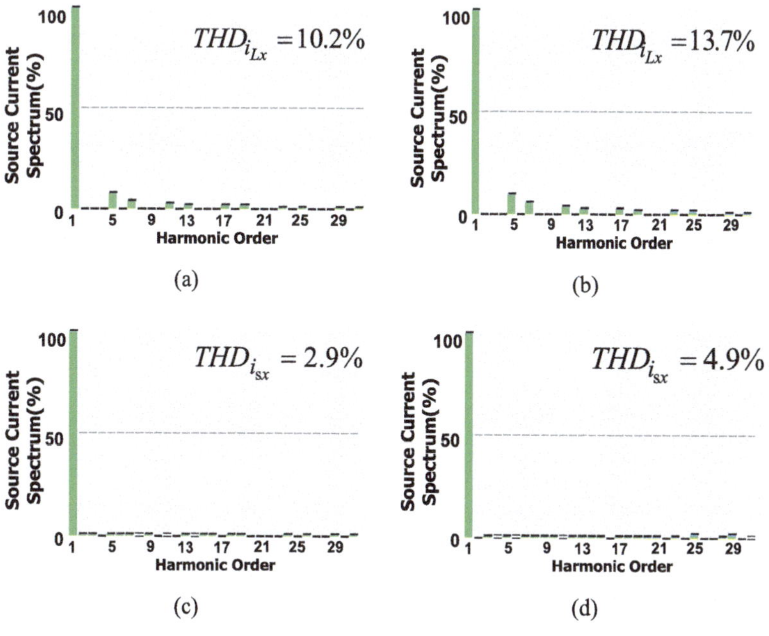

Fig. 2.17 Current harmonic spectrums of source current for **a** inductive loads before compensation, **b** capacitive loads before compensation, **c** inductive loads after compensation, and **d** capacitive loads current after compensation

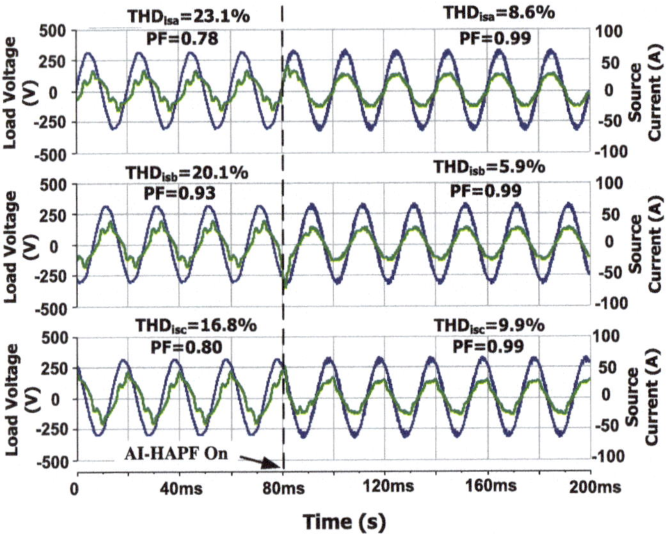

Fig. 2.18 Simulated waveforms of the load voltages and currents with the delta-connected AI-HAPF compensation under unbalanced harmonic loads

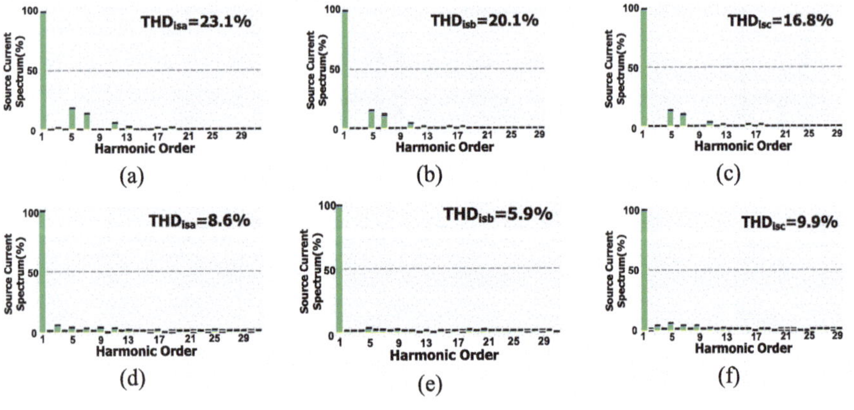

Fig. 2.19 Current harmonic spectrums of source currents: before compensation **a** phase *a*, **b** phase *b*, **c** phase *c*; after the AI-HAPF compensation **d** phase *a*, **e** phase *b*, **f** phase *c*

on Table 2.6, it can be seen that the current passing through the delta-connected AI-HAPF (I_{Cxy}) can be reduced to $1/\sqrt{3}$ times the compensating current (I_{Cx}). The DC-link voltage can be kept at a low level ($V_{DC} = 90$ V for 380 V system) for both inductive loads and capacitive load compensations.

From Fig. 2.18, it can be seen that after the delta-connected AI-HAPF compensation, the source PFs are compensated to 0.99 from the original 0.78, 0.93, and 0.80. In Figs. 2.18 and 2.19, the source current THD_{isx} is improved from 23.1 to 9.9% for

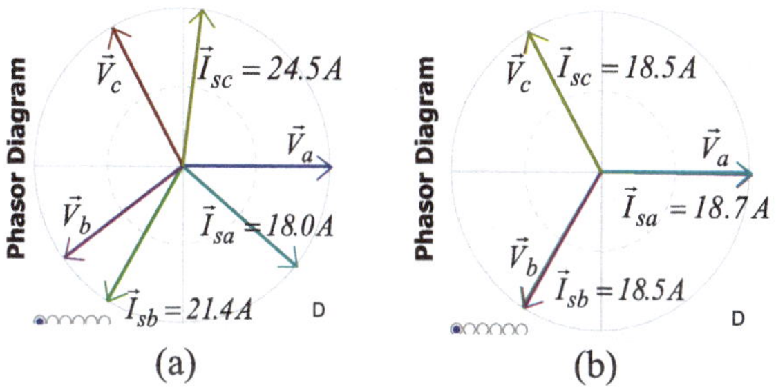

Fig. 2.20 Phasor diagrams of source voltages and currents: **a** before compensation and **b** after the AI-HAPF compensation under unbalanced harmonic loads

Table 2.6 Simulation results for different loads compensation before and after delta-connected AI-HAPF compensation

			Q_{sx} (var)	PF	I_{sx} (A)	I_{Cx}/I_{Cxy} (A)	THD_{isx} (%)	V_{DC} (V)
Balanced loads	Before comp.		2458	0.87	22.7	–	10.2	–
	AI-HAPF		2	1.00	19.9	11.6/6.7	2.9	90
	Before comp.		−2145	0.82	17.5	–	13.7	–
	AI-HAPF		1	1.00	14.6	10.2/5.8	4.9	90
Unbalanced loads	Before comp.	A	2443	0.78	18.2	–	23.1	–
		B	1727	0.93	21.3	–	20.1	
		C	3163	0.80	24.2	–	16.8	
	AI-HAPF	A	3	0.99	18.6	12.7/4.9	8.6	90
		B	5	0.99	18.6	8.6/6.1	5.9	
		C	4	0.99	18.6	14.8/10.4	9.9	

the worst phase. Based on Figs. 2.18 and 2.20, the source current i_{sx} and the load voltage v_x are compensated for being in phase with each other under the unbalanced loads' case. Table 2.6 shows that the current passing through the delta-connected AI-HAPF (I_{Cxy}) is smaller than the compensating current (I_{Cx}). The DC-link voltage can be kept at a low level ($V_{DC} = 90$ V) for unbalanced load compensation.

2.3.4 *Summary*

This Section proposes a new structure and control method of the delta-connected SVC-HAPF for reactive current, harmonic current, and unbalanced current compensation. The proposed delta-connected AI-HAPF has the desirable characteristics of a wide compensation range (from inductive to capacitive) with low DC-link operating voltage. Simulation results validate that the proposed delta-connected AI-HAPF can effectively compensate reactive current, harmonic current, and unbalanced current with low current passing through delta-connected AI-HAPF and low DC-link voltage ($V_{DC} = 90$ V for 380 V system).

2.4 Hybrid Structure of Static Var Compensator and Hybrid Active Power Filter (SVC//HAPF)

This Section proposes the structure, modeling, parameter design, and control method for a new hybrid structure of a static var compensator in parallel with a hybrid active power filter called SVC//HAPF. The SVC part of the SVC//HAPF is used to compensate for the reactive power dynamically. The hybrid active power filter as a HAPF part is used to provide harmonic power and a small amount of reactive power compensation. Due to the large fundamental voltage drop on the coupling LC part, the active inverter part of the SVC//HAPF has a low voltage rating. Meanwhile, the parallel connected SVC acts as a current divider to reduce the active inverter current. Therefore, the proposed SVC//HAPF shows great promise in compensating harmonic current and wide-range reactive power with a low (both) voltage and current rating active inverter part. To show the advantages of the proposed SVC//HAPF, simulation comparisons among the APF, HAPF, SVC, SVC-HAPF, and the proposed SVC//HAPF are provided. Finally, experimental results based on the laboratory-scaled hardware prototype are given to show the validity of the SVC//HAPF.

2.4.1 Circuit Configuration and Modeling of the Three-Phase SVC//HAPF

The circuit configuration of a three-phase, three-wire SVC//HAPF is provided in Fig. 2.21, where v_{sx}, v_x, and v_{invx} are the source voltage, load voltage, and inverter voltage, respectively. The subscript 'x' denotes phase $x = a, b, c$; i_{sx}, i_{Lx}, and i_{cx} are source current, load current, and compensating current, respectively. The SVC part of the SVC//HAPF consists of a coupling inductor L_c, a parallel capacitor C_{PF}, and a thyristor-controlled reactor (TCR) with an inductor L_{PF}. The SVC part can be both star and delta connections. The HAPF part consists of a coupling LC filter (L_p and C_p), a two-level voltage source inverter (VSI), and a DC-link capacitor C_{DC}. The VSI of the HAPF part has both a low voltage rating (due to a large fundamental voltage drop on the coupling LC and a low current rating due to the current divider between i_{cx1} and i_{cx2}.

The single-phase modeling for the SVC//HAPF is proposed in Fig. 2.22. At the fundamental frequency (Fig. 2.22a), the coupling LC filter is used to compensate a fixed amount of reactive power, while the SVC part impedance X_{SVCf} is used to dynamically compensate the reactive power difference between the LC filter and the loads by controlling its firing angle. Ideally, the fundamental inverter voltage V_{invxf}

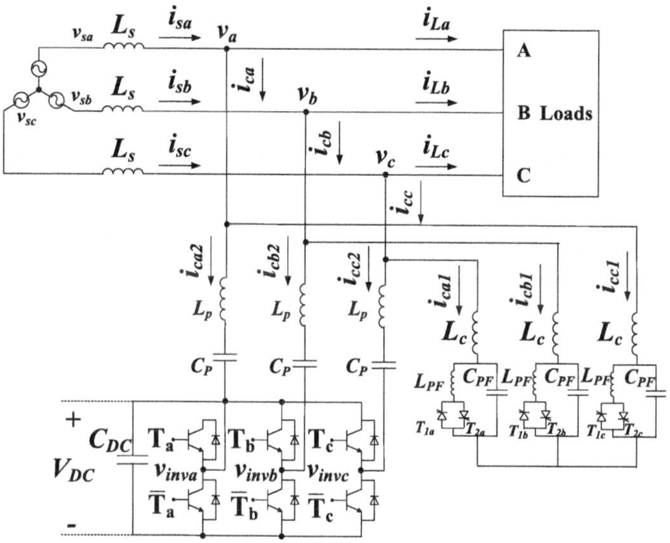

Fig. 2.21 Circuit configurations of the SVC//HAPF

can be designed close to zero, as it is not necessary to compensate the fundamental reactive power. At the harmonic frequency (Fig. 2.22b), the harmonic inverter voltage V_{invxn} is used to compensate the loading harmonic current. In the following, the steady-state compensation analysis and the parameter design of the SVC//HAPF are proposed according to Fig. 2.22.

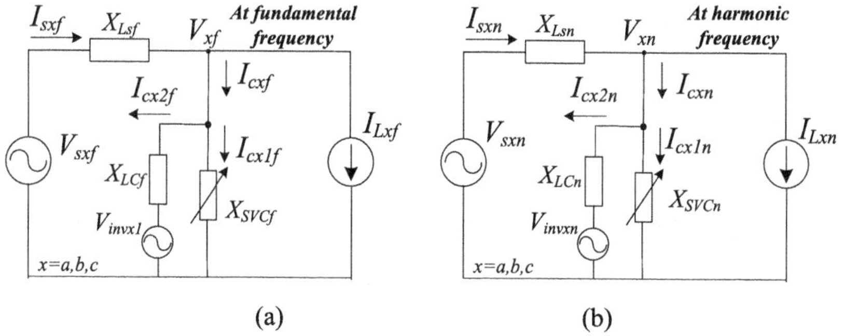

Fig. 2.22 The single-phase SVC//HAPF modeling: **a** at fundamental frequency and **b** at harmonic frequency

2.4.2 Parameter Design of SVC//HAPF

In this Section, a parameter design method is discussed and explained in three parts. In *Part A*, the parameter design of the C_{PF}, L_{PF} and C_P is proposed under the fundamental frequency consideration. In *Part B*, the parameter design of the dc-link voltage V_{DC} is proposed under harmonic frequency consideration. In *Part C*, the design of L_c and L_p is given.

A. *Design of C_{PF}, L_{PF} and C_P under fundamental frequency consideration*

At the fundamental frequency (Fig. 2.22a), the X_{SVCf} and X_{LCf} are used to compensate for the load-required fundamental reactive power. Therefore, the fundamental inverter voltage V_{invxf} can be kept minimized to zero because it is not necessary to compensate for the fundamental reactive power.

In Fig. 2.22a, the SVC//HAPF fundamental impedance (X_{SVCf} and X_{LCf}) can be calculated by applying Ohm's Law as:

$$\frac{X_{LCf}\cdot X_{SVCf}(\alpha_x)}{X_{LCf}+X_{SVCf}(\alpha_x)}=\frac{V_{xf}}{I_{cxf}}=\frac{V_{xf}^2}{Q_{Lx}} \tag{2.40}$$

where V_{xf}, I_{cxf} and Q_{Lx} are the fundamental load voltage, compensating current and the load reactive power, respectively, where x stands for phase *a*, *b* and *c*. X_{LCf} and X_{SVCf} are the fundamental impedance of the SVC part and the coupling *LC* part of the HAPF. And X_{LCf} and X_{SVCf} can be expressed as:

$$X_{LCf}=\omega\cdot L_p-1/\omega\cdot C_p \tag{2.41}$$

$$X_{SVCf}(\alpha_x)=\frac{\pi X_{L_{PF}}X_{C_{PF}}}{X_{C_{PF}}(2\pi-2\alpha_x+\sin 2\alpha_x)-\pi X_{L_{PF}}}+X_{L_c} \tag{2.42}$$

In (2.42), $\omega(=2\pi f)$ is the angular frequency, X_{Lc}, X_{CPF}, X_{LPF} are the fundamental impedances of L_c, C_{PF} and L_{PF}. α_x is the firing angle of the thyristor. The SVC part has two back-to-back connected thyristors (T_{1x}, T_{2x}), and they are triggered alternately in every half cycle. When $\alpha_x=180°$ (thyristors are opened for the whole cycle), the SVC part has the maximum capacitive SVC impedance X_{SVCf} ($\alpha_x=180°$)(<0). On the other hand, when the firing angle $\alpha_x=90°$ (one of the thyristors is closed for the whole cycle), the SVC part has the minimum inductive SVC impedance $X_{SVCf}(\alpha_x=90°)$. Therefore, X_{SVCf} ($\alpha_x=180°$) and X_{SVCf} ($\alpha_x=90°$) can be expressed as:

$$X_{SVCf}\left(\alpha_x=180^{\circ}\right)=X_{L_c}-X_{C_{PF}}=\omega L_c-1/\omega C_{PF} \tag{2.43}$$

$$X_{SVCf}\left(\alpha_x=90^{\circ}\right)=\frac{X_{L_{PF}}X_{C_{PF}}}{X_{C_{PF}}-X_{L_{PF}}}+X_{L_c}=\frac{\omega L_{PF}}{1-\omega^2 L_{PF}C_{PF}}+\omega L_c \tag{2.44}$$

Based on (2.40)–(2.44), the relationship between the load required inductive and capacitive reactive power and the SVC//HAPF components value can be expressed as:

$$\begin{aligned}\frac{V_{xf}^2}{Q_{Lx(\text{MaxCap})}} &= X_{SVCf}\left(\alpha_x = 90^{\circ}\right)//X_{LCf}\\ &\approx \frac{\omega L_{PF}}{1-\omega^2 L_{PF}(C_{PF}+C_P)}\end{aligned} \tag{2.45}$$

$$\frac{V_{xf}^2}{Q_{Lx(\text{MaxInd})}} = X_{SVCf}\left(\alpha_x = 180^{\circ}\right)//X_{LCf} \approx \frac{1}{\omega\left(C_{PF}+C_p\right)} \tag{2.46}$$

where $Q_{Lx(\text{MaxInd})}$ and $Q_{Lx(\text{MaxCap})}$ are the maximum load required for inductive and capacitive reactive power. In (2.45) and (2.46), the impedances of the L_c and L_p are ignored due to their small values (will discussed in part C). By satisfying the above (2.45)–(2.46), the C_{PF}, $L_{PF,}$ and C_p can be designed as:

$$C_{PF}+C_p = \frac{Q_{Lx(\text{MaxCap})}}{\omega V_{xf}^2} \tag{2.47}$$

$$L_{PF} = \frac{V_x^2}{\omega \cdot Q_{Lx(\text{MaxCap})} + \omega^2 V_{xf}^2\left(C_{PF}+C_p\right)} \tag{2.48}$$

The C_{PF}, $L_{PF,}$ and C_p can be designed from (2.47) and (2.48).

B. *Design of* V_{DC} *based on harmonic frequency analysis*

In the harmonic model given in Fig. 2.22b, the harmonic current I_{sxn} circulating in the source is provided as:

$$I_{Sxn} = \frac{X_{SVC//HAPFn}}{X_{Lsn}+X_{SVC//HAPFn}} \cdot I_{Lxn} \tag{2.49}$$

where the subscript "n" represents each harmonic order, $I_{sxn,}$ and I_{Lxn} are the source and load harmonic currents, $X_{SVC//HAPFn}$ is the harmonic equivalent impedance of SVC//HAPF which can be expressed as:

$$\begin{aligned}X_{SVC//HAPFn} &= X_{\text{SVC}n}//(X_{LCn}+V_{invxn}/I_{\text{cx}2n})\\ &= \frac{X_{SVCLn}\cdot[X_{LCn}+(V_{invxn}/I_{cx2n})]}{X_{SVCLn}+X_{LCn}+V_{inv\text{X}n}/I_{cx2n}}\end{aligned} \tag{2.50}$$

In (2.50), the purpose of the inverter voltage V_{invxn} is to adjust the equivalent harmonic impedance of the SVC//HAPF to be zero $\left(X_{SVC//HAPF_n} \approx 0\right)$ so that the load harmonic current will not pollute the source side. According to (2.50), the inverter voltage V_{invxn} at the harmonic frequency can be given as:

$$|V_{invxn}| = |X_{LCn} \cdot I_{cx2n}| = \left| \frac{X_{LCn} \cdot X_{SVCn}}{|X_{LCn}| + |X_{SVCn}|} \cdot I_{Lxn} \right| \tag{2.51}$$

In (2.51), the harmonic impedance of X_{LCn}, and X_{SVCn} at each n can be expressed as:

$$X_{LCn} = n\omega \cdot L_p - 1/n\omega \cdot C_p \tag{2.52}$$

$$|X_{SVCn}(\alpha_x)| = \left| \frac{\pi(n\omega L_{PF})}{(2\pi - 2\alpha_x + \sin 2\alpha_x) - \pi(n\omega)^2 \cdot L_{PF} C_{PF}} + n\omega L_c \right| \tag{2.53}$$

Finally, the dc-link voltage V_{DC} can be calculated as:

$$V_{DC} > \sqrt{6}\sqrt{\sum_{n=2}^{\infty} V_{invxn}^2} = \left(\sqrt{6}\sqrt{\sum_{n=2}^{\infty} V_{invan}^2}, \sqrt{6}\sqrt{\sum_{n=2}^{\infty} V_{invbn}^2}, \sqrt{6}\sqrt{\sum_{n=2}^{\infty} V_{invcn}^2} \right)_{\max} \tag{2.54}$$

where the V_{invxn} can be obtained from (2.51) designed based on worst-case I_{Lxn}, and the information on I_{Lxn} can be obtained by measuring the loading for a period of time before the design of SVC//HAPF. The final V_{DC} is designed to be larger than the maximum value among each phase (2.54).

C. *Design of L_p and L_c for harmonic order tuning*

The purpose of L_p is to provide a low LC impedance at the most dominated harmonic order so that the required dc-link voltage for harmonic compensation can be reduced. The value of L_p can be designed as:

$$L_p = \frac{1}{(2\pi f n_1)^2 C_P} \tag{2.55}$$

where n_1 is the dominated harmonic order (e.g. $n_1 = 5$), f is the system frequency.

The thyristors for each phase of the SVC part can be considered as a pair of bidirectional switches that generate low-order harmonic currents when the thyristors change states. The purpose of the design of L_c is to tune the possible generated harmonic current order to diverge from the characteristic harmonic orders $n_d = 6n \pm 1^{th}$ ($n = 1, 2, 3$) of a three-phase three-wire power system to avoid the SVC harmonic current generation problem. Based on the transient analysis, the harmonic orders generated by the SVC part can be given as:

$$n_2 = \frac{1}{2\pi f \sqrt{L_c C_{PF}}} \tag{2.56}$$

$$n_3 = \frac{1}{2\pi f}\sqrt{\frac{L_c + L_{PF}}{L_c L_{PF} C_{PF}}} \tag{2.57}$$

where the harmonic orders n_2 and n_3 mainly depend on the L_c, L_{PF} and C_{PF} values. And the L_c can be expressed as:

$$L_C = \frac{1}{(2\pi f n_2)^2 C_{PF}} \text{ .\& } L_C = \frac{1}{(2\pi f n_3)^2 C_{PF} - 1/L_{PF}} \tag{2.58}$$

For a three-phase three-wire system, there is no $3n^{\text{th}}$ order harmonic current and the common harmonic orders are $6n \pm 1 (\geq 5)$ with $n = 1, 2, 3 \ldots$. To avoid the harmonic currents injection by the SVC, it is suggested to tune n_2 and n_3 to be away from the $6n \pm 1^{th}$ order and/or close to the $3n^{\text{th}}$ order.

2.4.3 Control System

The control strategy of the SVC//HAPF is separated into three parts: (A) the SVC part and (B) the HAPF part, and (C) the overall control block (as Fig. 2.23).

A. *SVC part control*

The required SVC impedance can be calculated through X_{LCf}, Q_{Lx}, and V_{xf}.

$$X_{SVCf}(\alpha_x) = \frac{V_{xf}^2 \cdot X_{LCf}}{Q_{Lx} \cdot X_{LCf} - V_{xf}^2} \tag{2.59}$$

where X_{LCf} is the fundamental impedance of coupling LC in HAPF, and X_{LCf} is a fixed value. The Q_{Lx} and V_{xf} are the load reactive power and load voltage, which can be calculated in real-time as:

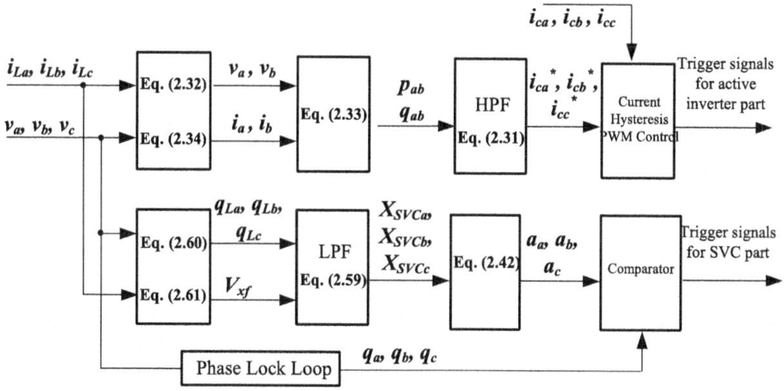

Fig. 2.23 The overall control block of SVC//HAPF

$$\begin{bmatrix} q_{La} \\ q_{Lb} \\ q_{Lc} \end{bmatrix} = \begin{bmatrix} v_a \cdot i_{La}^D - v_a^D \cdot i_{La} \\ v_b \cdot i_{Lb}^D - v_b^D \cdot i_{Lb} \\ v_c \cdot i_{Lc}^D - v_c^D \cdot i_{Lc} \end{bmatrix} \tag{2.60}$$

$$V_{xf} = \|v\|/\sqrt{3} = \sqrt{v_a^2 + v_b^2 + v_c^2}/\sqrt{3} \tag{2.61}$$

where ${v_x}^D$ and ${i_{Lx}}^D$ can be obtained by delaying v_x and i_{Lx} by a phase angle of *90°*. q_{Lx} is the phase instantaneous reactive power. Then $Q_{Lx} \approx -\overline{q}_{Lx}/2$ can be obtained by passing the q_{Lx} in (2.60) though low pass filters (LPF). $\|v\|$ is the norm of three-phase instantaneous load voltage.

After obtaining Q_{Lx} and V_{xf}, the required X_{SVCf} can be obtained from (2.59). With the calculated X_{SVCf}, the firing angle α_x can be calculated through (2.42). However, the (2.42) does not have a closed-form solution. A look-up table has been installed to directly obtain the firing angle α_x with the known X_{SVCf}. By comparing the firing angle α_x with the phase angle of the load voltage θ_x which can be obtained by using a phase lock loop.

B. *HAPF part control*

The instantaneous power theory is chosen to calculate the reference compensating current ${i_{cx}}^*$. The calculated ${i_{cx}}^*$ contains the harmonics, reactive power, and unbalanced power. By controlling the compensating current i_{cx} to track its reference ${i_{cx}}^*$, the active inverter part can compensate the load harmonic currents and improve the reactive power compensation ability and dynamic performance of the SVC part. The ${i_{cx}}^*$ can be calculated through the well-known instantaneous p-q theory as (2.31)–(2.34).

C. *Overall control block*

The overall control block is provided in Fig. 2.23. From Fig. 2.23, for the SVC part control, the (2.60) and (2.61) are used to calculate the load reactive power q_{Lx} and load voltage V_{xf}. Then, the required impedance of SVC part X_{SVCf} can be obtained from (2.59). In (2.59), the obtained X_{SVCf} compensates for the difference between the load reactive power and the reactive power provided by the coupling *LC* of the HAPF part. After that, the corresponding firing angle α_x of the SVC part can be obtained from (2.42). Finally, by comparing the firing angle α_x with the phase angle of the load voltage θ_x, the trigger signals for the thyristor of the SVC can be obtained.

The purpose of the HAPF part is to compensate the reactive power left by the SVC part and the harmonic current by limiting the compensating current i_{cx} to its reference ${i_{cx}}^*$. Through the (2.31)–(2.34), the ${i_{cx}}^*$ can be obtained. In this section, the current hysteresis PWM control is applied. Compared with the voltage control PWM methods, the current hysteresis PWM control has the advantage of fast dynamic response, simplicity of implementation, high robustness, etc. Through the current hysteresis PWM control method, the trigger signals of the HAPF can be generated by comparing the i_{cx} with ${i_{cx}}^*$.

2.4.4 Simulation Results

The simulation case studies are separately into two parts for discussion: *A*. Simulation case studies of PWM control, and *B*. Comparative simulation case studies of APF, HAPF, SVC-HAPF, and the proposed SVC//HAPF.

A. *Simulation case studies of PWM control*

The simulation comparisons of the proportional gain control and hysteresis band control for the SVC//HAPF are provided in Figs. 2.24 and 2.25. Based on Figs. 2.24 and 2.25, it can be seen that the proportional gain control can compensate the worst phase source current THD_{isx} to 3.3%, while the hysteresis band control can compensate THD_{isx} to 1.3% performance under the same average switching frequency $f_{sw(ave)} = 4\text{kHz}$.

Based on the simulation results, it can prove that the hysteresis control method (irregular switching frequency) can have better performance than the proportional gain control (relative regular switching frequency) under the same average switching frequency. Thus, the hysteresis band control is selected in this section.

B. *Comparative simulation case studies of APF, HAPF, SVC-HAPF, and the proposed SVC//HAPF*

The SVC//HAPF can obtain good compensation performances with a low voltage/current active inverter rating. In this section, the simulation case studies are provided to show the above advantages of the proposed SVC//HAPF in comparison with the conventional APF, HAPF, and SVC-HAPF. The parameters of the SVC//HAPF used in simulation and experimental case studies follow the proposed parameter design method. The summary of the SVC//HAPF parameter design is given in Table 2.7. The parameters of different compensators for simulation case studies are given in Table 2.8. In addition, for a fair comparison, the dc-link voltages of HAPF, SVC-HAPF, and the proposed SVC//HAPF are set to be the same ($V_{DC} = 3\text{kV}$) in simulation case studies.

Figure 2.26 shows the waveforms of load voltage, dc-link voltage, load current, source current, and compensating inverter current (i_{cx2}) for harmonic loads current compensation ($Q_{Lx} = 0$), and harmonic and reactive power loads compensation by using the APF, HAPF, SVC-HAPF and the proposed SVC//HAPF. Table 2.9 summarizes the corresponding simulation results.

From Fig. 2.26 and Table 2.9, it can be seen that the APF can compensate the power factor (PF) close to unity from the original 0.97 and 0.60 for the light harmonic loads and heavy inductive loads, respectively. Meanwhile, the source current THD_{isx} has been improved to lower than 6.5% from 28.8% of light loads and 6.5% of heavy loads. The required V_{DC} is 15kV, and the inverter compensating current increases to about 400A for heavy inductive loads compensation.

From Fig. 2.26a and Table 2.9, it can be seen that the APF can compensate the power factor (PF) close to unity from the original 0.97 and 0.60 for the light harmonic loads and heavy inductive loads, respectively. Meanwhile, the source current THD_{isx}

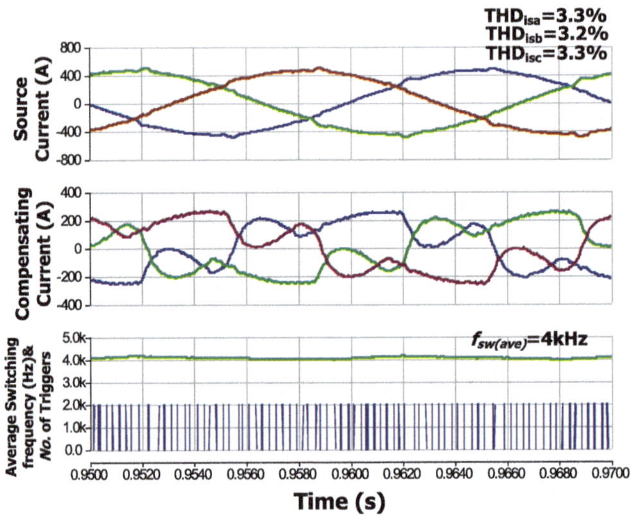

(a)

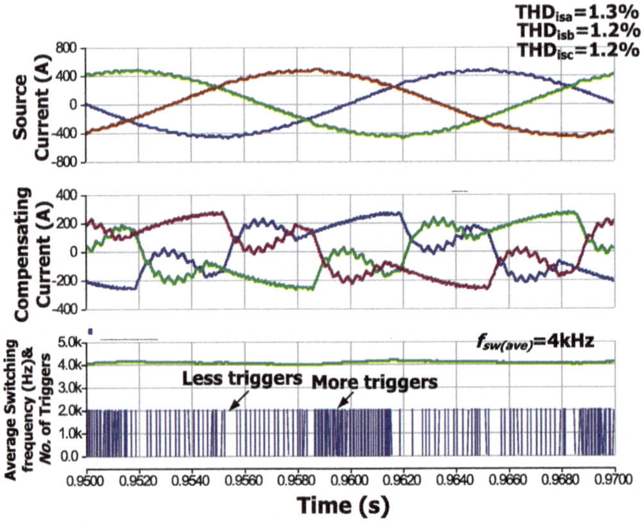

(b)

Fig. 2.24 The waveforms of source current, compensating current, average switching frequency, and trigger signals by using: **a** proportional gain control and **b** hysteresis band control

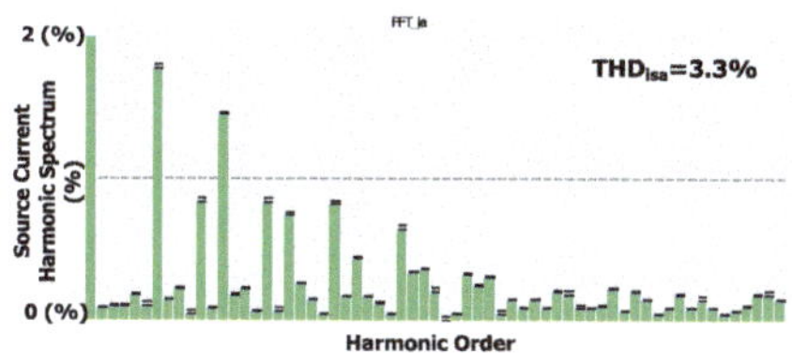

(a)

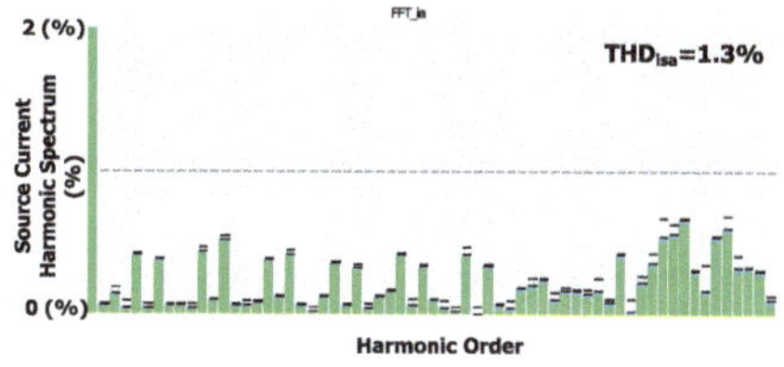

(b)

Fig. 2.25 The source current harmonic spectrum obtained by using **a** proportional gain control, and **b** hysteresis band control

Table 2.7 The parameter design of SVC//HAPF for simulation and experimental case studies

Parameter	Eqs	Load condition	Designed range	Design values
C_{PF}, L_{PF}, C_p	(2.8), (2.9)	$Q_{Lx1} = 30$ kvar $Q_{Lx2} = 2200$ kVar	-5.47 Mvar $< Q_{Lx} <$ 5.83 Mvar	$C_{PF} = 360\ \mu$F, $L_{PF} = 9$ mH, $C_p = 80\ \mu$F
L_p	(2.16)	Compensate dominated current harmonic (5th order)	$n_I = 5$ in (2.16)	$L_p = 5$ mH
L_c	(2.17)–(2.19)	Diverge from harmonic order $6n \pm 1(\geq 5)$	$n_2 \approx 2.4$, $n_3 \approx 2.9$ in (2.19)	$L_c = 5$ mH

Table 2.8 Parameters of APF, HAPF, SVC-HAPF and SVC//HAPF used for 10kV simulation case studies

	Parameters	Physical values
System parameters	V_{LL}, f, L_s	10 kV, 50 Hz, 0.2 mH
APF [3, 4]	L	5 mH
HAPF [6, 7]	L, C	5 mH, 80 μF
SVC-HAPF [10–18]	L_c, L_{PF}, C_{PF}	5 mH, 9 mH, 360 μF
SVC//HAPF	$L_c, L_{PF}, C_{PF}\ L_p, C_P$	5 mH, 9 mH, 360 μF, 5 mH, 80 μF

has been improved to lower than 6.5% from 28.8% of light loads and 6.5% of heavy loads. The required V_{DC} is 15kV, and the inverter compensating current increases to about 400A for heavy inductive loads compensation.

From Fig. 2.26b and Table 2.9, the HAPF only provides a fixed amount of capacitive reactive power, so the PF is getting worse for light harmonic load compensation. The PF has been compensated from 0.97 and 0.60 to 0.70 and 0.74 for the light harmonic loads and heavy inductive loads, respectively. The source current THD_{isx} has been compensated to 30.1 and 21.1%, which are not satisfactory. Even though the HAPF requires low V_{DC} (=3kV) and low fixed inverter compensating current (about 150A), the HAPF cannot obtain good compensation performance.

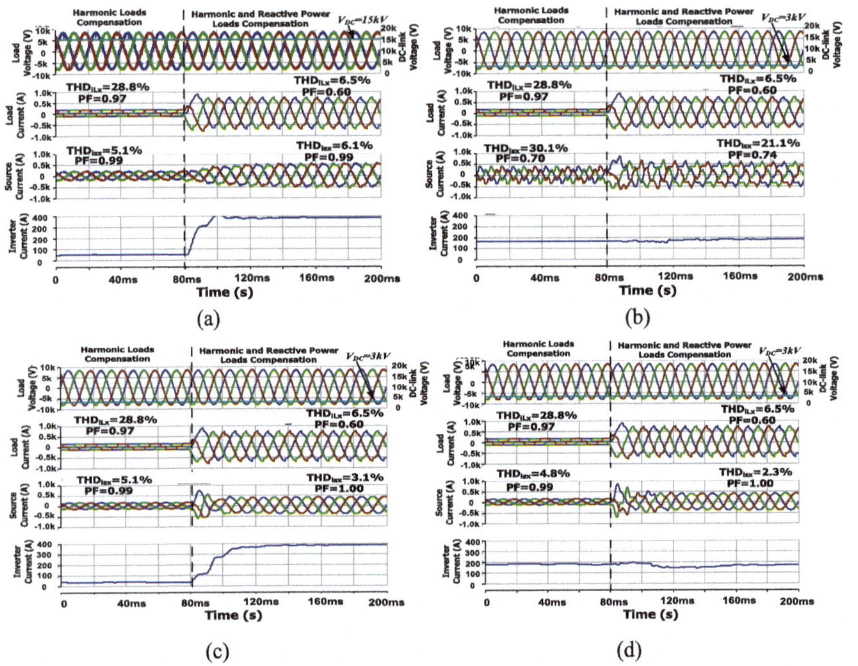

Fig. 2.26 Waveforms of load voltage, dc-link voltage, load current, source current, and compensating inverter current for harmonic loads current compensation (QLx = 0) and harmonic and reactive power loads compensation by using **a** APF, **b** HAPF, **c** SVC-HAPF and **d** the proposed SVC//HAPF

Table 2.9 Simulation results for different loads compensation before and after APF, HAPF, SVC-HAPF, and the proposed SVC//HAPF compensation

loads	Comp	Q_{sx} (var)	PF	i_{sx} (A)	THD_{vx} (%)	THD_{isx} (%)	V_{DC} (V)
Light harmonic loads	Before comp	30 k	0.97	110	1.0	28.8	–
	APF	7 k	0.99	120	0.6	5.1	15 k
	HAPF	−820 k	0.70	210	0.7	30.1	3 k
	SVC-HAPF	5 k	0.99	113	0.5	5.1	3 k
	SVC//HAPF	4 k	0.99	112	0.6	4.8	3 k
Heavy inductive loads	Before comp	2200 k	0.60	473	1.0	6.5	–
	APF	9 k	1.00	321	0.5	6.1	15 k
	HAPF	1290 k	0.74	342	0.7	21.1	3 k
	SVC-HAPF	6 k	1.00	274	0.6	3.1	3 k
	SVC//HAPF	4 k	1.00	289	0.6	2.3	3 k

Based on Fig. 2.26c and Table 2.9, the SVC-HAPF can improve the *PF* and THD_{isx} to about unity and lower than 5.5% for both light harmonic and heavy inductive loads. The required V_{DC} is 3 kV, and the inverter compensating current is increasing to about 400 A for heavy loads compensation.

From Fig. 2.26d and Table 2.9, the *PF* has been improved to about unity from 0.97 and 0.60 after SVC//HAPF compensation. The source current THD_{isx} has been improved to 4.8% and 2.3% from the original 28.8% for light harmonic loads and 6.5% for heavy inductive loads compensation, respectively. In addition, the SVC//HAPF requires low DC-link voltage (=3 kV) and a low fixed inverter compensating current (about 150 A).

Based on the above discussions and simulation results, the APF requires the highest V_{DC} to compensate different loads. For APF and SVC-HAPF, the required inverter current increases as the required compensating reactive power increases. On the other hand, even though HAPF requires the low V_{DC} and low fixed inverter current as the SVC//HAPF, it obtains the poorest compensation performance. Compared with the APF, HAPF, and SVC-HAPF, the proposed SVC//HAPF requires both low V_{DC} and low inverter current to obtain satisfactory performances. As the cost of the SVC part is much lower than that of the active inverter part, the reduction of the inverter rating (voltage and current) can thus lead to a decrease in the total cost of the SVC//HAPF system. Therefore, the SVC//HAPF is a cost-effective solution for medium voltage level heavy loads compensation.

2.4.5 Experimental Results

In this section, a 110 V–5 kVA experimental prototype of the SVC//HAPF with the proposed parameter and control design methods is built in the laboratory, as shown in Fig. 2.11. The digital control system of the SVC//HAPF is a digital signal processor (DSP) TMS320F2812, and the sampling frequency of the control system is 25 kHz. For every 1/25 kHz(s) period, the Timer will provide a signal to process analog to digital (A/D) conversion and the corresponding interrupt. After processing the control algorithm, the output PWM signal will be generated, and the maximum PWM switching frequency will be 12.5 kHz. For both simulation and experiment, the minimum switching frequency is about 0.7 kHz. The switching devices for the active inverter part are Mitsubishi IGBT intelligent power modules PM300DSA060. The switching devices for the SVC part are thyristors SanRex PK110FG160. Moreover, the experimental parameters of the SVC//HAPF are the same as shown in Table 2.8.

Figure 2.27 shows the waveforms of source voltage and current by using the SVC//HAPF for light nonlinear loading compensation and heavy inductive loading compensation. Figure 2.28 gives the harmonic spectrums of source current before and after compensation for different loads. Table 2.9 summarizes the corresponding experiment results.

From Fig. 2.27a, Fig. 2.28a and b, and Table 2.9, after SVC//HAPF compensation, the worst phase *PF* has been compensated from 0.96 to 0.99 for light harmonic loads

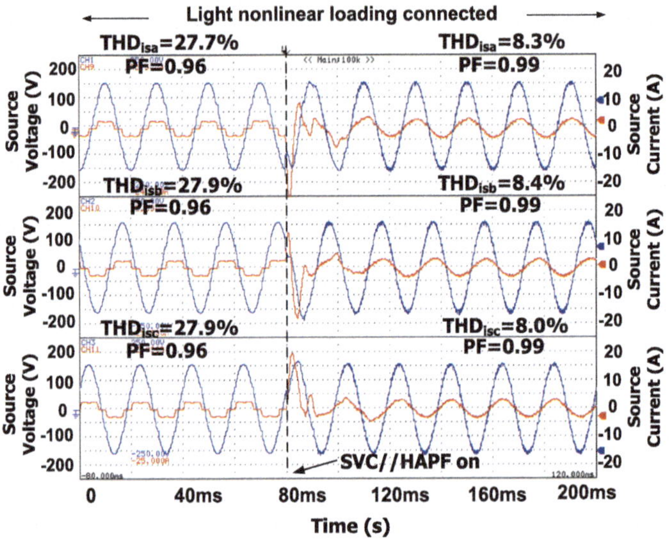

(a)

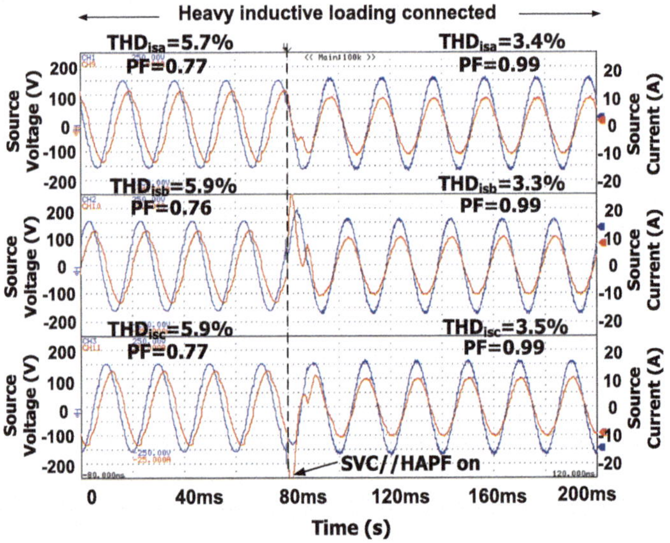

(b)

Fig. 2.27 Waveforms of v_{sx} and i_{sx} by using SVC//HAPF for: **a** light nonlinear loading compensation, **b** heavy inductive loading compensation

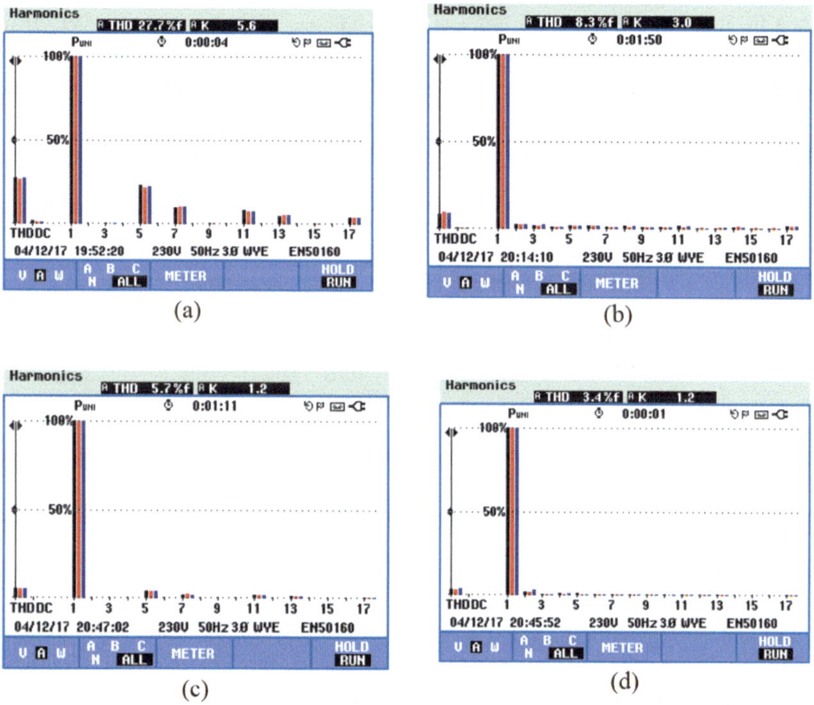

Fig. 2.28 The harmonic spectrums of source current for light harmonic loading compensation: **a** before SVC//HAPF compensation, **b** after SVC//HAPF compensation; for heavy inductive loading compensation: **c** before SVC//HAPF compensation, and **d** after SVC//HAPF compensation

compensation. The worst phase THD_{isx} has been compensated to 8.4% from the original 27.9% for light harmonic loads compensation. From Fig. 2.27b, Fig. 2.28c and d, and Table 2.9, the proposed SVC//HAPF can compensate the heavy loads situation. The fundamental source reactive power is reduced from 650 var to almost 10 var. The PF has been improved from 0.77 to 0.99. Also, the worst phase THD_{isx} has been compensated to 3.5% from the original 5.9%. Moreover, the source voltage and current are in phase with each other after SVC//HAPF compensation (Table 2.10).

2.4.6 Summary

In this Section, a new hybrid structure of SVC in parallel with HAPF (SVC//HAPF) in a three-phase power system has been proposed and discussed as a cost-effective compensator for medium voltage heavy loads compensation. The SVC part is used to compensate the reactive power dynamically, while the HAPF is used to provide harmonic and low fixed amounts of reactive power compensation. Moreover, the structure, modeling, operation principle, parameter design, and control method of

Table 2.10 Experimental results for different loads compensation before and after SVC//HAPF compensation

			Q_{sx} (var)	PF	i_{sx} (A)	THD_{isx} (%)	V_{DC} (V)
Light inductive loads	Before comp.	A	20	0.96	2.1	**27.7**	–
		B	10	0.96	2.0	**27.9**	
		C	20	0.96	2.1	**27.9**	
	SVC//HAPF	A	0	0.99	2.0	8.3	50
		B	10	0.99	2.0	8.4	
		C	10	0.99	2.0	8.0	
Heavy inductive loads	Before comp.	A	640	**0.77**	8.7	5.7	–
		B	650	**0.76**	8.8	5.9	
		C	650	**0.77**	8.7	5.9	
	SVC//HAPF	A	10	0.99	7.0	3.4	50
		B	10	0.99	6.9	3.3	
		C	10	0.99	7.0	3.5	

**Notes* the bold areas mean undesirable results

the SVC//HAPF are proposed and discussed. Finally, the representative simulation and experimental results are given to show that the SVC//HAPF has great promise in a wide reactive power compensation range with both low voltage and current inverter rating characteristics.

2.5 Summary

This chapter summarized the applications of adaptive impedance converters in distribution systems, including the related adaptive hybrid static synchronous compensator and adaptive hybrid active filter. This chapter proposed a centralized TCLC-STATCOM PTT in the smart substation for inductive and capacitive reactive power compensation, a delta-connected AI-HAPF for undesired current compensation, and a hybrid structure of SVC//HAPF as a high efficient compensator for medium voltage heavy loads compensation. The adaptive passive coupling impedance is used to effectively provide dynamic reactive power and reduce the DC side voltage of the converter, thereby significantly reducing the number of switches/voltage resistance and breaking through the bottleneck of low efficiency and high cost of large-capacity converters, with effectively compensating reactive current, harmonic current and unbalanced current.

References

1. C.W. Gellings, M. Samotyj, B. Howe, The future's smart delivery system [electric power supply]. IEEE Power Energy Mag. **2**(5), 40–48 (2004)
2. M. Moghbel, M.A.S. Masoum, A. Fereidouni, S. Deilami, Optimal sizing, siting and operation of custom power devices with STATCOM and APLC functions for real-time reactive power and network voltage quality control of smart grid. IEEE Trans. Smart Grid **9**(6), 5564–5575 (2018)
3. Q.-C. Zhong, Harmonic droop controller to reduce the voltage harmonics of inverters. IEEE Trans. Ind. Electron. **60**(3), 936–945 (2013)
4. J. Dixon, L. Moran, J. Rodriguez, R. Domke, Reactive power compensation technologies: state-of-the-art review. Proc. IEEE **93**(12), 2144–2164 (2005)
5. A. Hamadi, S. Rahmani, K. Al-Haddad, A hybrid passive filter configuration for VAR control and harmonic compensation. IEEE Trans. Ind. Electron. **57**(7), 2419–2434 (2010)
6. A. Kazemi, B. Badrzadeh, Modeling and simulation of SVC and TCSC to study their limits on maximum loadability point. Int. J. Electr. Power Energy Syst. **26**(8), 619–626 (2004)
7. S. Kim, P.N. Enjeti, A new hybrid active power filter (APF) topology. IEEE Trans. Power Electron. **17**(1), 48–54 (2002)
8. P. Rao, M.L. Crow, Z. Yang, STATCOM control for power system voltage control applications. IEEE Trans. Power Del. **15**(4), 1311–1317 (2000)
9. L. Wang, Y. Pang, C.-S. Lam, J.-Y. Deng, M.-C. Wong, Design and analysis of single-phase adaptive passive part coupling hybrid active power filter (HAPF), in *IECON 2018—44th Annual Conference of the IEEE Industrial Electronics Society, Washington, DC, USA* (2018), pp. 3615–3620
10. W.-H. Choi, C.-S. Lam, M.-C. Wong, Y.-D. Han, Analysis of DC-link voltage controls in three-phase four-wire hybrid active power filters. IEEE Trans. Power Electron. **28**(5), 2180–2191 (2013)
11. L. Wang, K.-W. Lao, C.-S. Lam, M.-C. Wong, Delta-connected static var compensator (SVC) based hybrid active power filter (SVC-HAPF) and its control method, in *IECON 2017—43rd Annual Conference of the IEEE Industrial Electronics Society, Beijing, China* (2017), pp. 6384–6388. https://doi.org/10.1109/IECON.2017.8217112
12. L. Wang, C.-S. Lam, M.-C. Wong, Modeling and parameter design of thyristor-controlled LC-coupled hybrid active power filter (TCLC-HAPF) for unbalanced compensation. IEEE Trans. Ind. Electron. **64**(3), 1827–1840 (2017)
13. L. Wang, C.-S. Lam, M.-C. Wong, Selective compensation of distortion, unbalanced and reactive power of a thyristor-controlled LC-coupling hybrid active power filter (TCLC-HAPF). IEEE Trans. Power Electron. **32**(12), 9065–9077 (2017)
14. O.P. Mahela, A.G. Shaik, A review of distribution static compensator. Renew. Sustain. Energy Rev. **50**, 531–546 (2015)
15. S. Hu, Z. Zhang, Y. Chen et al., A new integrated hybrid power quality control system for electrical railway. IEEE Trans. Ind. Electron. **62**(10), 6222–6232 (2015)
16. C. Wang, X. Yin, Z. Zhang, M. Wen, A novel compensation technology of static synchronous compensator integrated with distribution transformer. IEEE Trans. Power Del. **28**(2), 1032–1039 (2013)
17. Y. Li, L. Luo, C. Rehtanz, S. Rüberg, D. Yang, An industrial DC power supply system based on an inductive filtering method. IEEE Trans. Ind. Electron. **59**(2), 714–722 (2012)
18. Y. Li, F. Liu, T.K. Saha, O. Krause, Y. Cao, Hybrid inductive and active filtering method for damping harmonic resonance in distribution network with non-linear loads. IET Power Electron. **8**(9), 1616–1624 (2015)
19. Y. Li, Q. Liu, S. Hu et al., A virtual impedance comprehensive control strategy for the controllably inductive power filtering system. IEEE Trans. Power Electron. **32**(2), 920–926 (2017)
20. S. Hu, Y. Li et al., A Y-D multifunction balance transformer-based power quality control system for single-phase power supply system. IEEE Trans. Ind. Appl. **52**(2), 1270–1279 (2016)

21. B. Xie, Z. Zhang, S. Hu, Y. Li, L. Luo, S. Sun, YN/VD connected balance transformer-based electrical railway negative sequence current compensation system with passive control scheme. IET Power Electron. **9**(10), 2044–2051 (2016)
22. E. Lei, X. Yin, Z. Zhang, Y. Chen, An improved transformer winding tap injection DSTATCOM topology for medium-voltage reactive power compensation. IEEE Trans. Power Electron. **33**(3), 2113–2126 (2018)
23. Z. Bai, Y. Pang, Z. Xiang et al., A capacitive-coupling winding tap injection DSTATCOM integrated with distribution transformer for balance and unbalance operations. IEEE Trans. Ind. Electron. **70**(2), 1081–1093 (2023)
24. L. Wang, K.-W. Lao, C.-S. Lam, M.-C. Wong, Delta-connected static var compensator (svc) based hybrid active power filter (SVC-HAPF) and its control method, in *The 43rd Annual Conference of the IEEE Industrial Electronics Society (IECON 2017), Beijing, China* (2017)
25. Y. Yu, G. Konstantinou, C.D. Townsend, R.P. Aguilera, V.G. Agelidis, Delta-connected cascaded H-bridge multilevel converters for large-scale photovoltaic grid integration. IEEE Trans. Ind. Electron. **64**(11), 8877–8886 (2017)
26. P.H. Wu, H.C. Chen, Y.T. Chang, P.T. Cheng, Delta-connected cascaded H-bridge converter application in unbalanced load compensation. IEEE Trans. Ind. Appl. **53**(2), 1254–1262 (2017)
27. B. Wang, J. Shang, N. Dai, H. Chen, Harmonic reference currents balancing method for delta-connected static synchronous compensator. Electron. Lett. **51**(25), 2134–2136 (2015)
28. J. Dixon, Y. del Valle M. Orchard, M. Ortuzar, L. Moran, C. Maffrand, A full compensating system for general loads, based on a combination of thyristor binary compensator, and a PWM-IGBT active power filter. IEEE Trans. Ind. Electron. **50**(5), 982–989 (2003)
29. A. Luo, S. Peng, C. Wu, J. Wu, Z. Shuai, Power electronic hybrid system for load balancing compensation and frequency-selective harmonic suppression. IEEE Trans. Ind. Electron. **59**(2), 723–732 (2012)
30. L. Wang, C.-S. Lam, M.-C. Wong, Hybrid structure of static var compensator and hybrid active power filter (SVC//HAPF) for medium-voltage heavy loads compensation. IEEE Trans. Ind. Electron. **65**(6), 4432–4442 (2018)

Chapter 3
Adaptive Impedance Converters for Renewable Energy Integration and Power Quality

3.1 Introduction of Grid Connected Inverters

The grid is the interconnection of different power systems, from high-voltage transmission systems to low-voltage distribution systems, from centralized generation to distributed generation (DG), from utility control centers to end-user home-area networks, and from traditional energy resources to renewable generation energy resources [1]. However, the development of a smart grid brings many new challenges to power quality [2]. Especially in weak grid areas, the small power quality problem can lead to severe consequences. Therefore, the multi-functional grid-connected inverters have been paid more and more attention for their benefits on power quality improvement and renewable power integration.

Traditionally, the task of inductive grid-connected inverters (IGCIs) is used to transfer active power from renewable energy resources to the power grid. However, the integration of uncontrollable resources, such as wind and solar energy, can also cause power quality problems to the grid. Therefore, apart from the renewable integration function, many researchers suggest embedding the power quality enhancement function in the IGCIs. As the active power flow is orthogonal to the reactive power flow, it is economical to use the same inverter to transfer active and reactive power simultaneously. To attach reactive power and active power transfer, different control strategies have been proposed for IGCIs [3–8]. However, the harmonic and unbalanced power compensation abilities have not been included in the above control methods [3–8]. To include reactive, harmonic, and unbalanced compensation abilities, the more general control strategies have been proposed in [9–12], etc. The coupling impedance of IGCIs is a small inductive impedance, which is used to filter the ripple produced by power switches. However, the IGCIs require high dc-link voltage. The capacitive-coupling grid-connected inverters (CGCIs) [13, 14] combine active and passive components in order to reduce the power rate of power electronics devices, allowing high-power applications with reduced costs. However, the reactive

L. Wang et al., *Design and Applications of Adaptive Impedance Converters*,
https://doi.org/10.1007/978-981-97-5319-2_3

power compensation range is very narrow. To enlarge the reactive power compensation range, different hybrid inverter topologies have been proposed, including hybrid grid-connected inverter (HGCI) [15] and static var compensator (SVC) in parallel with a CGCI (SVC//CGCI) [16], etc. This Chapter will introduce the SVC//CGCI and HGCI in terms of circuit structure, control system, parameter design, and characteristic analysis.

The HGCI has a low voltage inverter rating and a wide range of reactive power compensation and active injection. The HGCI can simultaneously overcome the high DC-link voltage problem of the IGCIs and narrow the energy conversion range of the CGCIs for both active power transfer and power quality conditioning. Therefore, the HGCI can require low voltage rating switching devices and operate with fewer PV panels connected in series. With the advantage of low DC-link voltage, the power loss can be reduced compared with the conventional IGCI. In contrast, the SVC//CGCI has low voltage and current inverter ratings for a wide range of reactive power compensation but has poor active power injection ability.

A comparison table of IGCIs, CGCIs, HGCIs, and SVC//CGCIs is provided in Table 3.1. The IGCI has a wide range of reactive power, harmonic current and/or active power injection with low required dc-link voltage. The CGCI has a low dc-link voltage for a narrow range of reactive power compensation. However, to enlarge the range of active power injection, the dc-link voltage of CGCI and CGCI part of SVC//HAPF are significantly increasing. In contrast, the HGCI has a wide range of both reactive power compensation and active power injection.

In this Chapter, Sect. 3.1 presents the development of different grid-connected inverters. Then, the application of an adaptive hybrid grid-connected inverter for renewable energy generation with power quality conditioning is given in Sect. 3.2. After that, the multifunctional hybrid structure of SVC and capacitive grid-connected inverter for active power injection and nonactive power compensation is provided in Sect. 3.3. Finally, the Chapter summary is given in Sect. 3.4.

Table 3.1 Comparisons among different power quality Converters in terms of reactive power compensation range, inverter voltage rating, current rating, and power loss

	Reactive power Comp. range	Inverter current rating	Inverter voltage rating	Power loss
IGCI	Wide	High	High	High
CGCI	Narrow	Medium	Low	Low
HGCI	Wide	Medium	Low	Low
SVC//CGCI	Wide	Small	Low	Low

Notes The shaded area means the undesirable characteristic

3.2 Analysis, Control, and Design of Hybrid Grid-Connected Inverter for Renewable Energy Generation with Power Quality Conditioning

This chapter proposes a new type of DC/AC inverter named hybrid-coupling grid-connected inverter as HGCI for Photovoltaic (PV) active power generation with power quality conditioning, which consists of a full-bridge three-phase DC/AC inverter coupling to the power grid in series with a thyristor controlled LC (TCLC) filter. Compared with the conventional IGCI and CGCI, the proposed HGCI has distinct characteristics of a wide operational range and low DC-link operating voltage. Based on these prominent characteristics, the system cost and operational cost can be reduced. Moreover, it can simultaneously transfer the active power and compensate reactive, unbalanced, and harmonic power. The analysis of the structure, parameter design, and control method of the HGCI are proposed and presented. Finally, simulation and experimental results are provided to verify the effectiveness and performance of the proposed HGCI in comparison with the IGCI and CGCI.

3.2.1 *Circuit Configuration of the Proposed HGCI*

Figure 3.1 shows the circuit configuration of the proposed HGCI, in which the subscript "x" stands for phases a, b, and c in the following analysis. v_{sx} is the power grid source voltage; i_{sx}, i_{Lx}, and i_{cx} are the source, load, and injecting currents, respectively. L_s is the transmission line impedance. The HGCI consists of a TCLC and an active inverter part.

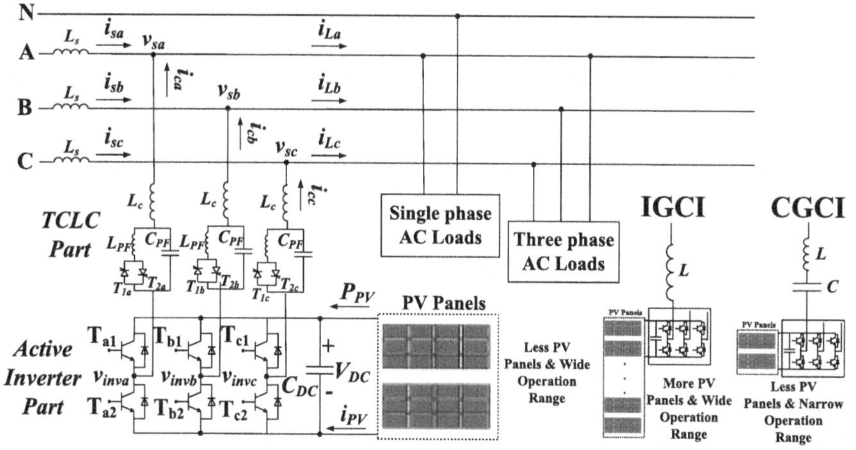

Fig. 3.1 The structure of the IGCI, CGCI, and the proposed HGCI for PV active power injection with power quality conditioning

The TCLC part is composed of a coupling inductor L_c, a parallel capacitor C_{PF}, and a thyristor-controlled reactor with L_{PF}. The TCLC part provides a wide and continuous inductive and capacitive reactive power compensation range that is controlled by controlling the firing angles α_x of the thyristors. The active inverter part is composed of a DC/AC voltage source inverter (VSI) with a DC-link capacitor C_{DC} and a DC-link voltage V_{DC}, and a small rating active inverter part is used to bear the PV active power injection, improve the performance of the TCLC part and also compensate the current harmonics. Thus, it is possible to connect fewer PV panels in series in comparison with the IGCI case for active power transfer and power quality conditioning. The HGCI can require low voltage rating switching devices of the active inverter part, and the cost and switching loss of the thyristor are much lower than those switching devices. Therefore, the initial cost and operational loss of the HGCI can be less than the IGCI. In addition, the coupling components of the conventional IGCI and CGCI are also presented in Fig. 3.1 for comparison.

3.2.2 Characteristic Study of the IGCI, CGCI, and the Proposed HGCI

As mentioned in the previous section, the grid-connected inverters are used to inject active power and reactive power to the grid at the fundamental frequency when they are used to integrate the PV energy generation system with power quality conditioning. The current injected into the grid by the VSI can be expressed as:

$$\mathrm{I}_{cxf} = \mathrm{I}_{cxfp} + j \cdot \mathrm{I}_{cxfq} \tag{3.1}$$

where the subscript "f" represents the fundamental frequency component, I_{cxfp} is the fundamental active current that injects or absorbs from the grid. I_{cxfq} is a fundamental reactive current that injects or absorbs into the grid. The fundamental output voltage of the VSI of the HGCI is given in (3.2).

$$V_{invxf} = \vec{V}_{sxf} + \mathrm{X}_{TCLC} \cdot \vec{I}_{cxf} = \vec{V}_{sxf} + \mathrm{X}_{TCLC}(\alpha_{\mathrm{x}}) \cdot (\vec{I}_{cxpf} + j\vec{I}_{cxqf}) \tag{3.2}$$

In (3.2), the TCLC part impedance $X_{TCLC}(\alpha_x)$ of the HGCI can be expressed as:

$$X_{TCLC}(\alpha_x) = \frac{X_{TCR}(\alpha_x) X_{C_{PF}}}{X_{C_{PF}} - X_{TCR}(\alpha_x)} + X_{L_c} = \frac{\pi X_{L_{PF}} X_{C_{PF}}}{X_{C_{PF}}(2\pi - 2\alpha_x + \sin 2\alpha_x) - \pi X_{L_{PF}}} + X_{L_c} \tag{3.3}$$

where X_{Lc}, X_{LPF}, and X_{CPF} are the fundamental impedances of L_c, L_{PF}, and C_{PF}, respectively. In (3.3), it is shown that the TCLC part impedance is controlled by the firing angle α_x. The minimum inductive and capacitive impedances (absolute value) of the TCLC part can be obtained by substituting the firing angles $\alpha_x = 90°$ and α_x

= *180°* into (3.3), respectively. In the following discussion, the minimum value for impedance stands for its absolute value. The minimum inductive ($X_{ind(min)} > 0$) and capacitive ($X_{Cap(min)} < 0$) TCLC part impedances can be expressed as:

$$X_{Ind(\min)}(\alpha_x = 90^\circ) = \frac{X_{L_{PF}} X_{C_{PF}}}{X_{C_{PF}} - X_{L_{PF}}} + X_{L_c} \tag{3.4}$$

$$X_{Cap(\min)}(\alpha_x = 180^\circ) = -X_{C_{PF}} + X_{L_c} \tag{3.5}$$

Based on the above (3.2)–(3.3), the injected active power and reactive power by the HGCI can be expressed as:

$$\begin{aligned}
S_{cxf} &= \vec{V}_{sxf} \vec{I}^{*}_{cxf} = \vec{V}_{sxf} \cdot \left(\frac{\vec{V}_{invxf} - \vec{V}_{sxf}}{jX_{TCLC}(\alpha_x)} \right)^{*} \\
&= V_{sxf} \cdot \left(\frac{(V_{invxf} \cos\delta + jV_{invxf} \sin\delta - V_{sxf}}{jX_{TCLC}(\alpha_x)} \right)^{*} \\
&= V_{sxf} \cdot \left(\frac{V_{invxf} \sin\delta}{X_{TCLC}(\alpha_x)} + j \frac{\vec{V}_{invxf} \cos\delta - \vec{V}_{sxf}}{X_{TCLC}(\alpha_x)} \right) \\
&= \frac{V_{sxf} V_{invxf} \sin\delta}{X_{TCLC}\alpha_x} + j \left(\frac{V_{sxf} V_{invxf} \cos\delta}{X_{TCLC}\alpha_x} - \frac{V^2_{sxf}}{X_{TCLC}\alpha_x} \right) \\
&= P_{cxf} + jQ_{cxf}
\end{aligned} \tag{3.6}$$

where * is the conjugate value, δ is the angle between the inverter voltage V_{invxf} and the source voltage V_{sxf}. Based on (3.6), the ratio of V_{invf}/V_{sxf} can be expressed as:

$$\frac{V_{invxf}}{V_{sxf}} = \sqrt{\left(\frac{P_{icxf}}{S_{baseH}} \right)^2 + \left(\frac{Q_{cxf}}{S_{baseH}} - 1 \right)^2} \tag{3.7}$$

where S_{baseH} is the base power of the HGCI, which can be expressed as:

$$S_{baseH} = \frac{V^2_{sxf}}{X_{TCLC}(\alpha_x)} \tag{3.8}$$

In (3.8), the $X_{TCLC}(\alpha_x)$ is controlled by the firing angle, which can be obtained from (3.3). In other words, the S_{baseH} is controlled by the firing angle of the thyristors. In order to get the minimum ratio of V_{invf}/V_{sxf} in (3.7), so as to achieve the minimum DC-link voltage requirement for the HGCI, the optimal S_{baseH} can be calculated by taking the derivative of V_{invf}/V_{sxf} with regard to S_{baseH} to zero.

$$\frac{\partial \left(V_{invxf} / V_{invxf} V_{sxf} \right)}{\partial S_{baseH}} = 0 \; S_{baseH} = \frac{P^2_{cxf} + Q^2_{cxf}}{Q_{cxf}} \tag{3.9}$$

If the deduced S_{baseH} in (3.9) is within the range $S_{baseH} \in [V_{sxf}{}^{2}/X_{TCLC}(\alpha_{x} = 180°), V_{sxf}{}^{2}/X_{TCLC}(\alpha_{x} = 90°)]$, the ratio of V_{invf}/V_{sxf} can be calculated by substituting the S_{baseH} in (3.9) into (3.7). Otherwise, the ratio of V_{invf}/V_{sxf} can be calculated with $S_{baseH} = V_{sxf}{}^{2}/X_{TCLC}(\alpha_{x} = 180°)$ or $S_{baseH} = V_{sxf}{}^{2}/X_{TCLC}(\alpha_{x} = 90°)$. Compared with the HGCI, the coupling impedances for the IGCIs and CGCIs are shown in Fig. 3.1, which are fixed as X_L and X_L-X_C. The base power of IGCI and CGCI can be expressed as:

$$\mathrm{S_{BaseI}} = \frac{V_{sxf}^{2}}{2\pi fL} \tag{3.10}$$

$$\mathrm{S_{BaseC}} = \frac{V_{sxf}^{2}}{2\pi fL - \frac{1}{2\pi fC}} \tag{3.11}$$

From (3.10) and (3.11), the base power of the IGCI and CGCI are fixed values. By substituting the (3.10) and (3.11) into (3.3) ($S_{baseH} = S_{baseI}$ and $S_{baseH} = S_{baseC}$), the ratio of V_{invf}/V_{sxf} in terms of P_{cxf} and Q_{cxf} can be plotted as in Fig. 3.2a, b. The plots of the IGCI in Fig. 3.2a, b are the same.

For IGCI, as shown in Fig. 3.2a, b, the $V_{invxf}/V_{sxf} > 1$, when the loading is inductive. In contrast, the required V_{invxf}/V_{sxf} is slightly smaller than 1 when the required reactive power is capacitive. Actually, the V_{invxf}/V_{sxf} is close to one due to the small value of the coupling inductor L.

For CGCI, as shown in Fig. 3.2a, the V_{invxf}/V_{sxf} is lower than 1 under a small inductive reactive power area. In contrast, the ratio of V_{invxf}/V_{sxf} increases rapidly when the Q_{cxf} moves away from S_{baseC}.

For the proposed HGCI as shown in Fig. 3.2b, if the deduced S_{baseH} in (3.9) is within the range $S_{baseH} \in [V_{sxf}{}^{2}/X_{TCLC}(\alpha_{x} = 180°), V_{sxf}{}^{2}/X_{TCLC}(\alpha_{x} = 90°)]$, the ratio of V_{invf}/V_{sxf} can always smaller or equal to 1. This phenomenon can be proved mathematically by substituting (3.9) with (3.7):

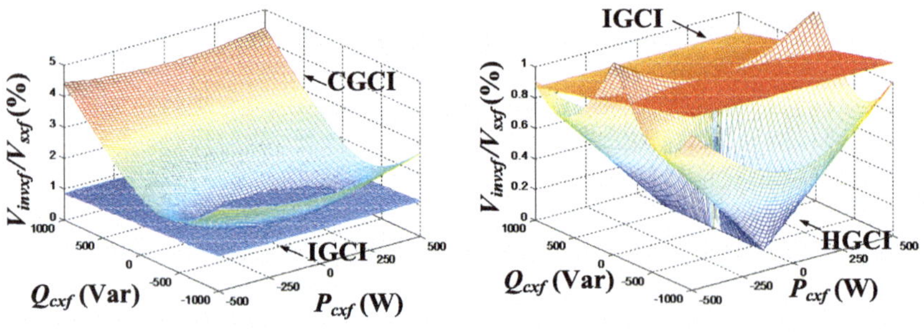

(a) (b)

Fig. 3.2 The ratio of V_{invf}/V_{sxf} in terms of P_{cxf} and Q_{cxf}: **a** IGCI and CGCI **b** IGCI and the proposed HGCI

$$\frac{V_{invxf}}{V_{sxf}} = \sqrt{\frac{\mathrm{P}_{cxf}^2}{\mathrm{S}_{baseH}} + \left(\frac{\mathrm{Q}_{\mathrm{cxf}}}{\mathrm{S}_{baseH}} - 1\right)^2} = \sqrt{\frac{\mathrm{P}_{cxf}^4 + \mathrm{P}_{cxf}^2\mathrm{Q}_{cxf}^2}{(\mathrm{P}_{cxf}^2 + \mathrm{Q}_{cxf}^2)^2}} \leq 1 \tag{3.12}$$

where the S_{baseH} is designed to be:

$$S_{baseH} = \frac{V_{sxf}^2}{\mathrm{X}_{TCLC}(\alpha_{\mathrm{x}})} = \frac{P_{cxf}^2 + Q_{cxf}^2}{Q_{\mathrm{cxf}}} \tag{3.13}$$

The basic function of the HGCI at the fundamental frequency is to inject P_{cxf} and compensate Q_{cxf} to the power grid simultaneously in the daytime, and it provides Q_{cxf} compensation at night, which makes the PV generation system work more cost-effectively instead of just injecting active power during sunshine hours in the daytime. Another desirable function of the HGCI is that V_{invf}/V_{sxf} can be as small as zero when it provides Q_{cxf} compensation only at nighttime. This phenomenon can be observed from Fig. 3.3b (with $P_{cxf} = 0$) and also can be proved by using (3.12) as:

$$\frac{V_{invxf}}{V_{sxf}} = \lim_{\mathrm{P_{cxf}} \to 0} \sqrt{\frac{\mathrm{P}_{cxf}^4 + \mathrm{P}_{cxf}^2\mathrm{Q}_{cxf}^2}{(\mathrm{P}_{cxf}^4 + 2\mathrm{P}_{cxf}^2\mathrm{Q}_{cxf}^2) + \mathrm{Q}_{cxf}^4}} = 0 \tag{3.14}$$

Compared with the IGCI and CGCI, the HGCI requires the lowest required fundamental inverter voltage V_{invf} to inject P_{cxf} and compensate Q_{cxf} to the power grid simultaneously in the daytime or compensate Q_{cxf} at nighttime. As the fundamental inverter voltage usually dominates the design of the DC-link voltage of the DC/AC VSI, the HGCI can obtain the lowest DC-link voltage requirement among the IGCI and CGCI.

3.2.3 *Parallel Resonance Prevention of HGCI*

The pure TCLC part operation may cause the resonance problem, but the active inverter part of HGCI can help alleviate the resonance problem. The TCLC part is controlled by thyristors, which can be modeled as a switch. When the switch is turned off, the TCLC part is the L_c in series with C_{PF} (*LC*-mode). When the switch is turned on, the TCLC part is the L_c in series with the paralleled combination of L_{PF} and C_{PF} (*LCL*-mode).

Figure 3.3 gives its equivalent circuit model of parallel HGCIs due to load harmonic current i_{Lxh}. The subscript 'h' denotes the harmonic component. In the harmonic model, as in Fig. 3.3, the inverters of the parallel HGCIs can be treated as a current-controlled voltage source, which can be expressed as:

$$v_{invxh1} = K_1 \cdot i_{sxh},\ v_{invxh2} = K_2 \cdot i_{sxh} \tag{3.15}$$

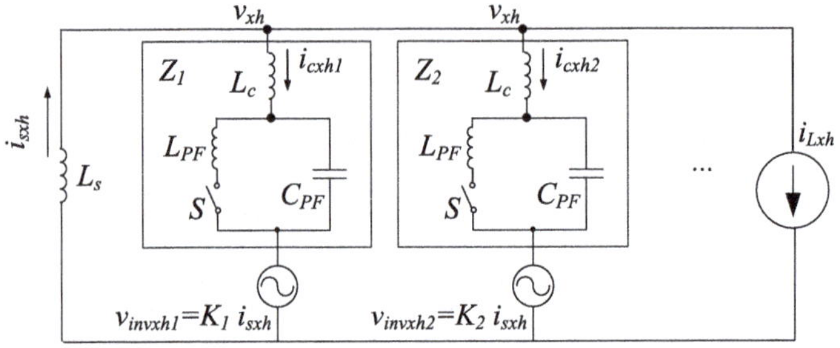

Fig. 3.3 Harmonic equivalent circuit model of parallel HGCI due to i_{Lxh}

where i_{sxh} is the source harmonic current, K_1 and K_2 are the control gains that depend on the inverter part of the parallel HGCIs. In Fig. 3.3, the ratio of i_{sxh} and i_{Lxh} under both *LC-mode* and *LCL-mode* can be expressed as:

$$K_{sh} = \frac{i_{sxh}}{i_{Lxh}} = \frac{Z_1//Z_2}{Z_{sh} + K_1 + K_2 + Z_1//Z_2} \tag{3.16}$$

where the Z_{sh} is the harmonic impedance of system L_s Z_{1h} and Z_{2h} are harmonic impedances of the TCLC part of the parallel HGCIs. The $K_1 = 0$ and $K_2 = 0$ are for the operation of the pure TCLC part (no inverter) of the parallel HGCIs, while $K_1 >$ 0 and $K_2 > 0$ are for the operation of the parallel HGCIs.

With the help of (3.16) and the bode diagrams as in Fig. 3.4, the possible parallel resonance problem of the parallel HGCIs can be analyzed. The existence of K_1 and K_2 can eliminate the original resonance condition of the pure TCLC part operation at $|Z_{sh} + Z_{1h}/Z_{2h}| \approx 0$ in (3.16), thus playing a similar damping role as the Z_{sh}. Hence, the pole of the transfer function shifts to a lower frequency. The $K_1 = K_2 = 0$ plot in Fig. 3.4 describes the corresponding transfer function with only pure TCLC part operation. The $K_1 = K_2 = 40$ and $K_1 = K_2 = 80$ plots indicate the transfer function with the parallel HGCIs operation. The amplitude-frequency response shows a damping characteristic due to the load harmonic current.

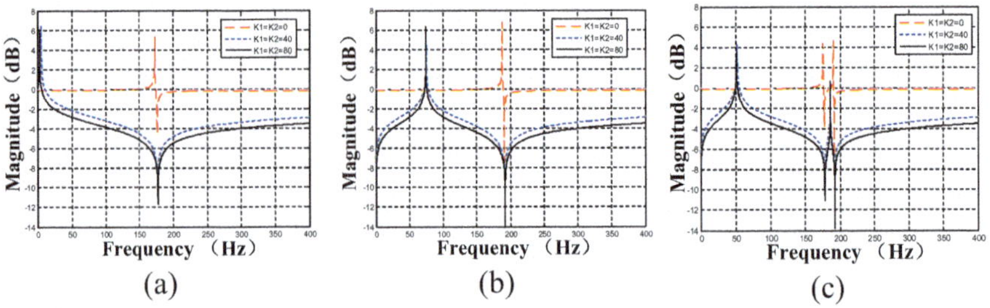

Fig. 3.4 Bode diagrams of parallel HGCIs under resonance analysis **a** both LC modes, **b** both LCL modes and **c** one LC mode and one LCL mode

3.2.4 Parameter Design of HGCI

The TCLC part of the proposed HGCI is designed based on the consideration of the injection of P_{cxf} and the compensation Q_{cxf} range (for L_{PF} and C_{PF}) and the limitation of the high-frequency switching current ripple (for L_c).

A. *Design of C_{PF} and L_{PF}*

The purpose of the TCLC part is to provide the optimum based power S_{BaseH} as in (3.9). In (3.9), when the $X_{TCLC}(\alpha_x) = X_{Cap(min)}(\alpha_x = 180°)$ or $X_{TCLC}(\alpha_x) = X_{Ind(min)}(\alpha_x = 90°)$, the TCLC part provides the maximum capacitive or inductive compensating reactive power $Q_{cxf(MaxCap)} > 0$ and $Q_{cxf(MaxInd)} < 0$, respectively. Therefore, the $S_{BaseH(MaxCap)}$ and $S_{BaseH(MaxCap)}$ via (3.13) can be expressed as:

$$S_{BaseH(MaxCap)} = \frac{P^2_{cxf(Max)} + Q^2_{cxf(MaxCap)}}{Q_{cxf(MaxCap)}} = \frac{V^2_{sxf}}{X_{Cap(min)}(\alpha_x = 180°)} = -\frac{V^2_{sxf}}{X_{C_{PF}} - X_{L_c}} \quad (3.17)$$

$$S_{BaseH(MaxInd)} = \frac{P^2_{cxf(Max)} + Q^2_{cxf(MaxInd)}}{Q_{cxf(MaxInd)}} = \frac{V^2_{sxf}}{X_{Ind(min)}(\alpha_x = 90°)} = \frac{V^2_{sxf}}{\frac{X_{L_{PF}}X_{C_{PF}}}{X_{C_{PF}}-X_{L_{PF}}} + X_{L_c}} \quad (3.18)$$

where the minimum inductive impedance $X_{Ind(min)}$ and the capacitive impedance $X_{Cap(min)}$ are obtained from (3.4) and (3.5), respectively. Based on (3.17) and (3.18), the parallel capacitor C_{PF} and inductor L_{PF} can be designed as:

$$C_{PF} = \frac{P^2_{cxf(Max)} + Q^2_{cxf(MaxCap)}}{\omega^2 L_c(P^2_{cxf(Max)} + Q^2_{cxf(MaxCap)}) - \omega V^2_{sxf} \cdot Q_{cxf(MaxCap)}} \quad (3.19)$$

$$L_{PF} = \frac{V^2_{sxf} Q_{cxf(MaxInd)} - \omega L_c\left(P^2_{cxf(Max)} + Q^2_{cxf(MaxInd)}\right)}{\omega(P^2_{cxf(Max)} + Q^2_{cxf(MaxInd)}) - \omega^3 L_c C_{PF}(P^2_{cxf(Max)} + Q^2_{cxf(MaxInd)}) + \omega^2 V^2_{sxf} C_{PF} Q_{cxf(MaxInd)}} \quad (3.20)$$

where $P_{cxf(Max)}$ is the designed maximum injected active power, $Q_{cxf(MaxCap)}$ and $Q_{cxf(MaxInd)}$ are the designed maximum compensated inductive and capacitive reactive power. ω is the fundamental angular frequency, and V_{sxf} is the root-mean-square (RMS) source voltage.

B. *Design of L_c*

A coupling inductor L_c is used to limit the high-frequency switching current ripple. The impedance of the coupling L_c can be designed to be larger or equal to 5% of the impedance of the coupling capacitor C_{PF} as:

$$L_c \geq \frac{0.05}{\omega^2 \cdot C_{PF}} \quad (3.21)$$

where C_{PF} is the TCLC part capacitor which can be found in (3.19).

3.2.5 Control System

In this part, the control strategy for the HGCI is proposed by coordinating the control of the TCLC part and the active inverter part so that the two parts can complement each other's disadvantages and the overall performance of the HGCI can be improved. The control block diagram of the HGCI is shown in Fig. 3.5. The control strategy of the HGCI is separated into two parts for discussion: *A.* TCLC part control and *B.* Active inverter part control.

A. *TCLC part control*

The TCLC part control is based on the instantaneous pq theory. The TCLC part is mainly used to instantaneously inject the reference $P_{cxf}^{\ *}$ (from the PV source) and compensate the load reactive power $Q_{cxf} = Q_{Lxf} \approx \overline{q}_{Lx}/2$. Referring to (3.13), to calculate the X_{TCLC} in real-time, the expression of X_{TCLC} can be expressed in terms of instantaneous values as:

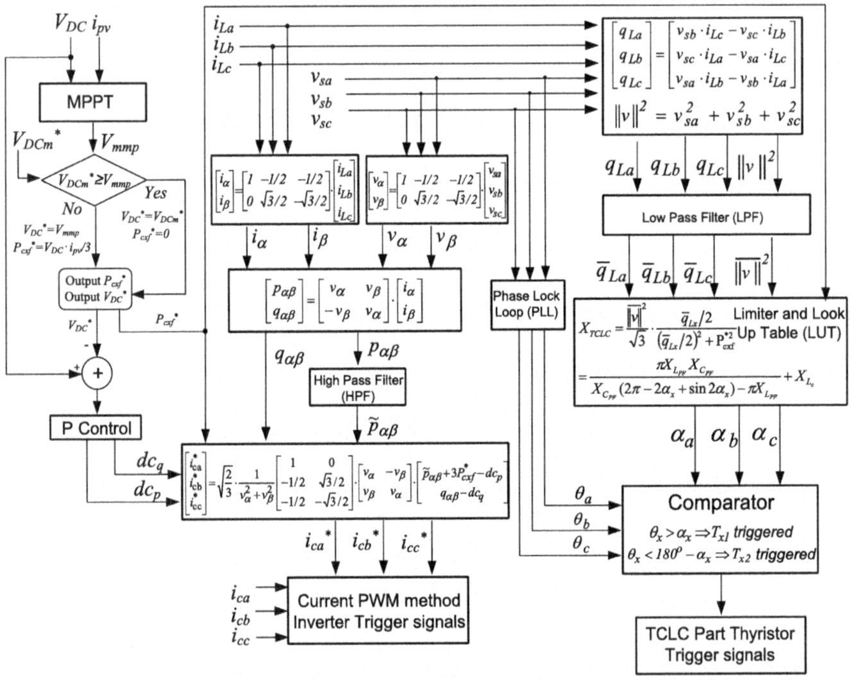

Fig. 3.5 The control block diagram of the proposed HGCI

$$X_{TCLC}\alpha_x = \frac{Q_{cxf}V_{sxf}^2}{P_{cxf}^2 + Q_{cxf}^2} = \frac{\overline{\|v\|}^2}{\sqrt{3}} \times \frac{\bar{q}_{Lx}/2}{\left(\bar{q}_{Lx}/2\right)^2 + P_{cxf}^{*2}} \tag{3.22}$$

where $\|v\|$ is the norm of the three-phase instantaneous source voltage v_{sx} and $\overline{q}_{Lx}$ is the DC component of the load reactive power. The expression of $\|v\|$ and $\overline{q}_{Lx}$ can be obtained by (3.25) and (3.26) with low-pass filters.

$$\|v\| = \sqrt{v_{sa}^2 + v_{sb}^2 + v_{sc}^2} \tag{3.23}$$

$$\begin{bmatrix} q_{La} \\ q_{Lb} \\ q_{Lc} \end{bmatrix} = \begin{bmatrix} v_{sb} \cdot i_{Lc} - v_{sc} \cdot i_{Lb} \\ v_{sc} \cdot i_{La} - v_{sa} \cdot i_{Lc} \\ v_{sa} \cdot i_{Lb} - v_{sb} \cdot i_{La} \end{bmatrix} \tag{3.24}$$

In (3.23) and (3.24), v_{sx} and q_{Lx} are the instantaneous load voltage and the load reactive power, respectively. By comparing the calculated X_{TCLC} in (3.22) and the X_{TCLC} expression in (3.3), the firing angle α_x can be determined. Because (3.3) is complicated, a look-up table between the firing angle α_x and the X_{TCLC} is installed inside the controller. The trigger signals to control the TCLC part can then be generated by comparing the firing angle α_x with θ_x, which is the phase angle of the source voltage v_{sx}. θ_x can be obtained by using a phase lock loop.

B. *Active inverter part control*

The active inverter part is explained in two parts. In part *B.1*, the DC-link voltage control with the maximum power point tracking (MPPT) is proposed and discussed. In Part B.2, the reference current calculation is introduced.

B.1 DC-link voltage control with the MPPT

Due to the photovoltaic power output usually being affected by the weather, the commonly used MPPT leads to active power changing in time, which can affect the active inverter part DC-link voltage V_{DC} of the HGCI.

For the HGCI system, the TCLC part is used to compensate the reactive power and unbalanced power. The active inverter part with DC-link voltage V_{DC} is used to compensate the harmonic power and inject active power (P_{cxf}). Therefore, a minimum DC-link voltage reference $V_{DCm}{}^*$ is still required and can be calculated based on harmonic current compensation when the reference active power injection $P_{cxf}{}^* = 0$. The $V_{DCm}{}^*$ can be expressed as:

$$\begin{aligned} V_{DCm}^* &= \max\left(\sqrt{6}\cdot V_{invah}, \sqrt{6}\cdot V_{invbh}, \sqrt{6}\cdot V_{invch}\right) \\ &= \max\left(\sqrt{6}\sqrt{\sum_{n=2}^{\infty}(X_{an}\cdot I_{Lan})^2}, \sqrt{6}\sqrt{\sum_{n=2}^{\infty}(X_{bn}\cdot I_{Lbn})^2}, \sqrt{6}\sqrt{\sum_{n=2}^{\infty}(X_{cn}\cdot I_{Lcn})^2}\right) \end{aligned} \tag{3.25}$$

where I_{Lxn} is the load harmonic current at each harmonic order n, X_{xn} is the harmonic impedance of the TCLC part, which can be expressed as:

$$|X_{xn}(\alpha)| = \left| \frac{\pi(n\omega L_{PF})}{(2\pi - 2\alpha + \sin 2\alpha) - \pi(n\omega)^2 \cdot L_{PF}C_{PF}} + n\omega L_c \right| \tag{3.26}$$

Once the MPPT method leads to the PV output voltage V_{mmp} lower than its $V_{DCm}{}^*$, the reference DC-link voltage $V_{DC}{}^*$ will be equal to $V_{DCm}{}^*$. In this situation, the HGCI is controlled to compensate the reactive power, unbalanced power, and harmonics power only (without active power injection $P_{cxf}{}^* = 0$), which is similar as its targeted operation at night. The HGCI can resume its active power injection once the MPPT-controlled PV output voltage V_{mmp} is higher than its $V_{DCm}{}^*$. When the $V_{mmp} > V_{DCm}{}^*$, $V_{DC}{}^* = V_{mmp}$, the HGCI with the photovoltaic power is used to inject active power ($P_{cxf}{}^* = V_{DC} \cdot i_{PV}/3$) and compensate reactive power, unbalanced power, and harmonics power simultaneously.

The HGCI can effectively control the DC-link voltage by feedback to the DC voltage-controlled signal. Based on the DC-link voltage control study, the feedback signals use both reactive and active components (dc_q, dc_p) with proportional (P) control:

$$dc_q = -k_q \cdot (V_{DC}^* - V_{DC}) \tag{3.27}$$

$$dc_p = k_p \cdot (V_{DC}^* - V_{DC}) \tag{3.28}$$

where k_q and k_p are the proportional gains of the DC-link voltage controller. dc_q and dc_p are the dc control signals related to the reactive and active current components. dc_q aims to step change the DC-link voltage under insufficient DC-link voltage, which can be effectively applied for the start-up process, while dc_p aims to maintain the DC-link voltage due to the system loss.

B.2 *Reference current calculation*

In the proposed control strategy, the instantaneous active and reactive power theory method is implemented for the active inverter part to inject active power ($3P_{cxf}{}^* = V_{DC} \cdot i_{PV}$) and compensate reactive power ($\overline{q}_{\alpha\beta}$) and harmonic current ($\tilde{p}_{\alpha\beta}$ and $\tilde{q}_{\alpha\beta}$). Specifically, the active inverter part is used to limit the compensating current i_{cx} to track its reference value $i_{cx}{}^*$. The $i_{cx}{}^*$ can be calculated as:

$$\begin{bmatrix} i_{ca}^* \\ i_{cb}^* \\ i_{cc}^* \end{bmatrix} = \sqrt{\frac{2}{3}} \cdot \frac{1}{v_\alpha^2 + v_\beta^2} \begin{bmatrix} 1 & 0 \\ -1/2 & \sqrt{3}/2 \\ -1/2 & -\sqrt{3}/2 \end{bmatrix} \cdot \begin{bmatrix} v_\alpha & -v_\beta \\ v_\beta & v_\alpha \end{bmatrix} \cdot \begin{bmatrix} \tilde{p}_{\alpha\beta} + 3P_{cxf}^* - dc_p \\ q_{\alpha\beta} - dc_q \end{bmatrix} \tag{3.29}$$

where V_{DC} and i_{PV} are the measured DC-link voltage and PV current, dc_q and dc_p are the dc control signals related to the reactive and active current components, which can be obtained from (3.27) and (3.28). $p_{\alpha\beta}$ and $q_{\alpha\beta}$ are the instantaneous active and reactive powers of the loading which include DC components $\overline{p}_{\alpha\beta}$ and $\overline{q}_{\alpha\beta}$, and ac components $\tilde{p}_{\alpha\beta}$ and $\tilde{q}_{\alpha\beta}$. $\overline{p}_{\alpha\beta}$ and $\overline{q}_{\alpha\beta}$ contain the fundamental active and reactive current components, respectively, while $\tilde{p}_{\alpha\beta}$ and $\tilde{q}_{\alpha\beta}$ contain harmonic currents and negative sequence components. $\tilde{p}_{\alpha\beta}$ is obtained by passing $p_{\alpha\beta}$ through a high pass filter (HPF). $p_{\alpha\beta}$ and $q_{\alpha\beta}$ can be obtained as:

$$\begin{bmatrix} p_{\alpha\beta} \\ q_{\alpha\beta} \end{bmatrix} = \begin{bmatrix} v_\alpha & v_\beta \\ -v_\beta & v_\alpha \end{bmatrix} \cdot \begin{bmatrix} i_\alpha \\ i_\beta \end{bmatrix} \tag{3.30}$$

In (3.29) and (3.30), the voltages (v_α and v_β) and currents (i_α and i_β) in α-β plane are transformed from a-b-c frames by:

$$\begin{bmatrix} v_\alpha \\ v_\beta \end{bmatrix} = \begin{bmatrix} 1 & -1/2 & -1/2 \\ 0 & \sqrt{3}/2 & -\sqrt{3}/2 \end{bmatrix} \cdot \begin{bmatrix} v_{sa} \\ v_{sb} \\ v_{sc} \end{bmatrix} \text{ and } \begin{bmatrix} i_\alpha \\ i_\beta \end{bmatrix} = \begin{bmatrix} 1 & -1/2 & -1/2 \\ 0 & \sqrt{3}/2 & -\sqrt{3}/2 \end{bmatrix} \cdot \begin{bmatrix} i_{La} \\ i_{Lb} \\ i_{Lc} \end{bmatrix} \tag{3.31}$$

where v_{sx} and i_{Lx} are source voltage and load current signals. Finally, by comparing compensating current i_{cx} with the reference value $i_{cx}{}^*$ through the hysteresis pulse-width modulation (PWM) control method, the triggers for the active inverter part can be generated.

3.2.6 *Simulation Case Studies of HGCI*

In this section, simulation case studies are provided to verify that the proposed HGCI can perform active power injection, reactive power, and harmonic current compensation under different loading conditions (balanced/unbalanced) in comparison with the IGCI and CGCI. With the based value of $V_{base} = 110$ V, $I_{base} = 10$ A, and $X_{base} = V_{base}/I_{base} = 11\ \Omega$, Table 3.2 gives the parameters used in both simulations and experiments. The values of L_{PF} and C_{PF} in HGCI are designed based on the reactive power/current compensation and active power injection range. A small rating HGCI (110 V–10 A) experimental prototype is built in the laboratory for testing purposes. The L_{PF} and C_{PF} are selected to be 30 mH and 160 μF due to the small rating HGCI. For practical industrial applications with a wider reactive power compensation range and active power injection, a much smaller L_{PF} value can be designed. Table 3.3 summarizes all the above simulation results by using the IGCI, CGCI, and HGCI. The comparative performances of IGCI, CGCI, and HGCI are given in Part A. The electromagnetic interference (EMI) impact of the HGCI is provided in Part. B. The HGCI for unbalanced loads compensation is presented in Part C. Finally, the HGCI with the proposed MPPT control method during the night time and day time is discussed in Part D.

Part A: The comparative performances of IGCI, CGCI and HGCI

Table 3.2 Simulation and experimental parameters for the IGCI, CGCI, and HGCI

	Parameters	Physical values
System parameters	V_{sx}, f, L_s	*110 V, 50 Hz,0.1 mH*
IGCI	L	*5 mH* (0.14 p.u)
CGCI	L, C	*5 mH*(0.14 p.u), *80* μF(3.6 p.u)
HGCI	L_c, L_{PF}, C_{PF}	*5 mH* (0.14 p.u), *30 mH* (0.87 p.u),*160* μF(1.8 p.u)

Table 3.3 Simulation results for different loads compensation before and after the IGCI, CGCI, and HGCI compensation

			I_{sx} (A)	Q_{sx} (var)	P_{sx} (W)	PF	THD_{isx}(%)	V_{DC} (V)
Linear load ($P_{cxf}^* = 200$ W)	Before		7.6	540	630	0.76	0.1	–
	IGCI Comp		4.2	8	468	1.00	2.1	300
	CGCI Comp		4.0	18	433	1.00	2.3	300
	HGCI Comp		4.1	15	443	1.00	2.5	100
Non-Linear load ($P_{cxf}^* = 200$ W)	Before		9.3	540	864	0.85	16.2	–
	IGCI Comp		7.3	10	700	1.00	10.7	300
	IGCI Comp		6.4	11	680	1.00	3.7	400
	CGCI Comp		6.2	20	650	1.00	2.4	300
	HGCI Comp		6.2	28	674	1.00	4.3	100
Unbalanced loads ($P_{cxf}^* = 100$ W)	Before Comp	A	6.2	486	457	0.68	27.4	–
		B	6.2	280	608	0.91	27.3	
		C	8.1	508	708	0.81	20.7	
	After HGCI Comp	A	4.6	1	498	1.00	3.0	100
		B	4.6	10	498	1.00	4.7	
		C	4.6	5	499	1.00	4.6	

Figures 3.6 and 3.7 show the performance of the IGCI, CGCI, and HGCI for linear loading and nonlinear loading compensation, in which they are switched on at 60 ms. The IGCI, CGCI, and HGCI provide the load-required reactive power $Q_{Lxf} = Q_{cxf}$, and inject the active power to its reference value $P_{cxf}^* = P_{cxf}$. Based on Fig. 3.6 and Table 3.3, the IGCI and CGCI require about $V_{DC} = 300$ V to compensate the $Q_{Lxf} = 540$ var and inject $P_{cxf}^* = 200$ W to the linear loads power system, while the HGCI just requires $V_{DC} = 100$ V for the same Q_{Lxf} compensation and P_{cxf} injection.

As shown in Fig. 3.6a, the IGCI with $V_{DC} = 300$ V cannot inject the same $P_{cxf}^* = 200$ W ($P_{cxf} = 120$ W) when the loads have the harmonic current ($THD_{isx} = 16.2\%$). However, when the V_{DC} is increased to 400 V, the IGCI can fully compensate the reactive power and current harmonics problem and inject the $P_{cxf}^* = 200$ W, as shown in Table 3.3. Based on Fig. 3.7b, c, the CGCI and HGCI require $V_{DC} = 300$ V and $V_{DC} = 100$ V for fully reactive power ($Q_{cxf} = Q_{Lxf} = 540$ var) and current harmonics compensation and also active power injection ($P_{cxf}^* = P_{cxf} = 200$ W).

Fig. 3.6 Simulation results for balanced linear loading compensation before and after **a** IGCI, **b** CGCI and **c** HGCI operation

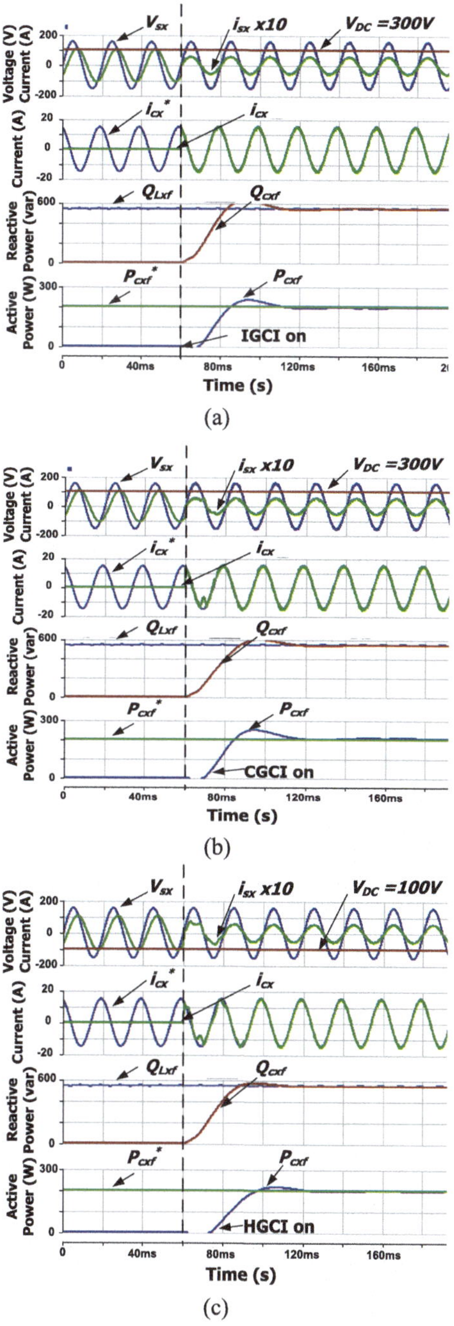

Fig. 3.7 Simulation results for balanced non-linear loading compensation before and after **a** IGCI, **b** CGCI and **c** HGCI operation

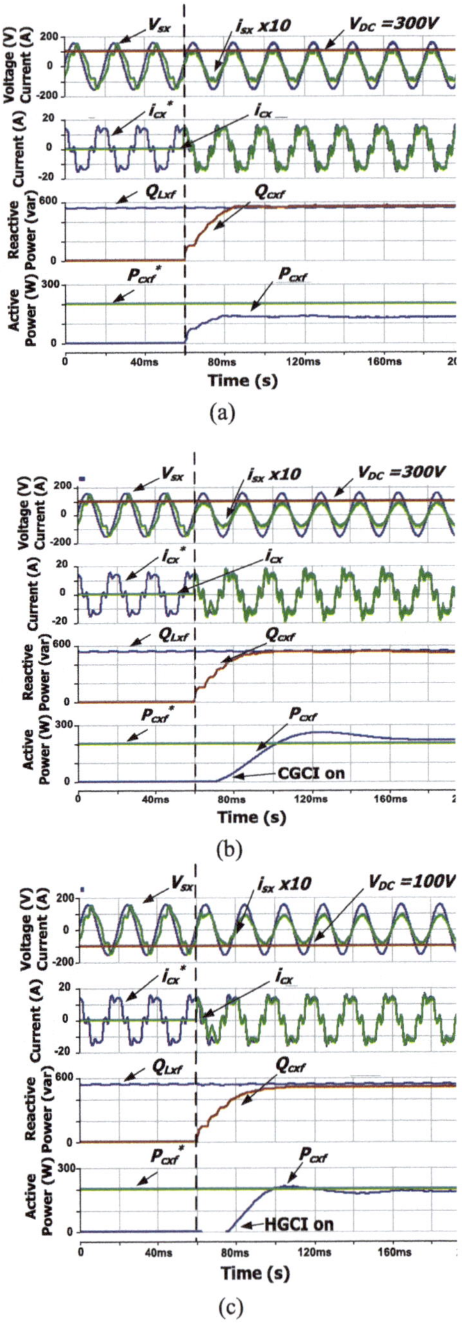

Part B: EMI impact of the HGCI

The high switching frequency will introduce high-frequency voltage and current noise components to the system, which is one of the main reasons for EMI noise. In order to reduce the EMI noise of the HGCI, the hysteresis PWM method can be used. The operational principle of hysteresis current PWM can spread the energy concentrated around the harmonics of switching frequency to a wider frequency range to reduce the EMI peak value. The following simulation case study is provided to explain the hysteresis PWM control method. From Figs. 3.8 and 3.9, it can be seen that the hysteresis PWM method can spread the switching frequency energy in a wider frequency range to reduce the EMI peak value.

Moreover, compared with the conventional high DC-link voltage controlled IGCIs, the low DC-link voltage characteristic of the HGCI can also help to reduce the EMI noise. The lower the DC-link voltage, the lower the high-frequency power (voltage). Thus, lowering the high-frequency power (voltage) can lead to the EMI reduction. Besides, the EMI filter for the power supply of the controller, transducers,

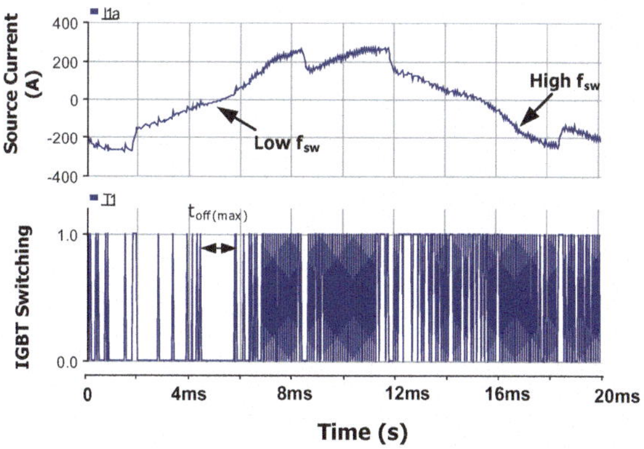

Fig. 3.8 Waveforms of inverter compensating current and corresponding switching signal of IGBT

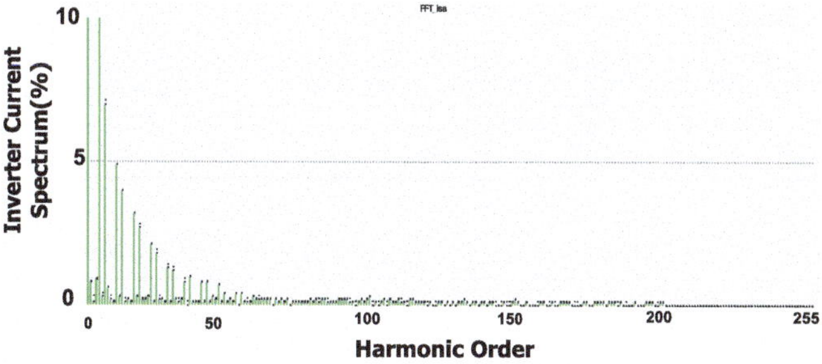

Fig. 3.9 Spectrum of the inverter compensating current

and drivers of the IGBT and thyristors has been installed in the hardware prototype to reduce the EMI problem.

Part C: The HGCI for unbalanced loads operation

Figure 3.10 gives the performance of the HGCI for unbalanced load compensation, in which it is switched on at 80 ms. From Fig. 3.10 and Table 3.3, by using the HGCI with $V_{DC} = 100$ V and the proposed control method, the three-phase system source power factors (*PFs*) have been compensated to unity from the original 0.68 (showing the worst phase). The source reactive powers are compensated close to zero from the original 486 var, 280 var, and 508 var. The average value of source active power is 591 W ($P_{saf} = 457$ W, $P_{sbf} = 608$ W, and $P_{scf} = 708$ W) before compensation. After compensation, the source active power is balanced to around 498 W with $P_{cxf}{}^* = 100$ W injection for each phase. Moreover, from Table 3.3, the source current *THD* is compensated to lower than 5% from 27.4% (showing the worst phase).

Part D: The HGCI with the proposed MPPT control method during night time and day time

Figure 3.11 illustrates the performance of the HGCI with the proposed MPPT control method during the nighttime and daytime. From Fig. 3.11, it can be seen that the reference DC-link voltage $V_{DC}{}^*$ is keeping at the minimum DC-link voltage reference $V_{DCm}{}^* \approx 45$ V (when $P_{cxf} \approx 0$) during the night time for harmonic and reactive power compensation. During the daytime, the HGCI can resume its active power injection once the MPPT method leads to the PV output voltage V_{mmp} higher than its $V_{DCm}{}^*$

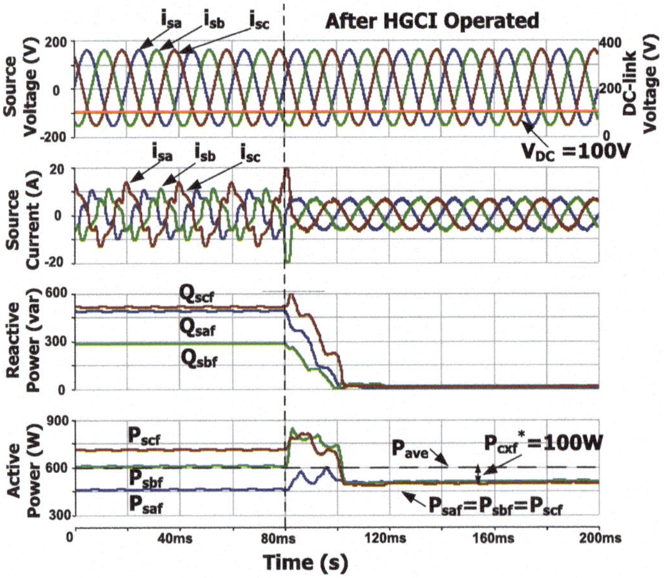

Fig. 3.10 Simulation results for unbalanced loading compensation before and after HGCI operation

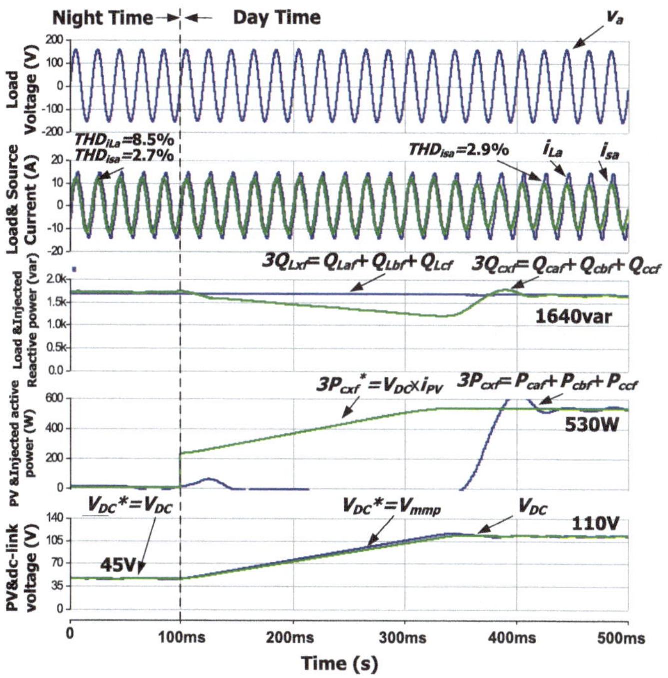

Fig. 3.11 Simulated waveforms of load voltage, load and source current, three-phase load and injected reactive power, PV output and three-phase injected active power, PV output voltage and DC-link voltage during HGCI with MPPT control

≈45 V. When the $V_{mmp} > V_{DCm}{}^{*}$, $V_{DC}{}^{*} = V_{mmp}$, the HGCI with the photovoltaic power are used to inject active power ($P_{cxf}{}^{*} = V_{DC} \cdot i_{PV}/3$) and compensate reactive power, unbalanced power and harmonics power simultaneously.

From Fig. 3.11, the THD_{isx} has been improved to 2.7% during the nighttime and 2.9% during the day time from the original 8.5%. The injected reactive power can fully compensate the load reactive ($Q_{cxf} = Q_{Lxf} = 1640$ var/3) during both nighttime and daytime. By using the MPPT method, the V_{DC} increases from 45 V to about 110 V, while the three-phase injected active power increases from zero to 530 W.

3.2.7 Experimental Results

A three-phase, three-wire 110 V–10 A HGCI experimental prototype is designed and constructed in the laboratory. The performances of the proposed HGCI for balanced and unbalanced loads compensation are provided below in Part A and Part B. Figures 3.12, 3.13, 3.14, 3.15, 3.16, 3.17, 3.18 and 3.19 show the waveforms of v_{sx}, i_{sx} Q_{sxf} and P_{sxf} for balanced and unbalanced reactive power and harmonic

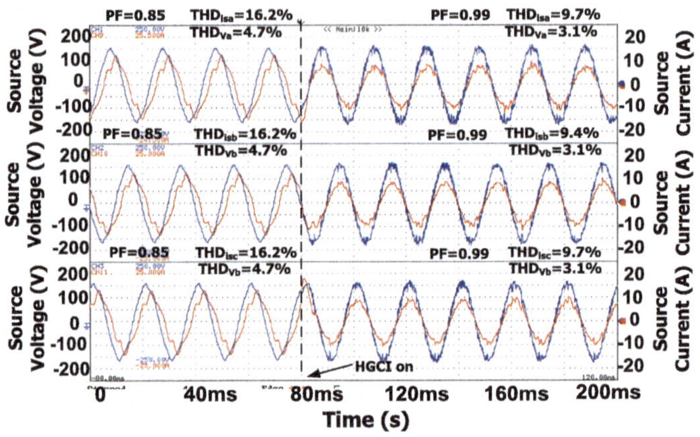

Fig. 3.12 Waveforms of v_{sx} and i_{sx} for balanced inductive reactive power and harmonic current compensation and active power injection by using the HGCI

current compensation and active power injection by using the HGCI. And Table IV summarizes the experimental results.

Part A: *HGCI for balanced loads compensation*

From Fig. 3.12 and Table 3.4, it can be seen that the HGCI can compensate the THD_{isx} to 9.7% and improve the power factor (PF) to 0.99 from the original 16.2% and 0.85. The voltage THD_{Vx} has been improved from 4.7% to 3.2% for the worst phase among phases *a, b, and* c. As shown in Fig. 3.13 and Table 3.4, the three-phase reactive power has been compensated to 0.14 kvar from 1.74 kvar. With the active power injection function ($P_{cxf}{}^{*} = 200$ W) during the daytime, the three-phase active power $3P_{sxf}$ has been reduced to 1.63 kW from the original 2.19 kW by using the HGCI.

Part B: HGCI for unbalanced loads compensation

As shown in Fig. 3.14 and Table 3.4, the HGCI can improve the source current THD_{isx} from 18.7% to 10.7% and source voltage THD_{Vx} from 4.7% to 3.2% (showing the worst phase among phases *a, b, and c*). The worst phase PF is improved from 0.59 to 0.99. From Fig. 3.15, the HGCI can compensate the Q_{sxf} of each phase (0.56 kvar, 0.28 kvar, and 0.58 kvar) close to zero. The different P_{sxf} of each phase (0.37 kW, 0.57 kW, and 0.64 kW) has been compensated to approximately the same. Moreover, the total three-phase $3P_{sxf}$ has been reduced to 1.37 kW from the original 1.58 kW. In addition, as shown in Fig. 3.16, the source voltage and current have been compensated to be in phase with each other for the compensation of unbalanced loads.

Table 3.4 Experimental results for different loads compensation before and after the proposed HGCI compensation

			I_{sx} (A)	Q_{sx} (var)	P_{sx} (W)	PF	THD_{Vx} (%)	THD_{isx} (%)	V_{DC} (V)
Nonlinear loads (HGCI Comp.)	Before Comp	A	7.8	0.58 k	0.73 k	0.85	4.7	16.2	–
		B	7.8	0.58 k	0.73 k	0.85	4.7	16.2	
		C	7.8	0.58 k	0.73 k	0.85	4.7	16.2	
	After Comp	A	4.3	0.05 k	0.54 k	0.99	3.1	9.7	100
		B	4.3	0.05 k	0.55 k	0.99	3.1	9.2	
		C	4.3	0.04 k	0.54 k	0.99	3.1	9.7	
Unbalanced loads (HGCI Comp.)	Before Comp	A	5.7	0.56 k	0.37 k	0.59	4.6	18.7	–
		B	5.6	0.28 k	0.57 k	0.88	4.7	17.4	
		C	7.6	0.58 k	0.64 k	0.77	4.6	13.4	
	After Comp	A	4.4	0.02 k	0.43 k	0.99	3.1	10.5	100
		B	4.8	0.04 k	0.50 k	0.99	3.1	10.6	
		C	4.4	0.09 k	0.44 k	0.99	3.2	10.7	

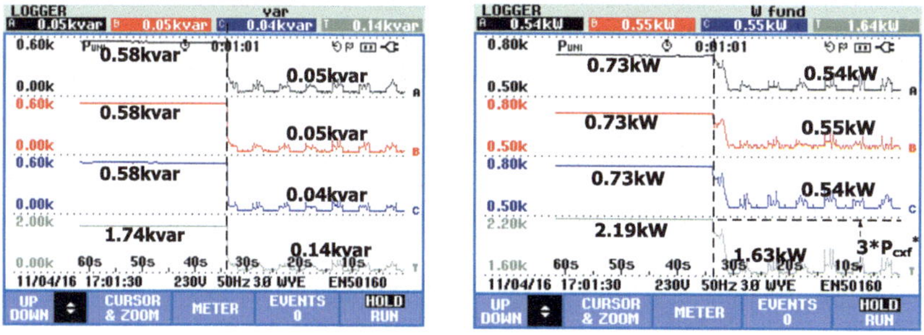

(a) (b)

Fig. 3.13 Dynamic balanced inductive loads compensation by using the HGCI **a** Q_{sxf} (reactive power compensation) and **b** P_{sxf} (active power injection)

3.2.8 Section Summary

In this Section, the analysis of the structure, parameter design and control method of a new type DC/AC inverter HGCI for PV power generation with power quality conditioning is proposed. Compared with the conventional IGCI and CGCI, the proposed HGCI has distinct characteristics of a wide operational range and low DC-link operating voltage. Thus, the system and operational costs can be reduced. Moreover, the HGCI can inject active power and provide reactive power, unbalanced power, and harmonic power compensation at the same time. Simulation and experimental results are presented to verify the effectiveness and performance of the proposed HGCI, which also shows a much lower DC-link operating voltage than the IGCI and CGCI.

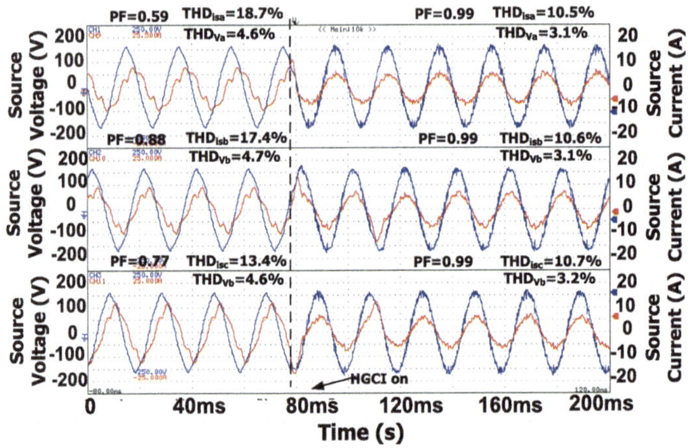

Fig. 3.14 Waveforms of v_{sx} and i_{sx} for unbalanced inductive reactive power and harmonic current compensation and active power injection by using the HGCI

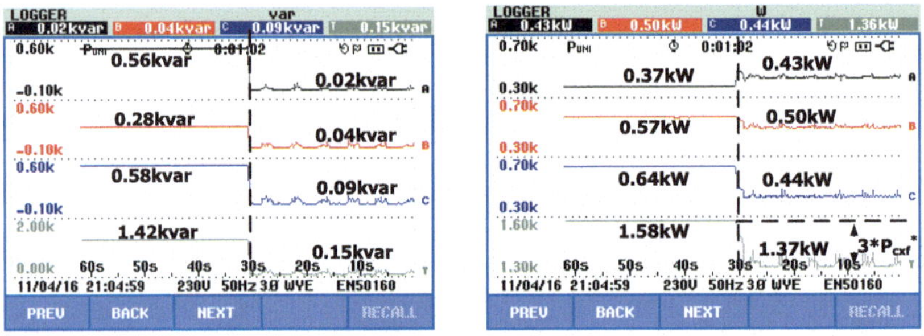

(a) (b)

Fig. 3.15 Dynamic unbalanced inductive loads compensation by using the HGCI **a** Q_{sxf} (reactive power compensation) and **b** P_{sxf} (active power injection)

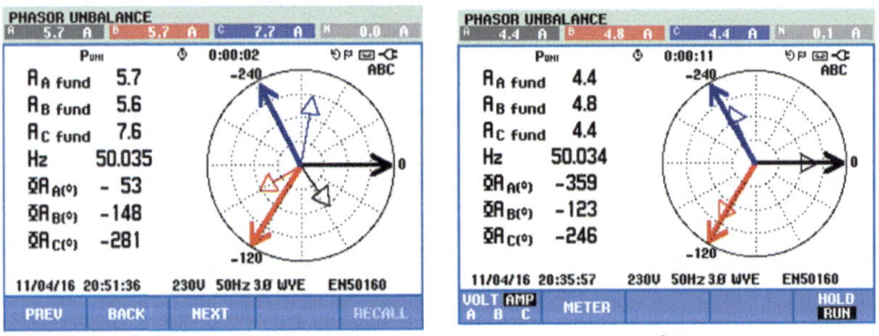

(a) (b)

Fig. 3.16 Experimental phasor diagrams of v_{sx} and i_{sxf} for unbalanced inductive loads: **a** before compensation and **b** after HGCI compensation

3.3 Multifunctional Hybrid Structure of SVC and Capacitive Grid-Connected Inverter (SVC//CGCI) for Active Power Injection and Nonactive Power Compensation

This Section proposes the structure, coordinate control method, and parameter design of a hybrid system for active power injection and non-active power (reactive, harmonic, and unbalanced power) compensation. The proposed hybrid system consists of SVC//CGCI. In SVC//CGCI, the SVC part is used to dynamically compensate the reactive power and unbalanced power, while the low-rating CGCI part is used to inject active power, provide harmonic power, and a fixed amount of capacitive reactive power. Compared with conventional inductive-coupling grid-connected inverters (IGCI), the CGCI can provide active, reactive, and harmonic power with a low rating of active inverter parts. The SVC//CGCI can be a high-efficient solution for active power injection and non-active power compensation. Finally, simulation and experimental results are provided to show the advantages and validity of the proposed SVC//CGCI compared to the conventional IGCI and SVC//IGCI.

3.3.1 Circuit Configuration of the Three-Phase SVC//CGCI

Figure 3.17 shows the system topology of a three-phase three-wire SVC//CGCI. In this paper, the subscript '*x*' denotes phase $x = a, b, c$; v_{sx}, $v_{x,}$, and v_{invx} are the source voltage, load voltage, and inverter voltage, respectively. i_{sx}, $i_{Lx,}$ and i_{cx} are source current, load current, and compensating current, respectively. The SVC part of the SVC//CGCI consists of a coupling inductor L_c, a parallel capacitive C_{PF}, and a thyristor-controlled reactor (TCR) with an inductor L_{PF}. The CGCI part consists of a coupling *LC* filter (L_p and C_p), a two-level voltage source inverter (VSI), and a dc-link capacitor C_{DC}. The CGCI part has both 1) *a* low voltage rating due to a large voltage drop on *LC* and 2) *a* low current rating due to the current divider between i_{cx1} and i_{cx2}.

3.3.2 Operation Principle and Coordinated Control of SVC//CGCI

In this Section, the operation principle and coordinated control of SVC//CGCI are discussed and explained in three parts: *Part A*: SVC part control, *Part B:* CGCI part control, and *Part C:* the overall control block.

A. *The Operation Principle and Control of SVC Part*

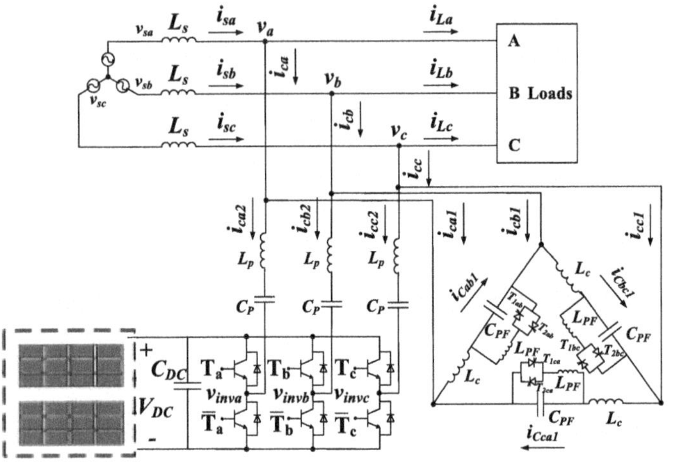

Fig. 3.17 Circuit configurations of the SVC//CGCI

The SVC part of SVC//CGCI is used to compensate the fundamental load imbalance and reactive power dynamically by controlling its firing angles. The key concept of SVC part control can be described in *Part A.1*. And the controls of SVC part in delta connection are explained in *Part A.2*.

A.1 The operation principle and concept of SVC part

In this paper, the study of the proposed SVC//CGCI is based on a balanced three-phase source voltage assumption. For the three-phase, three-wire system, the unbalanced source active power can be automatically controlled to be balanced when the unbalanced reactive power is compensated to be balanced. Figure 3.18 shows the power flow analysis of SVC//CGCI compensation.

The phase fundamental source apparent power is defined as: $S_{sx} = P_{sx} + jQ_{sx} = \vec{V}_{sx} \cdot \vec{I}^{*}_{sxf}$, where the note "*" denotes the complex conjugate. In Fig. 3.18, the sum of $\vec{I}^{*}_{sxf}$ can be expressed as:

Fig. 3.18 Power low analysis for unbalanced power compensation

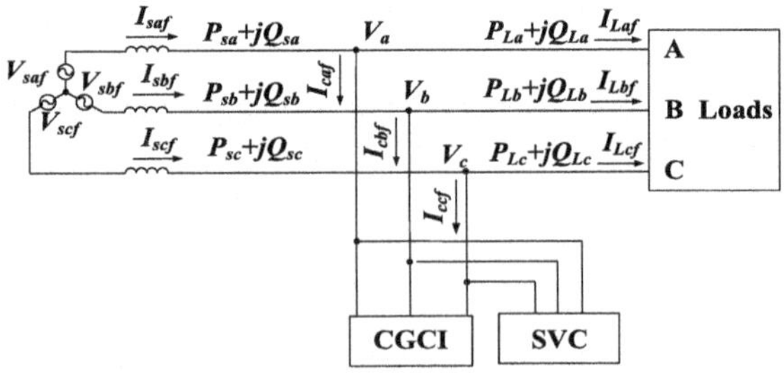

$$\vec{I}^{*}_{saf} + \vec{I}^{*}_{sbf} + \vec{I}^{*}_{scf} = \frac{P_{sa} + jQ_{sa}}{\vec{V}_{saf}} + \frac{P_{sb} + jQ_{sb}}{\vec{V}_{sbf}} + \frac{P_{sc} + jQ_{sc}}{\vec{V}_{scf}} = 0 \tag{3.32}$$

where $\vec{I}_{saf}$, $\vec{I}_{sbf}$ and $\vec{I}_{scf}$ are the fundamental source current phasors. $\vec{V}_{saf}$, $\vec{V}_{sbf}$ and $\vec{V}_{scf}$ are the fundamental source voltage phasors. $\vec{V}_{saf}$ is set to be the reference phasor, so $\vec{V}_{saf} = V_{sxf}\angle 0^{o}$, $\vec{V}_{sbf} = V_{sxf}\angle -120^{o}$ and $\vec{V}_{scf} = V_{sxf}\angle 120^{o}$, where $1\angle -120^{o} = -1/2 - j\sqrt{3}/2$ and $1\angle 120^{o} = -1/2 + j\sqrt{3}/2$, V_{sxf} is the root mean square value of the source voltage. Simplifying (3.32), one can get:

$$\begin{aligned} &\left(2P_{sa} - P_{sb} - \sqrt{3}Q_{sb} - P_{sc} + \sqrt{3}Q_{sc}\right) + \\ &j\left(2Q_{sa} - Q_{sb} + \sqrt{3}P_{sb} - Q_{sc} - \sqrt{3}P_{sc}\right) = 0 \end{aligned} \tag{3.33}$$

In (3.33), both the real and the imaginary parts are equal to zero. Thus, the following relationship can be obtained:

$$\begin{aligned} 2P_{sa} - P_{sb} - \sqrt{3}Q_{sb} - P_{sc} + \sqrt{3}Q_{sc} = 0 \\ 2Q_{sa} - Q_{sb} + \sqrt{3}P_{sb} - Q_{sc} - \sqrt{3}P_{sc} = 0 \end{aligned} \tag{3.34}$$

If the source reactive power is compensated to be balanced by the SVC part as (3.34), the source active power can be balanced as (3.35).

$$Q_{sa} = Q_{sb} = Q_{sc} = Q_{LC} \tag{3.35}$$

where Q_{sx} is source reactive power, Q_{LC} is the fixed value which is provided by the coupling *LC* of the CGCI part. By substituting (3.34) into (3.33), the relationship of the three-phase source active power can be obtained as:

$$P_{sa} = P_{sb} = P_{sc} \tag{3.36}$$

Based on the analysis of (3.32)-(3.36), it can be concluded that the source active power can become balanced once the reactive power is compensated to be balanced.

A.2 The SVC part control

The purpose of this part is to obtain the firing angle (α_{xy}) of the SVC part to compensate the phase source reactive power Q_{sx} to equal Q_{LC}, so that the small left Q_{LC} can be compensated by CGCI. Moreover, the SVC part can help to balance the source's active power. The compensating reactive power passing through the SVC part can be expressed as:

$$Q_{c1x} = -(Q_{Lx} - Q_{LC}) \tag{3.37}$$

where the Q_{c1x} is the phase compensating reactive power. And the line-to-line reactive power passing through the SVC part can be expressed as:

$$\begin{bmatrix} Q_{c1ab} \\ Q_{c1bc} \\ Q_{c1ca} \end{bmatrix} = \begin{bmatrix} Q_{c1c} - Q_{c1a} - Q_{c1b} \\ Q_{c1a} - Q_{c1b} - Q_{c1c} \\ Q_{c1b} - Q_{c1c} - Q_{c1a} \end{bmatrix} = \begin{bmatrix} Q_{La} + Q_{Lb} - Q_{Lc} - Q_{LC} \\ Q_{Lb} + Q_{La} - Q_{La} - Q_{LC} \\ Q_{Lc} + Q_{La} - Q_{Lb} - Q_{LC} \end{bmatrix} \quad (3.38)$$

where Q_{LC} is the small and fixed reactive power provided by LC part of CGCI, which can be expressed as:

$$Q_{LC} = \frac{\overline{V}_x^2}{X_{L_P} - X_{C_P}} \quad (3.39)$$

where $\overline{V}_x$ is the RMS value of load voltage which can be obtained in real time as $\overline{V}_x = \|v_L\| \Big/ \sqrt{3} = \sqrt{v_a^2 + v_b^2 + v_c^2} \Big/ \sqrt{3}$. X_{Lp} and X_{Cp} are the impedances of L_P and C_P.

The purpose of the SVC part is to compensate the fundamental reactive and unbalanced power by providing different α_{xy} for each phase. To obtain α_{xy}, the SVC part impedance X_{xy} in real-time is calculated as:

$$\begin{bmatrix} X_{ab} \\ X_{bc} \\ X_{ca} \end{bmatrix} = \begin{bmatrix} \frac{V_{ab}^2}{Q_{c1ab}} \\ \frac{V_{bc}^2}{Q_{c1bc}} \\ \frac{V_{ca}^2}{Q_{c1ca}} \end{bmatrix} = \begin{bmatrix} \frac{3 \cdot \overline{V}_x^2}{Q_{c1ab}} \\ \frac{3 \cdot \overline{V}_x^2}{Q_{c1bc}} \\ \frac{3 \cdot \overline{V}_x^2}{Q_{c1ca}} \end{bmatrix} = \begin{bmatrix} \frac{3 \cdot \overline{V}_x^2}{Q_{La}+Q_{Lb}-Q_{Lc}-Q_{LC}} \\ \frac{3 \cdot \overline{V}_x^2}{Q_{Lb}+Q_{La}-Q_{La}-Q_{LC}} \\ \frac{3 \cdot \overline{V}_x^2}{Q_{Lc}+Q_{La}-Q_{Lb}-Q_{LC}} \end{bmatrix} \quad (3.40)$$

where V_{xy} and V_x are the RMS values of the line-to-line and phase load voltage, respectively, Q_{Lx} is the phase load reactive power. The V_x and $Q_{Lx} \approx -\overline{q}_{Lx}/2$ can be calculated in real-time as:

$$\begin{bmatrix} q_{La} \\ q_{Lb} \\ q_{Lc} \end{bmatrix} = \begin{bmatrix} v_a \cdot i_{La}^D - v_a^D \cdot i_{La} \\ v_b \cdot i_{Lb}^D - v_b^D \cdot i_{Lb} \\ v_c \cdot i_{Lc}^D - v_c^D \cdot i_{Lc} \end{bmatrix} \quad (3.41)$$

$$V_x = \|v\| \Big/ \sqrt{3} = \sqrt{v_a^2 + v_b^2 + v_c^2} \Big/ \sqrt{3} \quad (3.42)$$

where ${v_x}^D$ and ${i_{Lx}}^D$ are obtained by delaying v_x and i_{Lx} by a phase angle of *90°*. $q_{Lx}$is the phase instantaneous reactive power. Then, $Q_{Lx} \approx -\overline{q}_{Lx}/2$ is obtained by passing q_{Lx} in (3.41) through low pass filters (LPFs). $\|v\|$ is the norm of the three-phase instantaneous load voltage. After obtaining the X_{xy} through (3.40), the α_{xy} is determined by solving the equation below.

$$\begin{bmatrix} X_{ab}(\alpha_{ab}) \\ X_{bc}(\alpha_{bc}) \\ X_{ca}(\alpha_{ca}) \end{bmatrix} = \begin{bmatrix} \frac{\pi X_{L_{PF}} X_{C_{PF}}}{X_{C_{PF}}[2\pi - 2\alpha_{ab} + \sin(2\alpha_{ab})] - \pi X_{L_{PF}}} + X_{L_c} \\ \frac{\pi X_{L_{PF}} X_{C_{PF}}}{X_{C_{PF}}[2\pi - 2\alpha_{bc} + \sin(2\alpha_{bc})] - \pi X_{L_{PF}}} + X_{L_c} \\ \frac{\pi X_{L_{PF}} X_{C_{PF}}}{X_{C_{PF}}[2\pi - 2\alpha_{ca} + \sin(2\alpha_{ca})] - \pi X_{L_{PF}}} + X_{L_c} \end{bmatrix} \tag{3.43}$$

where X_{L_c}, $X_{C_{PF}}$ and $X_{L_{PF}}$ are the impedances of the coupling inductor, the SVC capacitor, and the inductor, respectively. With the calculated X_{xy} from (3.42)-(3.38), the α_{xy} can be determined by solving (3.43). The details can be found in *Part C.*

B *The Operation Principle and Control of CGCI Part*

The operation principle and power range of the CGCI part are provided as *Part B.1.* The control of CGCI is explained in *Part B.2.*

B.1 The Operation principle and power range of the CGCI part

The inverter of CGCI is used to inject active power and reactive power at the fundamental frequency. The injected current from the CGCI and the output voltage of the VSI are given as:

$$I_{cx2f} = I_{cx2fp} + j \cdot I_{cx2fq} \tag{3.44}$$

$$V_{invxf} = \vec{V}_{xf} + X_{LC} \cdot \vec{I}_{cx2f} = \vec{V}_{xf} + X_{LC} \cdot (\vec{I}_{cx2fp} + j\vec{I}_{cx2fq}) \tag{3.45}$$

where I_{cx2fp} and I_{cx2fq} are the fundamental active and reactive currents that inject or absorb from the grid. Based on the above (3.44) and (3.45), the injected active power and reactive power by the CGCI part can be expressed as:

$$\begin{aligned} S_{cx2f} &= \vec{V}_{xf}\vec{I}^{*}_{cx2f} = \vec{V}_{xf} \cdot \left(\frac{\vec{V}_{invxf} - \vec{V}_{xf}}{jX_{LC}}\right)^{*} = V_{xf} \cdot \left(\frac{(V_{invxf} + jV_{invxf} - V_{xf}}{jX_{LC}}\right)^{*} \\ &= V_{xf} \cdot \left(\frac{V_{invxf}}{X_{LC}} + j\frac{\vec{V}_{invxf} - \vec{V}_{xf}}{X_{LC}}\right) \\ &= \frac{V_{xf}V_{invxf}}{X_{LC}} + j \cdot \left(\frac{V_{xf}V_{invxf}}{X_{LC}} - \frac{V_{xf}^2}{X_{LC}}\right) \\ &= P_{cx2f} + jQ_{cx2f} \end{aligned} \tag{3.46}$$

where * is the conjugate value, V_{invxf} and V_{xf} are the fundamental inverter voltage and the load voltage. Based on (3.46), the ratio of V_{invxf}/V_{xf} can be expressed as:

$$\frac{V_{invxf}}{V_{xf}} = \sqrt{\frac{P^2_{cx2f}}{S_{base}} + \left(\frac{Q_{cx2f}}{S_{base}} - 1\right)^2} \tag{3.47}$$

where S_{base} is the base power of CGCI, which can be expressed as:

Fig. 3.19 The ratio of V_{invxf}/V_{xf} in terms of P_{cxf}/S_{base} and Q_{cx2f}/S_{base}

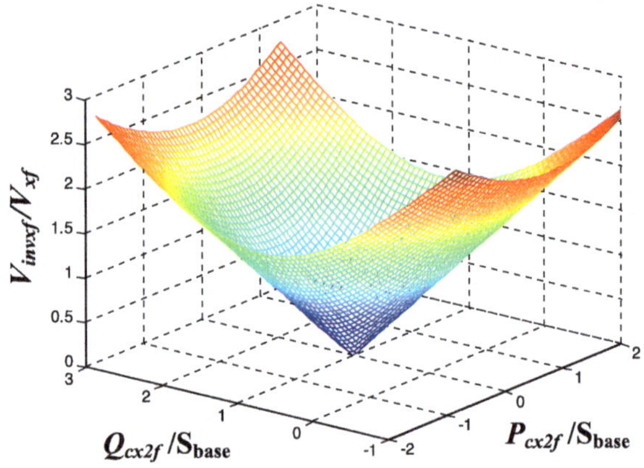

$$S_{base} = \frac{V_{xf}^2}{X_{LC}} \tag{3.48}$$

where X_{LC} is the fundamental coupling impedance of CGCI. Based on (3.47) and (3.48), the ratio of V_{invxf}/V_{xf} in terms of P_{cx2f} and Q_{cx2f} can be plotted as Fig. 3.19. In Fig. 3.19, at the fundamental frequency, the coupling *LC* of the CGCI part is used to compensate the fixed amount of reactive power ($Q_{cx2f} = S_{base}$), and the V_{invxf} is used to inject active power.

B.2 The CGCI part control

The purpose of CGCI part is to inject active power $P_{cx}^* = P_{cx2f}$, and compensate the harmonic power and reactive power left by the SVC part. However, to avoid the mistuning problem of SVC part, the CGCI is controlled to limit the compensating current i_{cx} (instead of inverter current i_{cx2}) to the reference value i_{cx}^*. The i_{cx}^* can be calculated through the well-known instantaneous p-q theory as:

$$\begin{bmatrix} i_{ca}^* \\ i_{cb}^* \\ i_{cc}^* \end{bmatrix} = \sqrt{\frac{2}{3}} \cdot \frac{1}{v_\alpha^2 + v_\beta^2} \begin{bmatrix} 1 & 0 \\ -1/2 & \sqrt{3}/2 \\ -1/2 & -\sqrt{3}/2 \end{bmatrix} \cdot \begin{bmatrix} v_\alpha & -v_\beta \\ v_\beta & v_\alpha \end{bmatrix} \cdot \begin{bmatrix} \tilde{p}_{\alpha\beta} + 3P_{cx}^* \\ q_{\alpha\beta} \end{bmatrix} \tag{3.49}$$

where P_{cx}^* is reference phase injected active power, and the total reference active power is $3P_{cx}^*$for three-phase system, $\tilde{p}_{\alpha\beta}$ is three phase instantaneous harmonic active power,$q_{\alpha\beta}$ is the three-phase instantaneous reactive power which includes harmonic component $\tilde{q}_{\alpha\beta}$ and dc component $\overline{q}_{\alpha\beta}$. As the reference compensating current i_{cx}^* is calculated from $\tilde{p}_{\alpha\beta}$ and $q_{\alpha\beta}$ (included $\tilde{q}_{\alpha\beta}$ component), the three-phase $\tilde{p}_{\alpha\beta}$ and $\tilde{q}_{\alpha\beta}$ fluctuation can be suppressed by limiting the compensating current i_{cx} to track its reference i_{cx}^*.The major part of $\overline{q}_{\alpha\beta}$ is shared by SVC part, while the remaining amount of $\overline{q}_{\alpha\beta}$ for three phase is compensated by *LC* part of CGCI. The $p_{\alpha\beta}$,$q_{\alpha\beta}$, v_α,v_β,i_α, and i_β can be obtained by (2.32)-(2.34).

C. *The Overall Control Block of SVC//CGCI*

The overall control block of the proposed SVC//CGCI is provided in Fig. 3.20. According to Fig. 3.20, and the SVC part control in *Part A*, the (3.49) and (2.33) are used to calculate the load reactive power q_{Lx} and load voltage V_x. Then, SVC part impedance X_{xy} can be obtained from (3.43) with the help of $Q_{Lx} \approx -\overline{q}_{Lx}/2$ and reactive power of the coupling *LC* of CGCI part in (3.42). After that, by comparing the impedance X_{xy} in (3.48) and (2.34), the firing angle α_{xy} can be obtained. As (2.34) does not have a closed-form solution, a look-up table has been installed to directly obtain the firing angle α_{xy} with known X_{xy}. By comparing the pre-calculated α_{xy} of the load voltage θ_{xy}, the trigger signals for the thyristor of the SVC can be obtained by using a phase lock loop.

Based on Fig. 3.20 and the CGCI part control in *Part B*, the load voltage and current (v_x and i_{Lx}) in the *a-b-c* frame are transformed to α–β frame through the (2.32) and (2.34). By using the instantaneous p-q theory, the instantaneous active power and reactive power inα–β frame can be calculated through (2.33). With the help of the HPF and (3.49), the load reactive power and harmonic active power can be transferred to i_{cx}^*. Through the current hysteresis PWM control method, the trigger signals of the CGCI can be generated by comparing the i_{cx} with i_{cx}^*.

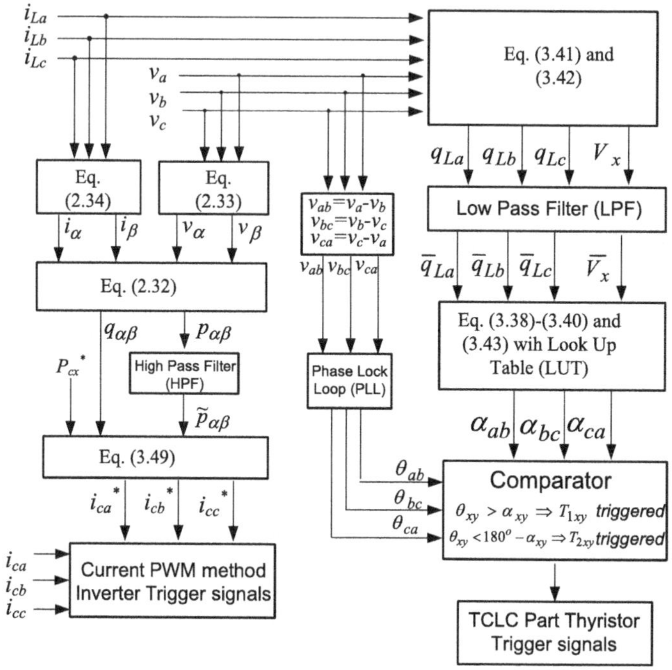

Fig. 3.20 The overall control block of the proposed SVC//CGCI

3.3.3 Parameter Design of Proposed SVC//CGCI

In this section, a parameter design method is discussed and proposed in two parts. In *Part A*, the parameter design of the C_P and L_p is proposed for the CGCI part. In *Part B*, the design of L_c, C_{PF}, and L_{PF} is given for the SVC part.

A. *Design of C_P and L_P*

According to Fig. 3.19 and (3.46), the injected active power can be used to determine the X_{LC}. Based on (3.46), the X_{LC} is designed as:

$$X_{LC} = \frac{V_{xf} V_{invxf}}{P_{cx2f(Max)}} \approx \frac{V_{xf} V_{DC}}{\sqrt{6} \cdot P_{cx2f(Max)}} \tag{3.50}$$

where $P_{cx2f(max)}$ is the maximum injected active power that can be provided by CGCI, V_{xf} is the load voltage, V_{DC} is the designed dc-link voltage. From another perspective, the X_{LC} can be expressed as the impedance of coupling components of CGCI:

$$X_{LC} = X_{L_p} - X_{C_p} = \omega \cdot L_p - {}^{1}/_{\omega \cdot C_p} \tag{3.51}$$

where ω is the angular frequency $\omega = 2\pi f$. In (3.51), the impedance of X_{Cp} is much larger than X_{Lp} ($X_{Cp} >> X_{Lp}$). Therefore, based on (3.50) and (3.51), the C_p can be designed as:

$$C_p \approx \frac{\sqrt{6} \cdot P_{cxf(Max)}}{2\pi f \cdot V_{xf} V_{DC}} \tag{3.52}$$

The coupling inductor L_P of CGCI is used to reduce the dominant harmonic order n_I of the loads. By tuning the zero impedance point of $X_{LC}(n_I\omega) = 0$ in (3.51) at the dominant harmonic order n_I, the L_p can be expressed as:

$$L_p \approx \frac{1}{(2\pi f n_1)^2 \cdot C_p} \tag{3.53}$$

where n_I is the dominant harmonic order which equals to 5th ($n_I = 5$) for three-phase, three-wire system.

B. *Design of L_{PF}, C_{PF} and L_c*

The impedance of the SVC part is used to compensate the fundamental reactive power left by the CGCI part. The SVC part impedance can be expressed as:

$$\begin{aligned} X_{xy} &= \frac{V_{xy}^2}{Q_{c1xy}} = \frac{3 \cdot \overline{V}_x^2}{Q_{c1xy}} = \frac{3 \cdot \overline{V}_x^2}{Q_{Lx} - Q_{LC}} \\ &= \frac{\pi X_{L_{PF}} X_{C_{PF}}}{X_{C_{PF}}[2\pi - 2\alpha_{xy} + \sin(2\alpha_{xy})] - \pi X_{L_{PF}}} + X_{L_c} \end{aligned} \tag{3.54}$$

In (3.54), X_{Lc}, X_{CPF}, and X_{LPF} are the fundamental impedances of L_c, $C_{PF,}$ and L_{PF}. When $\alpha_x = 180°$ (thyristors are opened for the whole cycle), the SVC part has the maximum capacitive SVC impedance $X_{xy}(\alpha_{xy} = 180°)(<0)$. In this situation, the SVC part can provide maximum capacitive reactive power $Q_{cx(MaxCap)}$ for maximum inductive loads reactive power compensation $Q_{Lx(MaxInd)}$ (=-$Q_{cx(MaxCap)}$).On the other hand, when $\alpha_{xy} = 90°$ (one of the thyristors is closed for the whole cycle), the SVC part has the minimum inductive SVC impedance $X_{xy}(\alpha_x = 90°)$. The SVC part can provide maximum inductive reactive power $Q_{cx(MaxInd)}$ for maximum capacitive loads reactive power compensation $Q_{Lx(MaxCap)}$ (=-$Q_{cx(MaxInd)}$). Based on the above discussions and (3.54), the relationship among $Q_{Lx(MaxCap)}$, $Q_{Lx(MaxInd)}$, and coupling SVC impedance can be given as:

$$\frac{3 \cdot \overline{V}_x^2}{Q_{Lx(MaxInd)} - Q_{LC}} = X_{xy}(\alpha_{xy} = 180^o) = -\frac{1}{X_{C_{PF}} - X_{L_c}} \quad (3.55)$$

$$\frac{3 \cdot \overline{V}_x^2}{Q_{Lx(MaxCap)} - Q_{LC}} = X_{xy}(\alpha_{xy} = 90^o) = \frac{X_{L_{PF}} X_{C_{PF}}}{X_{C_{PF}} - X_{L_{PF}}} + X_{L_c} \quad (3.56)$$

Based on (3.55) and (3.56), the parallel capacitor C_{PF} and inductor L_{PF} can be designed as:

$$C_{PF} = \frac{Q_{Lx(MaxInd)} - Q_{LC}}{\omega^2 L_c (Q_{Lx(MaxInd)} - Q_{LC}) + 3 \cdot \omega \overline{V}_x^2} \quad (3.57)$$

$$L_{PF} = \frac{3\overline{V}_x^2 + \omega L_c (Q_{Lx(MaxCap)} - Q_{LC})}{-\omega (Q_{Lx(MaxCap)} - Q_{LC}) + \omega^3 L_c C_{PF} (Q_{Lx(MaxCap)} - Q_{LC}) + 3\omega^2 \overline{V}_x^2 C_{PF}} \quad (3.58)$$

where Q_{LC} can be found from (3.39).

The purpose of the L_c design is to mitigate the harmonic currents generated by the thyristors. The thyristors of the SVC part can be considered as a pair of bidirectional switches, which would generate low-order harmonic currents when the switches change their states. Through the harmonic currents rejection analysis, the L_c can be expressed as:

$$L_c = \frac{1}{(2\pi f n_2)^2 C_{PF}} \& L_c = \frac{1}{(2\pi f n_3)^2 C_{PF} - 1/L_{PF}} \quad (3.59)$$

where the harmonic orders n_2 and n_3 mainly depend on the L_c, $L_{PF,}$ and C_{PF} values. For a three-phase three-wire system, there is no 3nth order harmonic current and the common harmonic orders are $6n \pm 1 (\geq 5)$ with $n = 1, 2, 3 \ldots$. To avoid the harmonic currents injection by the SVC part, it is suggested to tune n_2 and n_3 away from the $6n \pm 1$ th order and/or close to the 3th order.

3.3.4 Simulation Results

In this section, the simulation case studies are provided to show the advantages of the proposed SVC//CGCI in comparison with the conventional IGCI and SVC//IGCI. The control scheme of the IGCI is based on the instantaneous reactive power theory, which is valid for sinusoidal or non-sinusoidal (harmonic), balanced or unbalanced three-phase power systems. The parameters of IGCI, SVC//IGCI, and the proposed of the SVC//CGCI and simulation setups are given in Table 3.5. The selections of the DC-link voltage of IGCI and CGCI are provided as follows.

The coupling fundamental impedances for the traditional IGCI part and CGCI part are fixed as X_L(>0) and X_{Lp}-X_{Cp}(<0). The relationships among the V_x, V_{invxf}, I_{cx2f}, and the coupling impedance of the traditional IGCI and CGCI can be expressed as:

$$V_{invxf\,(IGCI)} = V_x + X_L \cdot I_{cx2f} \tag{3.60}$$

$$V_{invxf\,(CGCI)} = V_x + (X_{L_p} - X_{C_p}) \cdot I_{cx2f} \tag{3.61}$$

where X_L(>0) and X_{Lp}-X_{Cp}(<0) are the coupling impedance of IGCI and CGCI. I_{cx2f} is the compensating current passing through the coupling impedance. The X_L(>0) is a small inductive impedance, and X_{Lp}-X_{Cp}(<0) is a large capacitive impedance. In (3.60), if the I_{cx2f} is the capacitive, the $V_{invxf(IGCI)}$ for IGCI is larger than V_x. On the other hand, as (3.61), the $V_{invxf(CGCI)}$ for CGCI is smaller than V_x. due to a large voltage drop of coupling *LC*. The fundamental dc-link voltage can be expressed as:

$$V_{DCf} \approx \sqrt{6} \cdot V_{invxf} > \sqrt{6} \cdot V_x \tag{3.62}$$

In (3.62), the scale of $\sqrt{6}$ (=$\sqrt{2} \cdot \sqrt{3}$) can be explained by the following two reasons: *A.* to guarantee the sufficient V_{DCf}, the peak value of fundamental inverter voltage need to be considered $V_{invxf\,(p)} > \sqrt{2} \cdot V_{invxf}$ *B.* to transfer the phase voltage V_{invxf} to line-to-line dc-link voltage, the scale of $\sqrt{3}$ is required. According to (3.62), the dc-link voltage of IGCI for the 110 V system $V_{DCf} > \sqrt{6} \cdot 110 = 270$ V. Also, the required DC-link voltage of CGCI can be smaller than IGCI. In this section,

Table 3.5 The Parameters of IGCI and SVC//CGCI used for Simulation Case Studies

	Parameters	Physical values
System parameters	V_{LL}, f, L_s	*191 V, 50 Hz,0.2 mH*
IGCI	L	*5 mH*
SVC//IGCI	$L_c, L_{PF}, C_{PF} L_p$	*5 mH, 30 mH,160 uF, 5 mH*
SVC//CGCI	$L_c, L_{PF}, C_{PF} L_p, C_P$	*5 mH, 30 mH,160 uF, 5 mH, 80 uF*

the dc-link voltage of the IGCI and CGCI parts are set to be 300 V and 120 V, respectively.

From Fig. 3.21a and Table 3.6, it can be seen that the conventional IGCI can compensate the power factor (PF) to close to unity from the original 0.97 and 0.71 for the light harmonic loads and heavy harmonic loads, respectively. Meanwhile, the THD_{isx} has been improved to lower than 5.4% for light harmonic loads and 4.1% for heavy harmonic loads. The required V_{DC} and I_{cx2} are about 300 V and 7.4 A, respectively.

Based on Fig. 3.21b and Table 3.6, the proposed SVC//CGCI can also compensate the PF to unity for the light harmonic loads (PF = 0.97) and heavy harmonic loads (PF = 0.71), respectively. Meanwhile, the source current THD_{isx} has been improved to lower than 6.1% and 5.1% for light loads with 28.8% and heavy loads with 10.7%, respectively. The required V_{DC} and I_{cx2} are 120 V and 3.2A, respectively.

During the unbalanced loading case, from Fig. 3.22a and Table 3.6, the source reactive power Q_{sx} can be compensated to about zero, and the P_{sx} can compensated to be balanced (about 562 W) with $P_{cx}{}^{*} = 70$ W. From Fig. 3.6a, the *PF* and THD_{isx} are improved to 0.99 and 15.6% (worst phase) for unbalanced loads after conventional IGCI compensation.

From Fig. 3.22b and Table 3.6, the Q_{sx} can be compensated to about 18 var and the P_{sx} can be balanced to 560 W with $P_{cx}{}^{*} = 70$ W. From Fig. 3.22b, the *PF* and THD_{isx} have been compensated to unity and 4.7% (worst phase).

On the other hand, from Fig. 3.22c and Table 3.6, the Q_{sx} can be compensated from 586 var to about 12 var, and the P_{sx} can be balanced to 555 W with $P_{cx}{}^{*} = 70$ W. From Fig. 3.22b, the *PF* and THD_{isx} have been compensated to unity and 3.5%

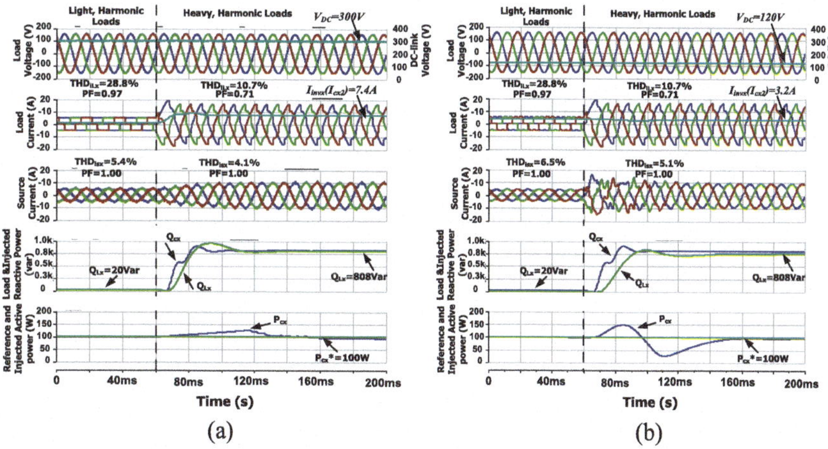

Fig. 3.21 Waveforms of load voltage, dc-link voltage, load current, inverter current, source current, load and injected reactive power, reference, and injected active power for light and heavy harmonic loads compensation by using **a** IGCI and **b** proposed SVC//CGCI

Table 3.6 Simulation Results for Different Loads Compensation Before and After IGCI and the Proposed SVC//CGCI Compensation

Loads	Compensation		PF	$THD_{isx}(\%)$	P_{sx}(W)	Q_{sx}(var)	i_{invx}(A)	V_{DC}(V)
Light Harmonic Loads	Before comp		0.97	28.3	442	20	–	–
	IGCI		1.00	5.4	308	9	1.3	300
	SVC//CGCI		1.00	6.1	308	7	3.0	120
Heavy Harmonic, Loads	Before comp		0.71	10.7	813	808	–	–
	IGCI		1.00	4.1	681	9	7.4	300
	SVC//CGCI		1.00	5.1	680	8	3.6	120
Unbalanced Harmonic Loads	Before comp	a	0.69	20.2	558	586	–	–
		b	0.84	25.7	546	546		
		c	0.86	18.6	759	349		
	IGCI	a	0.99	15.6	558	12	5.4	300
		b	0.99	3.5	560	14	4.8	
		c	0.99	1.9	567	13	3.4	
	SVC// IGCI	a	1.00	4.4	563	16	3.8	300
		b	1.00	3.8	560	17	3.7	
		c	1.00	4.7	557	18	3.9	
	SVC// CGCI	a	1.00	3.4	558	12	3.8	120
		b	1.00	3.5	550	11	3.5	
		c	1.00	1.9	557	10	3.9	

(worst phase). Compared with IGCI, the proposed SVC//CGCI has better THD_{isx} compensating performance.

Compared with the compensation performance of IGCI ($THD_{isx} > 15.0\%$) in Fig. 3.22a, the SVC//IGCI in Fig. 3.22b can obtain better THD_{isx} performance ($THD_{isx} < 5.0\%$) due to the compensating current sharing by SVC part of the SVC//IGCI. Among the three compensators, the SVC//CGCI in Fig. 3.22c obtains the best THD_{isx} (<4.0%) performance. In addition, the required DC-link voltage of SVC//CGCI (V_{DC} = 120 V) is smaller than IGCI (V_{DC} = 300 V) and SVC//IGCI (V_{DC} = 300 V).

3.3.5 Experimental Results

In this section, a 110 V-5kVA experimental prototype of IGCI and SVC//CGCI with the proposed design method is built in the laboratory as Fig. 2.11. Moreover, the experimental parameters of the SVC//CGCI are approximately the same as the simulation study, which is shown in Table 3.2. During the startup process, the SVC part is turned on first. Until the SVC part goes to a steady state, the CGCI part is

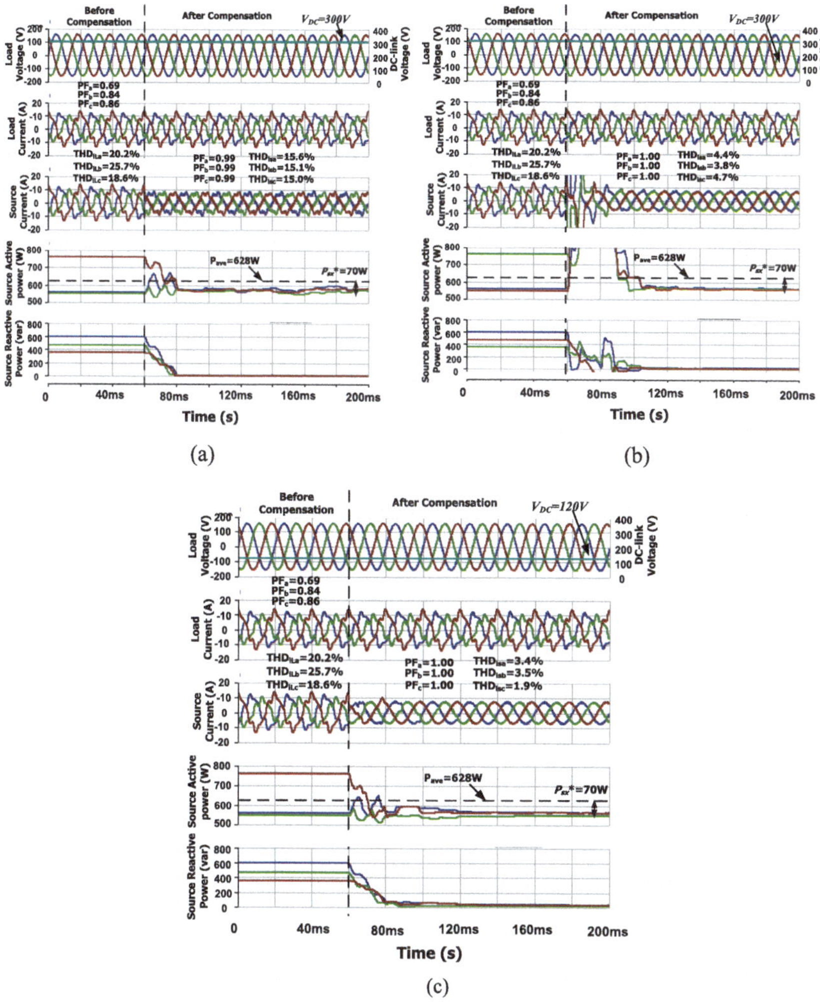

Fig. 3.22 Waveforms of load voltage, DC-link voltage, load current, source current, source reactive and active powers for unbalanced harmonic loads compensation by using **a** IGCI, **b** SVC//CGCI, and **c** proposed SVC//CGCI

connected to the system with the protection of series-connected resistors. After the CGCI goes to the steady state, the protect resistors are bypassed.

Figure 3.23 illustrates the waveforms of voltage and current by using the conventional IGCI. From Fig. 3.23a and Table 3.7, for the inductive loading, the worst phase PF and THD_{isx} have been improved from the original 0.96 and 24.5% to 0.99 and 13.5% after IGCI compensation. From Fig. 3.23b and Table 3.7, for the unbalanced loads compensation, the worst case PF and THD_{isx} have been compensated to 0.98

and 15.4% after IGCI compensation. And the DC-link voltage of IGCI is slightly larger than 300 V.

From Figs. 3.24, 3.25, and Table 3.7, after SVC//CGCI compensation with V_{DC} = 120 V, the worst phase PF has been improved from 0.76 to 0.99., the worst phase THD_{isx} has been kept to about 5.5% (worst phase). Since the proposed SVC//CGCI is

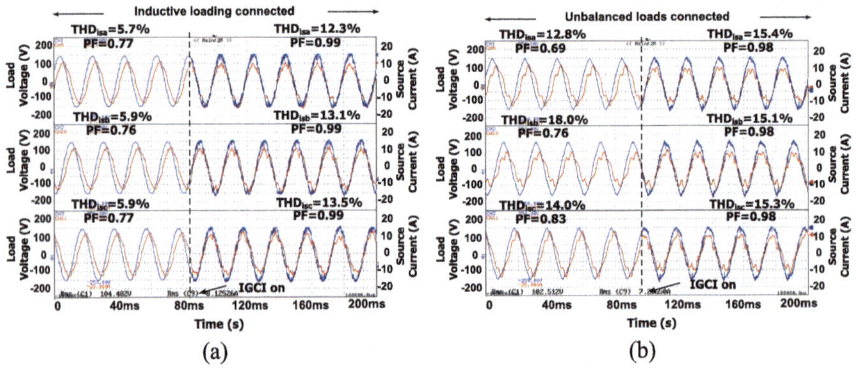

Fig. 3.23 Waveforms of v_x and i_{sx} by using IGCI for reactive power and harmonic compensation: **a** during inductive loads and **b** during unbalanced loads

Table 3.7 Experimental Results for Different Loads Before and After SVC//CGCI Compensation

			Q_{sx}(var)	P_{sx}(W)	PF	THD_{isx}(%)	V_{DC}(V)
Inductive loads	Before Comp	A	650	700	0.77	5.7	–
		B	650	740	0.76	5.9	
		C	650	740	0.77	5.9	
	IGCI	A	70	690	0.99	12.3	320
		B	80	700	0.99	13.1	
		C	70	700	0.99	13.5	
	SVC//CGCI	A	60	650	0.99	5.3	120
		B	50	690	0.99	5.4	
		C	80	680	0.99	5.5	
Unbalanced loads	Before Comp	A	600	590	0.69	12.8	–
		B	480	560	0.76	18.0	
		C	520	780	0.83	14.0	
	IGCI	A	70	640	0.98	15.4	320
		B	90	630	0.98	15.3	
		C	80	630	0.98	15.4	
	SVC//CGCI	A	10	620	0.99	8.4	120
		B	80	620	0.99	6.9	
		C	20	600	0.99	9.8	

connected to a low-rating isolation transformer with inductive impedance, the load voltage is increased after reactive power compensation. From Fig. 3.26a, b, after compensation, the load voltage and load active power slightly increase due to the inductive impedance of the isolation transformer. Compared with load active power (710 W, 780 W, and 780 W) in Fig. 3.26b, the source active power is reduced to 650 W, 690 W, and 680 W after SVC//CGCI compensation as Fig. 3.26c with $P_{cx}{}^{*} = 100$ W. From Fig. 3.26d, it can be seen that reactive power are compensated from 650 to 80 var for the worst phase.

From Figs. 3.27, 3.28, and Table 3.7, after SVC//CGCI compensation with $V_{DC} = 120$ V, the PF has been compensated from 0.69, 0.76, and 0.83 to 0.99 for three

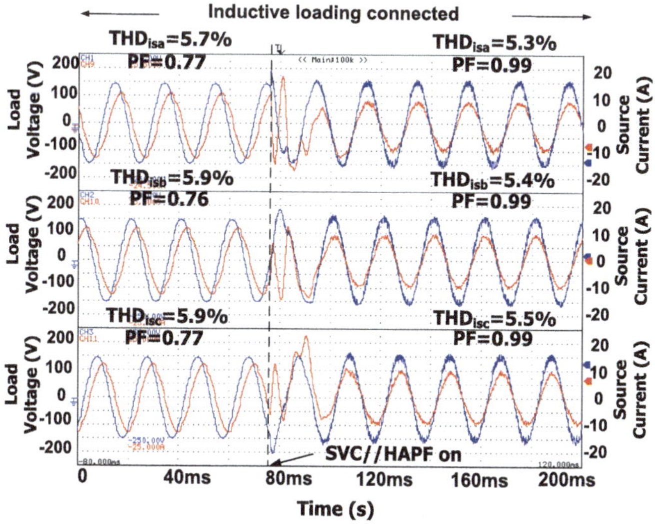

Fig. 3.24 Waveforms of v_x and i_{sx} by using SVC//CGCI for active power injection and non-active power compensation under inductive loads

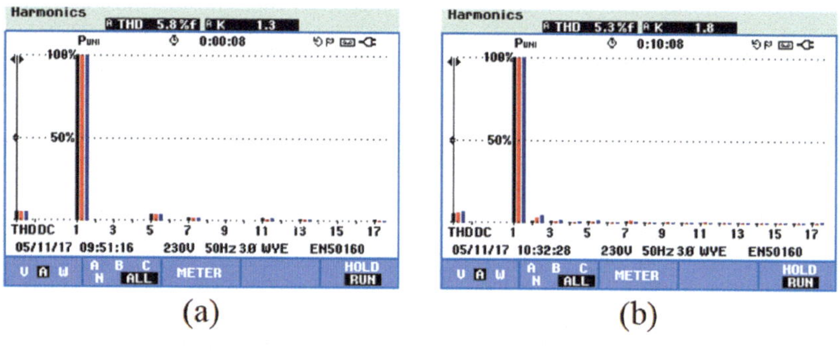

Fig. 3.25 The i_{sx} harmonic spectrums of SVC//CGCI for harmonic loads compensation: **a** before compensation, **b** after compensation

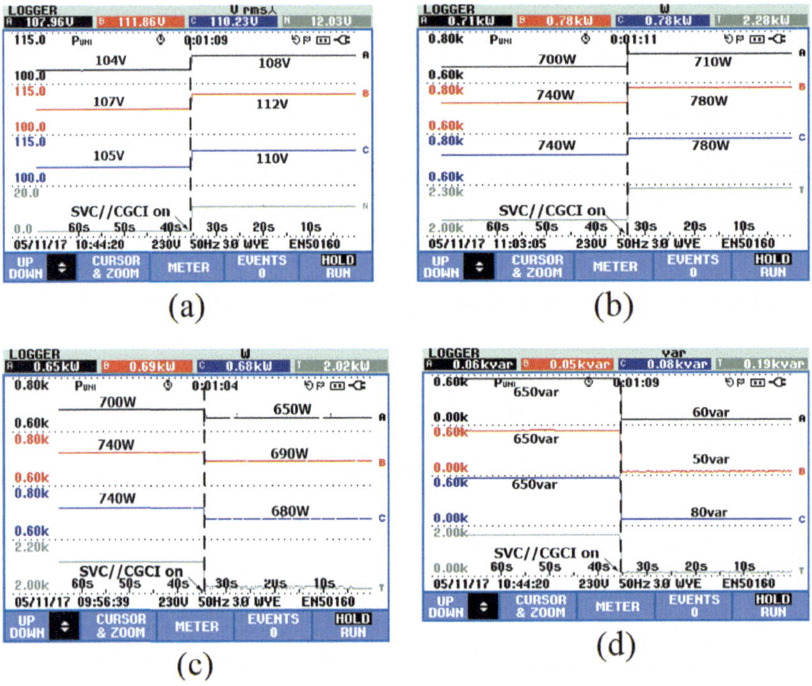

Fig. 3.26 Waveforms before and after SVC//CGCI compensation for harmonic loads compensation: **a** the load voltage, **b** load active power, **c** source active power, **d** source reactive power

phases under unbalanced loads compensation. The source current THD_{isx} has been improved to 9.8% for the worst phase. Figure 3.29 shows that the load voltage and source current are in phase with each other after compensation, even under unbalanced loads. From Fig. 3.30b, c, after SVC//CGCI compensation, the source active power of three phases have been balanced to 620 W, 620 W, and 600 W. Compared with the load active power (660 W, 650 W, and 840 W), the total source active power is reduced due to active power injection by the CGCI part with $P_{cx}{}^{*}$ = 70 W. Meanwhile, the source reactive power has been compensated from 600 var, 480 var, and 520 var to 10 var, 80 var, and 20 var, as shown in Fig. 3.30d.

Compared to the experiments results of the IGCI with the proposed SVC//CGCI, the SVC//CGCI obtains better compensation performances (PF and THD_{isx}) with lower dc-link voltage.

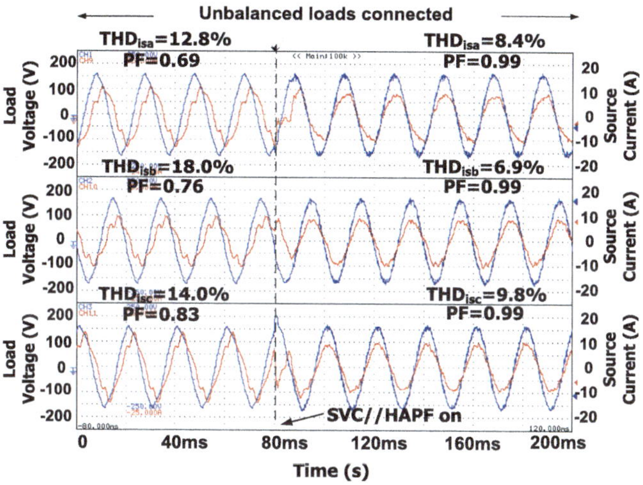

Fig. 3.27 Waveforms of v_x and i_{sx} by using SVC//CGCI for active power injection and non-active power compensation under unbalanced loads

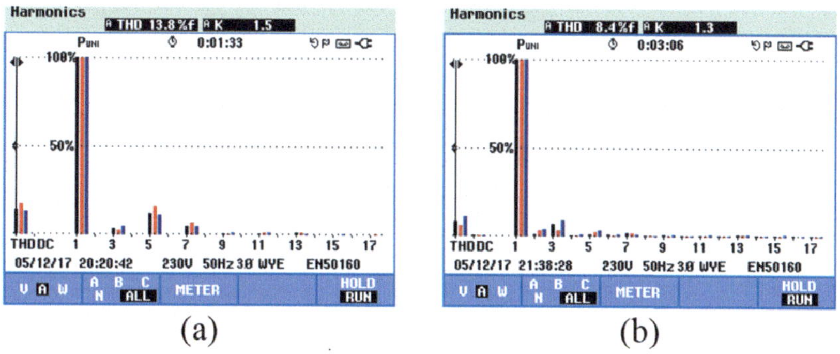

Fig. 3.28 The i_{sx} harmonic spectrums of SVC//CGCI for unbalanced harmonic loads compensation: **a** before compensation, **b** after compensation

3.3.6 Section Summary

In this chapter, a new structure of SVC in parallel with CGCI(SVC//CGCI) in a three-phase power system has been proposed and discussed as a cost-effective solution for active power injection and non-active power (harmonic, reactive and unbalanced power) compensation. The SVC part is used to compensate the reactive power dynamically and balance the source active power. And the CGCI is used to compensate the harmonic power and provide a fixed amount of reactive power (coupling LC). Finally, representative simulation and experimental results are provided to show that

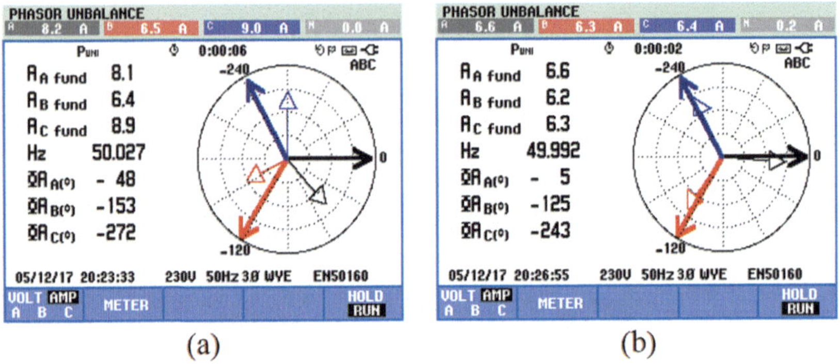

Fig. 3.29 The vector diagram of SVC//CGCI for unbalanced loads: **a** before compensation, **b** after compensation

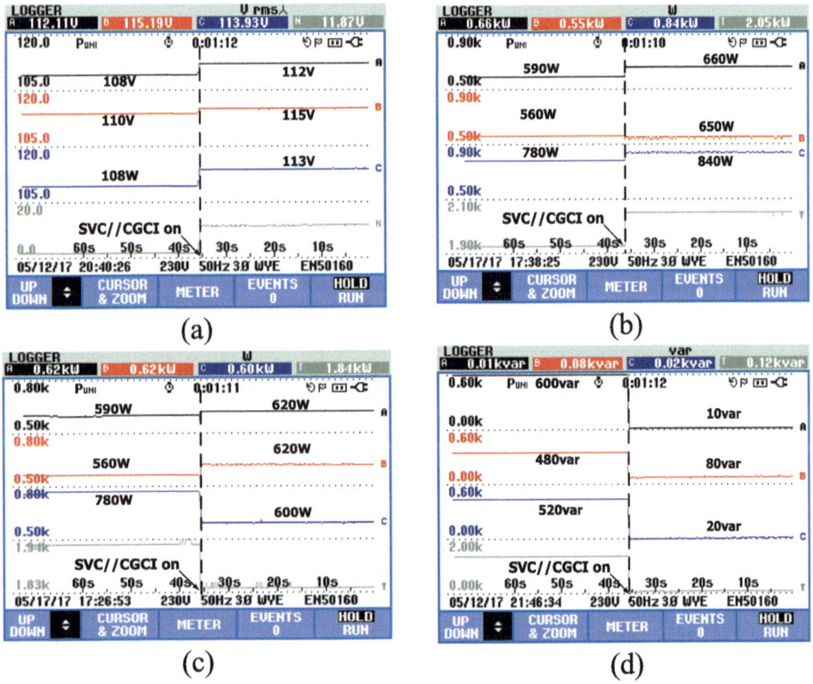

Fig. 3.30 Waveforms before and after SVC//CGCI compensation for unbalanced loads: **a** load voltage, **b** load active power, **c** source active power, **d** source reactive power

the SVC//CGCI has great promise for a wide operation range with both low voltage and current rating of active inverter.

3.4 Summary

This chapter summarized the applications of adaptive impedance converters in grid-connected systems for renewable energy integration and power quality compensation. This chapter proposed an adaptive hybrid grid-connected inverter topology and multifunctional SVC in parallel with a CGCI topology to enlarge the reactive power compensation range and reduce operational power loss. Compared with the conventional IGCI and CGCI, the proposed HGCI has distinct characteristics of a wide operational range and low DC-link operating voltage. Thus, the system and operational costs can be reduced. Moreover, the HGCI can inject active power and provide reactive power, unbalanced power, and harmonic power compensation at the same time. The SVC, in parallel with a CGCI topology, has great promise for a wide operation range with both low voltage and current rating of active inverter.

References

1. E. Santacana, G. Rackliffe, L. Tang, X. Feng, Gettingsmart. IEEE Power Energy Mag. **8**(2), 41–48 (2010)
2. Z. Shuaietal, Microgridstability: classificationandareview. Renew. Sustain. Energy Rev. **58**, 167–179 (2016)
3. H. Han, X. Hou, J. Yang, J. Wu, M. Su, J.M. Guerrero, Review of power sharing control strategies for islanding operation of ac microgrids. IEEE Trans. Smart. Grid **7**(1), 200–215 (2016)
4. B. Wang, L. Xian, V.R.K. Kanamarlapudi, K.J. Tseng, A. Ukil, H.B. Gooi, A digital method of power-sharing and cross-regulation suppres- sion for single-inductor multiple-input multiple-output dc–dcconverter. IEEE Trans. Ind. Electron. **64**(4), 2836–2847 (2017)
5. Z. Shuai, Y. Hu, Y. Peng, C. Tu, Z.J. Shen, Dynamic stability analysis of synchronverter-dominated microgrid based on bifurcation theory. IEEE Trans. Ind. Electron. **64**(9), 7467–7477 (2017)
6. Z. Shuai, W. Huang, C. Shen, J. Ge, Z.J. Shen, Characteristicsandre- straining method of fast transient inrush fault currents in synchronverters. IEEE Trans. Ind. Electron. **64**(9), 7487–7497 (2017)
7. H. Komurcugil, N. Altin, S. Ozdemir, I. Sefa, Lyapunov-function and proportional-resonant-based control strategy for single-phase grid- connected VSI with LCL filter. IEEE Trans. Ind. Electron. **63**(5), 2838–2849 (2016)
8. Q.C. Zhong, Y. Wang, B. Ren, UDE-basedrobustdroopcontrolof inverters in parallel operation. IEEE Trans. Ind. Electron. **64**(9), 7552–7562 (2017)
9. S. Dasgupta, S.N. Mohan, S.K. Sahoo, S.K. Panda, Applicationof four-switch-based three-phase grid-connected inverter to connect renew- able energy source to a generalized unbalanced microgrid system. IEEE Trans. Ind. Electron. **60**(3), 1204–1215 (2001)
10. A. Shayestehfard, S. Mekhilef, H. Mokhlis, IZDPWM-based feed- forward controller for grid-connected inverters under unbalanced and distorted conditions. IEEE Trans. Ind. Electron. **64**(1), 14–21 (2017)

11. M. Singh, A. Chandra, Real-time implementation of and is control for renewable interfacing inverter in 3p4w distribution network. IEEE Trans. Ind. Electron. **60**(1), 121–128 (2013)
12. R. Chilipi, N. Alsayari, K.H. AlHosani, A.R. Beig, Adaptivenotch filter based multipurpose control scheme for grid-interfaced three-phase four-wire DG inverter. IEEE Trans. Ind. Appl. **53**(4), 4015–4027 (2017)
13. T. Ye, N. Dai, C.S. Lam, M.C. Wong, J.M. Guerrero, Analysis, design, and implementation of a quasi-proportional-resonant controller for a multifunctional capacitive-coupling grid-connected inverter. IEEE Trans. Ind. Appl. **52**(5), 4269–4280 (2016)
14. N.Y. Dai, W.C. Zhang, M.C. Wong, J.M. Guerrero, C.S. Lam, Anal- ysis, control and experimental verification of a single-phase capacitive- coupling grid-connected inverter. IET Power Electron. **8**(5), 770–782 (2015)
15. L. Wang, C.S. Lam, M.C. Wong, Analysis, control and design of hybrid grid-connected inverter for renewable energy generation with power quality conditioning. IEEE Trans. Power Electron. **33**(8), 6755–6768 (2018)
16. L. Wang, C.S. Lam, M.C. Wong, Multifunctional hybrid structure of SVC and capacitive grid-connected inverter (SVC//CGCI) for active power injection and nonactive power compensation. IEEE Trans. Ind. Electron. **66**(3), 1660–1670 (2019)

Chapter 4
Adaptive Impedance Interlink Converters for AC/DC Microgrid

4.1 Introduction of Interlink Converters

The respective advantages of AC and DC microgrids lead to the blooming development of the hybrid AC/DC microgrid, which consists of AC and DC microgrids tied by an interlinking converter [1–3]. Nowadays, microgrids have become an attractive option for distributed generation systems with renewable energy sources (RES), storage systems, and different AC/DC loads. AC microgrids are rapidly developing since AC power systems are still mainly utilized [4–6]. On the other hand, DC microgrids have some attractive advantages, such as (1) higher efficiency for DC output sources and loads, (2) no reactive and unbalanced power problems, and (3) much fewer voltage problems due to the stored energy of DC capacitors. Hybrid AC/DC microgrids have been proposed to combine the advantages of both AC microgrids and DC microgrids in [7–14]. A hybrid AC/DC microgrid consists of three main parts: AC microgrid, DC microgrid, and power electronics interlinking converters (ICs). The IC is one of the most important elements in the hybrid AC/DC microgrids. To explore and improve the system performance, power rating, and application areas, the researchers are continuously developing the different structures and controls of ICs. A comparison of different interlinking converters is provided in Table 4.1.

The conventional structure of the interlinking converters is the interlinking converters with inductively coupled impedance (ICICs) [1–3]. Some improved topologies of ICICs, using dual power converters (Dual ICICs) or even more parallel connected modular converters, are proposed in [4–6]. Moreover, to reduce the cost and control complexity of paralleled ICICs, the auxiliary ICIC paralleled with the low-cost uncontrollable diode rectifier is proposed in [7]. Furthermore, the back-to-back connected ICICs have been proposed in [8, 9] to connect the unity grid, DC microgrid, and AC microgrid. However, all the above-mentioned ICICs are used to connect between AC microgrid and medium or high voltage DC microgrid since the

L. Wang et al., *Design and Applications of Adaptive Impedance Converters*,
https://doi.org/10.1007/978-981-97-5319-2_4

Table 4.1 Comparison of Different Interlinking Converters

Stage	Structure	Voltage level	Control	Control complexity	Converter power rating	Number of switches	Switching loss	Cost
Single-stage	Single ICIC	AC:200/ 380/400 V DC 400/ 600/800 V	P/Q exchange or voltage control, or power quality	Low	High	6	Medium	Medium
	Parallel multiply ICICs	AC:380/ 400 V DC:600/ 800 V	P/Q exchange and harmonic compensation	High	High	6N	High	High
	Auxiliary ICIC + paralleled diode rectifier	AC:200 V DC 370 V	As active filter	Medium	Low	6 (+6 Diode)	Medium	Medium
Multi-stages	Back to Back Connected ICIC	AC:3.5 kV DC 3.5 kV	P/Q flow + voltage stability	High	High	12	High	High
	ICIC + Boost converter	AC:180/ 310 V DC 300/ 380 V	P/Q exchange + voltage control	High	High	8	High	Medium
	ICIC + qZSI	AC:400 V DC 420 V	Voltage boost + unbalance condition control	High	High	7	High	Medium
	ICIC + DC transformer	AC:400 V DC 400 V	AC/DC bus voltage control	High	High	14	High	High
Single-stage	Proposed HCIC (droop/selective droop control)	AC:200 V DC 200 V	P/Q exchange + voltage control + power quality	High	Low	6 IGBTs + 6 thyristors	Low	Medium

peak voltage in AC microgrid is required to be lower than the voltage in DC microgrid. Thus, the conventional ICICs cannot be directly applied to the low-voltage DC (LVDC) microgrid.

To connect the LVDC microgrid, different two-stage converter structures such as ICIC + Boost converter [10, 11], ICIC + quasi-Z-source inverters (ICIC + qZSI) [12], and ICIC + DC transformer [13, 14], have been proposed. However, in these structures, the additional DC converter, quasi-Z-source inverters, and transformer (or DC/DC) can lead to an increase in the power switch capacity, control complexity, and cost. Therefore, to solve the above problems, a hybrid coupled interlinking converter (HCIC) is proposed for the hybrid AC/LVDC microgrid. The key features of the proposed HCIC system are that the coupled impedance is considered as one type of static var compensator (SVC) with controllable impedance, and the SVC part of HCIC enables the voltage drop between AC microgrid and LVDC microgrid.

With regards to the control of the interlinking converters, droop control is a well-known strategy for ICICs. It has the advantages of no communication requirements between converters, good power-sharing, high reliability, expandability and modularity, and low bandwidth, etc. However, as mentioned in various literature [15–23], the conventional droop control suffers many drawbacks, such as low accuracy, lack of the ability of harmonic compensation, slow response, etc. To enhance the current or power sharing accuracy, the improved droop control scheme has been proposed in [15, 16]. In some applications (e.g., fuel cells and batteries), the corresponding harmonic current is detrimental, for it may shorten the lifetime of devices. Therefore, the harmonic control functions [17–19] have been added to the original droop methods. In [20, 21], the main goals of the control system are to achieve satisfactory current sharing and maintain the microgrid stability. In addition, other factors like generation costs, efficiencies, and emission penalties at different load demands have been considered in [22, 23]. However, all the above state-of-the-art droop controls [15–23] are effectively applied to the conventional ICICs, but they cannot be directly applied to the proposed HCIC. In [24, 25], a similar structure of the hybrid active power filters and hybrid grid-connected inverter are proposed for reactive and harmonic and/or active power compensation. However, the instantaneous *PQ* control method is only adopted in [24, 25]. Moreover, the hybrid active power filters in [24] only act as power quality compensators and lack of active power transfer capability. Additionally, the hybrid grid-connected inverter in [25] is not suitable for microgrid applications, for its PQ control cannot affect the system voltage and frequency. Therefore, in the Chapter, a new droop control method for the HCIC is first proposed in the hybrid AC/LVDC microgrid applications. Generally, the large coupling capacitance in HCIC reduces the converter capacity. Thus, the practical over-capacity situation may be more likely to occur in HCIC. The over-capacity situation can deteriorate the performance of HCIC. Therefore, in the Chapter, a selective power droop control method for the HCIC is secondly proposed in the hybrid AC/LVDC microgrid applications. The proposed control strategy enables the HCIC to act as a PV inverter, load converter, and power quality compensator. Moreover, the proposed HCIC can be beneficial to the grids to unify the frequencies and voltages in microgrids. Thus, the proposed HCIC can connect with other ICICs and HCICs in parallel and is suitable for microgrid applications. Thus, the first main objectives of this Chapter are: (1) to propose a new structure of interlinking converter, namely, HCIC, composed of a converter in series with an SVC for the hybrid AC/LVDC microgrid; (2) to develop a new droop and selective droop control for the converter part of HCIC for the control of fundamental active and reactive power flow in the hybrid AC/DC microgrids; (3) to present a fast harmonic control for converter part of HCIC for the instantaneous harmonic power compensation.

In remote areas, the distribution networks are usually underdeveloped, and it is difficult to obtain a power supply. To reduce remote areas' investment cost and engineering difficulties, power supplies normally draw from the medium-voltage three-phase power lines. For example, in remote residential areas along the railway, the railway electric distribution power lines (27.5 kV) or other three-phase medium-voltage transmission power grid lines (10 kV) can support power for these areas [26]. However, there are always some major power quality problems in the traction power supply systems and other three-phase medium-voltage grids, namely voltage fluctuation, system unbalance, reactive power, and harmonics [27–29]. Therefore, these power quality problems must be solved when the traction power supply systems are used as residential power supply. In detail, the power transfer among different grids causes power quality issues. In particular, a small power quality problem can result in severe consequences in remote residential areas with microgrid grids [27–29]. Thus, the applicational demands and obstacles of the remote area residential power supply can be summarized as (1) power quality problems in both grid and load side elimination and (2) unified interface connections with unidirectional or bidirectional AC/DC loads. Therefore, a centralized adaptive impedance converter-based three-port power hub is proposed for the dedicated loads in bidirectional or unidirectional, and at the same time, isolate the power quality problems generated by the three-phase grid side and single-phase AC grid side. Thus, the second main objectives of this Chapter are: (1) to propose an adaptive impedance converter-based three-port power hub for improving power transfer and power quality compensation; (2) to develop the corresponding operation principle, power flow analysis, DC-link voltage design, and control strategy of the proposed power hub.

Section 4.1 presents the development of different interlink converters for AC/DC microgrids. Then, the application of adaptive impedance interlink converters and their new droop control and selective droop control are proposed in Sect. 4.2. After that, a centralized adaptive impedance converter-based three-port power hub is proposed in Sect. 4.3 to the dedicated loads in bidirectional or unidirectional, and at the same time, isolate the power quality problems generated by the three-phase grid side and single-phase AC grid side. Finally, the Chapter summary is drawn in Sect. 4.4.

4.2 A Hybrid Coupled Interlinking Converter for Hybrid AC/LVDC Microgrid

4.2.1 Circuit Configuration and Basic Control Block of the Proposed HCIC

The circuit configurations of the conventional ICIC and the proposed HCIC are illustrated in Fig. 4.1. Here, the subscript 'x' denotes phase $x = a, b, c$; v_{sx}, v_{x}, and v_{conx} are the source voltage, load voltage, and converter voltage, respectively. i_{sx}, i_{Lx}, and i_{cx} are source current, load current, and compensating current, respectively. The SVC part of HCIC consists of a coupled inductor L_c, a parallel capacitive C_{PF}, and a thyristor-controlled reactor (TCR) with an inductor L_{PF}, as seen in Fig. 4.1.

The control block of HCIC is provided in Fig. 4.1. A P-Q-α droop control is proposed to control the SVC part of HCIC. The purpose of the P-Q-α droop control is to compensate for the reactive power and provide a low impedance for injecting active power from the converter part. The firing angle α_x of the SVC part can be different for each phase so that the unbalanced power is compensated. For the converter part, there are two control loops, namely, the fundamental droop control and harmonic control. Based on the controllable coupled impedance, new P-δ, and Q-P-V droop controls are proposed for the fundamental power injection. A fast harmonic control of the converter is developed for the instantaneous harmonic power compensation.

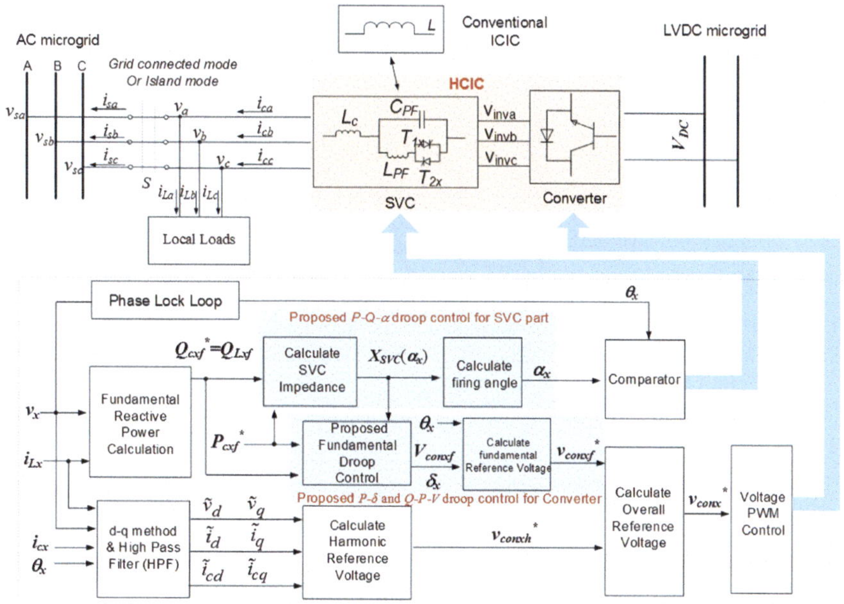

Fig. 4.1 Circuit configuration of HCIC and its proposed control block

4.2.2 Droop Control Method for HCIC

In this section, the control strategy of the HCIC, which includes three parts, is proposed. In *Part A*, the fundamental droop control is developed for the proposed HCIC, and the comparison between the control of the conventional ICICs and that of the proposed HCIC is made. Then, in *Part B*, a harmonic control method is given for HCIC. Finally, in *Part C,* the overall control strategy of the proposed HCIC is summarized.

A. *Fundamental Droop Control of the Proposed HCIC*

A.1 *Proposed P-δ and Q-P-V droop control for converter part*

The power flow between the converter and the grid can be calculated as below:

$$P^*_{cxf} = \left(\frac{V_x V_{conxf}}{X}\cos\delta_x - \frac{V_x^2}{X}\right)\cos\varphi + \frac{V_x V_{conxf}}{X}\sin\delta_x\sin\varphi \tag{4.1}$$

$$Q^*_{cxf} = \left(\frac{V_x V_{conxf}}{X}\cos\delta_x - \frac{V_x^2}{X}\right)\sin\varphi + \frac{V_x V_{conxf}}{X}\sin\delta_x\cos\varphi \tag{4.2}$$

where the subscript "*f*" represents the fundamental frequency component, V_{conxf} is the fundamental output voltage of the converter, and δ_x represents the phase angle between V_x and V_{conxf}. φ is related to the coupled impedance, X is the coupled impedance, and $\varphi = -90^\circ$ and $\varphi = 90^\circ$ are for capacitive and inductive coupled impedances, respectively. The variation of the phase angle δ_x operation region is small. Thus, (4.1) and (4.2) can be simplified into (4.3) and (4.4) for the capacitive impedance as:

$$P^*_{cxf} = -\frac{V_x V_{conxf}}{X_{SVC}(\alpha_x)}\sin\delta_x \approx -\frac{V_x V_{conxf}}{X_{SVC}(\alpha_x)}\delta_x \tag{4.3}$$

$$Q^*_{cxf} = -\left(\frac{V_x V_{conxf}}{X_{SVC}(\alpha_x)}\cos\delta_x - \frac{V_x^2}{X_{SVC}(\alpha_x)}\right) \tag{4.4}$$

For the fundamental converter voltage can be controlled by:

$$\delta_x = -\frac{X_{SVC}(\alpha_x)}{V_x \cdot V_{conxf}}P^*_{cxf} = -m_x \cdot P^*_{cxf} \tag{4.5}$$

$$\begin{aligned} V_{conxf} &= \sqrt{\left(\frac{X_{SVC}(\alpha_x)}{V_x}P^*_{cxf}\right)^2 + \left(V_x - \frac{X_{SVC}(\alpha_x)}{V_x}Q^*_{cxf}\right)^2} \\ &= \sqrt{\left(n_x P^*_{cxf}\right)^2 + \left(V_x - n_x Q^*_{cxf}\right)^2} \end{aligned} \tag{4.6}$$

where m_x and n_x can be expressed as:

$$m_x = n_x \big/ V_{conxf} \tag{4.7}$$

$$n_x = X_{SVC}(\alpha_x) \big/ V_x \tag{4.8}$$

From above (4.3)–(4.8), the $X_{SVC}(\alpha_x)$ can be given as:

$$\begin{aligned} X_{SVC}(\alpha_x) &= \frac{X_{TCR}(\alpha_x) X_{C_{PF}}}{X_{C_{PF}} - X_{TCR}(\alpha_x)} + X_{L_c} \\ &= \frac{\pi X_{L_{PF}} X_{C_{PF}}}{X_{C_{PF}}(2\pi - 2\alpha_x + \sin 2\alpha_x) - \pi X_{L_{PF}}} + X_{L_c} \end{aligned} \tag{4.9}$$

where X_{Lc}, X_{LPF}, and X_{CPF} are the fundamental impedances of L_c, L_{PF}, and C_{PF}, respectively. Thus, the fundamental component $v_{conxf}{}^*$ is expressed as:

$$v_{conxf}^* = \sqrt{2} V_{conxf} \sin(\omega t + \vartheta_x + \delta_x) \tag{4.10}$$

where δ_x and V_{conxf} are from (4.5) and (4.6), respectively. $\theta_x = \omega t + \vartheta_x$ is the phase of grid voltage which is directly obtained through the phase lock loop. Based on previous discussions, the comparisons between the conventional droop control for ICIC and the proposed droop control for HCIC are provided in Table 4.2.

In Table 4.2, the droop control strategy modified its form for converters with different types of coupled impedance. Compared with fixed inductive coupled ICIC, the SVC part of HCIC can be dynamically controlled by firing angle α_x, which can be inductive or capacitive. In particular, the droop control strategies for inductive coupled inverter and for capacitive coupled inverter are in the opposite way. Thus, the conventional droop control for ICICs does not work for the controllable coupled

Table 4.2 Comparisons between the conventional droop control for ICIC and the proposed droop control for HCIC

	ICIC	Proposed HCIC	
Coupled impedance	Inductive $\varphi = 90°$	Inductive $\varphi = 90°$	Capacitive $\varphi = -90°$
Active power	$P_{cxf}^* \approx \frac{V_x V_{conxf}}{X_L} \delta_x$	$P_{cxf}^* \approx \frac{V_x V_{conxf}}{X_{SVC}(\alpha)} \delta_x$	$P_{cxf}^* \approx -\frac{V_x V_{conxf}}{X_{SVC}(\alpha)} \delta_x$
Reactive power	$Q_{cxf}^* = \frac{V_x V_{conxf}}{X_L} \cos \delta_x - \frac{V_x^2}{X_L}$	$Q_{cxf}^* = \frac{V_x V_{conxf}}{X_{SVC}(\alpha)} \cos \delta_x - \frac{V_x^2}{X_{SVC}(\alpha)}$	$Q_{cxf}^* = -\left(\frac{V_x V_{conxf}}{X_{SVC}(\alpha)} \cos \delta_x - \frac{V_x^2}{X_{SVC}(\alpha)}\right)$
m in (4.5)	$m_x = \Delta\delta_x \big/ \Delta P_{cxf\,\max}^*$	$m_x = n_x \big/ V_{conxf}$	$m_x = n_x \big/ V_{conxf}$
n in (4.6)	$n_x = \Delta V_{conxf} \big/ \Delta Q_{cxf\,\max}^*$	$n_x = X_{SVC}(\alpha) \big/ V_x$	$n_x = X_{SVC}(\alpha) \big/ V_x$

HCIC. However, the proposed droop control strategy enables it to work for the controllable coupled HCIC system.

A.2 *Proposed P-Q-α droop control for SVC part*

Besides the droop control applied in the converter part of HCIC, the SVC part of HCIC is controlled by *P-Q-α* droop control. As shown in (4.6), the V_{conxf} and $X_{SVC}(\alpha_x)$ are operated together to determine the injected active and reactive power, and if the V_{conxf} in (4.11) fixed, the injected P_{cxf} and Q_{cxf} power range of the HCIC can reach the maximum at the minimum value of $X_{\text{SVC}}(\alpha_{\text{x}})$. The $X_{SVC}(\alpha_x)$ in (4.6) can be calculated by setting the derivative of V_{conxf} equal to zero.

$$\frac{\partial V_{conxf}}{\partial X_{SVC}(\alpha_x)} = 0, \quad X_{SVC}(\alpha_x) = \frac{V_x^2 \cdot Q_{cxf}^*}{\left(P_{cxf}^*\right)^2 + \left(Q_{cxf}^*\right)^2} \tag{4.11}$$

By comparing the calculated $X_{SVC}(\alpha_x)$ in (4.9) and (4.11), the firing angle α_x can be determined. Considered (4.9) is complicated, a look-up table between the firing angle α_x and the $X_{SVC}(\alpha_x)$ is created inside the controller. Besides, *P-Q-*α_x droop control of the SVC part is plotted in Fig. 4.2, which presents the relationship between the active and reactive powers and the firing angle, and the active and reactive powers can be controlled by α_x. In Fig. 4.2, the reactive power can be positive (inductive reactive power) and negative (capacitive reactive power) power.

It is necessary to point out that here, the reactance $X_{\text{SVC}}(\alpha_x)$ is affected by the active and reactive power passing through the HCIC, and the *P-Q-α* droop control is proposed for the SVC part of the HCIC. Figure 4.3 shows the relationship between

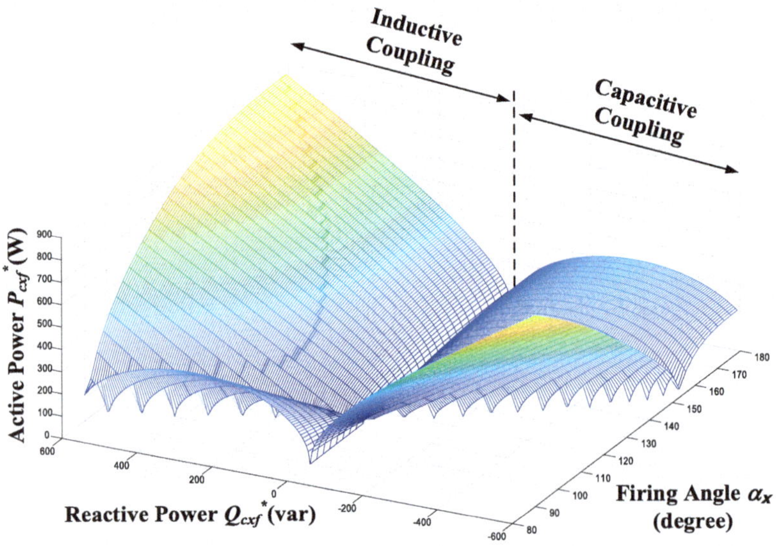

Fig. 4.2 *P-Q-*α_x droop control of SVC part

SVC reactance and the firing angle, and the reactance of $X_{SVC}(\alpha_x)$ is controlled by firing angle α_x. Therefore, the $X_{SVC}(\alpha_x)$ reactance is not constant when the active and reactive power are not fixed. While under the steady-state (fixed active power P and reactive power Q), the $X_{svc}(\alpha_x)$ reactance is fixed.

B. *Harmonic Control of the Proposed HCIC*

The converter is used to improve the filtering characteristics of the SVC part, suppress the harmonic power, and inject active power. The fast harmonic control is compared with the fundamental droop control of HCIC. The proposed P-δ and Q-P-V droop controls in the converter are used to control the fundamental active and reactive power, and the control period is above one cycle. On the other hand, the purpose of the fundamental P-Q-α droop control is to determine SVC impedance, and the SVC part has two back-to-back connected thyristors T_{1x} and T_{2x}, for each phase, and they are triggered alternately in every half cycle, as shown in Fig. 4.4.

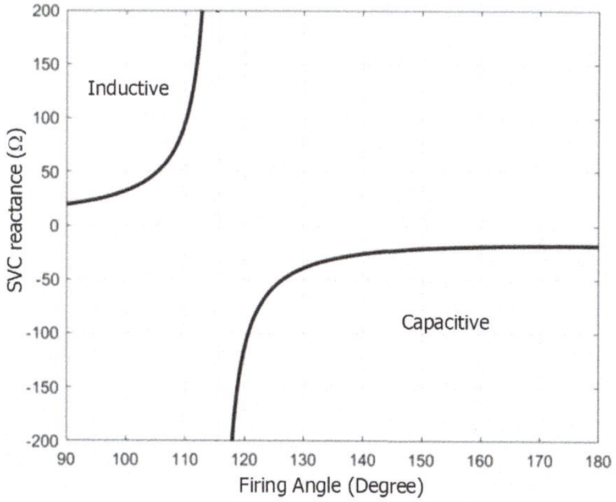

Fig. 4.3 Relationship between SVC reactance and firing angle

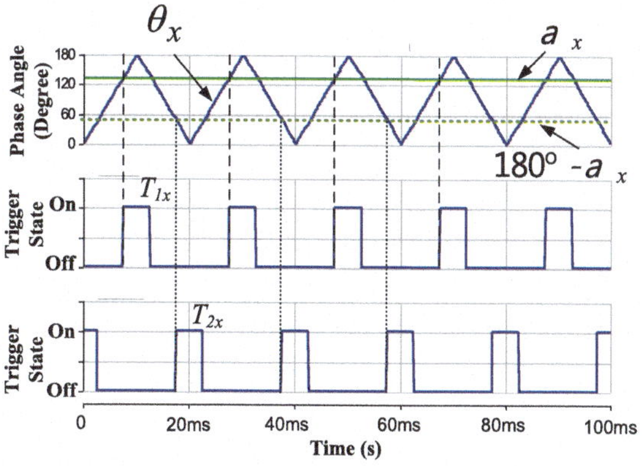

Fig. 4.4 Generation of trigger signals for the thyristors of the SVC part

Compared with the relatively slow fundamental droop control (0.02 s), the fast harmonic droop control (above 1/(5 kHz)) can enhance the overall control speed of the HCIC. The detailed control time of the proposed HCIC is provided in Table 4.3.

To achieve the above purposes, the reference converter voltage $v_{conx}{}^{*}$ needs to be calculated. Using Kirchhoff's voltage law, one can be written as:

$$v_{conx} = v_x - L_c \frac{di_{cx}}{dt} - v_{CPFx} \tag{4.12}$$

$$v_{CPFx} = L_{PF} \frac{di_{cx}}{dt} - L_{PF} C_{PF} \frac{d^2 v_{CPFx}}{dt^2} \tag{4.13}$$

The system of (4.13) is transformed into the synchronous orthogonal frame (d-q method) with the simplifications:

$$\begin{aligned} v^{*}_{conhd} &= \frac{1}{1 - L_{PF}C_{PF}\omega^2} \begin{bmatrix} -\omega\left(L_c\left(1 - L_{PF}C_{PF}\omega^2\right) + L_{PF}\right) \cdot \tilde{i}_{cq} + \\ \left(1 - L_{PF}C_{PF}\omega^2\right)v_d - \left(L_c\left(1 - L_{PF}C_{PF}\omega^2\right) + L_{PF}\right)u_d \end{bmatrix} \\ &= K_{p1} \cdot \tilde{i}_{cq} + K_{p2} \cdot v_d + K_{p3} \cdot u_d \end{aligned} \tag{4.14}$$

$$\begin{aligned} v^{*}_{conhq} &= \frac{1}{1 - L_{PF}C_{PF}\omega^2} \begin{bmatrix} -\omega\left(L_c\left(1 - L_{PF}C_{PF}\omega^2\right) + L_{PF}\right) \cdot \tilde{i}_{cd} + \\ \left(1 - L_{PF}C_{PF}\omega^2\right)v_q - \left(L_c\left(1 - L_{PF}C_{PF}\omega^2\right) + L_{PF}\right)u_q \end{bmatrix} \\ &= K_{p1} \cdot \tilde{i}_{cd} + K_{p2} \cdot v_q + K_{p3} \cdot u_q \end{aligned} \tag{4.15}$$

in which,

$$K_{P1} = \frac{\omega\left(L_c\left(1 - L_{PF}C_{PF}\omega^2\right) + L_{PF}\right)}{1 - L_{PF}C_{PF}\omega^2} \tag{4.16}$$

$$K_{P2} = \frac{1 - L_{PF}C_{PF}\omega^2}{1 - L_{PF}C_{PF}\omega^2} \tag{4.17}$$

$$K_{P3} = \frac{L_c\left(1 - L_{PF}C_{PF}\omega^2\right) + L_{PF}}{1 - L_{PF}C_{PF}\omega^2} \tag{4.18}$$

Table 4.3 Control time of the proposed HCIC

	SVC part	Converter part		
Control	Fundamental droop control		Fast harmonic control	Total transient response
	P-Q-α droop	*P-δ* and *Q-P-V droop*		
Time	20 ms	20 ms	1/(5 kHz)*	10 ms

* Note that the 5 kHz is the switching frequency

where K_{p1}, K_{p2}, and K_{p3} are the proportional gains of the controller. With the practical parameters of the proposed HCIC in Table 4.4, the K_{p1}, K_{p2}, and K_{p3}, which are 19.45, 1.00, and 0.06, respectively, are calculated. In (4.14) and (4.15), i_{cd}, i_{cq}, v_d, v_q, u_d, and u_q can be expressed as:

$$\begin{bmatrix} i_{cd} \\ i_{cq} \end{bmatrix} = \sqrt{\frac{2}{3}} \begin{bmatrix} \cos\theta_a & \cos\left(\theta_a - \frac{2\pi}{3}\right) & \cos\left(\theta_a - \frac{4\pi}{3}\right) \\ -\sin\theta_a & -\sin\left(\theta_a - \frac{2\pi}{3}\right) & -\sin\left(\theta_a - \frac{4\pi}{3}\right) \end{bmatrix} \begin{bmatrix} i_{ca} \\ i_{cb} \\ i_{cc} \end{bmatrix} \tag{4.19}$$

$$\begin{bmatrix} v_d \\ v_q \end{bmatrix} = \sqrt{\frac{2}{3}} \begin{bmatrix} \cos\theta_a & \cos\left(\theta_a - \frac{2\pi}{3}\right) & \cos\left(\theta_a - \frac{4\pi}{3}\right) \\ -\sin\theta_a & -\sin\left(\theta_a - \frac{2\pi}{3}\right) & -\sin\left(\theta_a - \frac{4\pi}{3}\right) \end{bmatrix} \begin{bmatrix} v_a \\ v_b \\ v_c \end{bmatrix} \tag{4.20}$$

$$\begin{bmatrix} u_d \\ u_q \end{bmatrix} = \begin{bmatrix} \left(L_c\left(1 - L_{PF}C_{PF}\omega^2\right) + L_{PF}\right) \cdot \frac{di_{cd}}{dt} \\ \left(L_c\left(1 - L_{PF}C_{PF}\omega^2\right) + L_{PF}\right) \cdot \frac{di_{cq}}{dt} \end{bmatrix} = \begin{bmatrix} K_p \cdot \left(\tilde{i}_{Ld} - \tilde{i}_{cd}\right) \\ K_p \cdot \left(\tilde{i}_{Lq} - \tilde{i}_{cq}\right) \end{bmatrix} \tag{4.21}$$

where $\tilde{i}_{Lq}$, $\tilde{i}_{Ld}$, $\tilde{i}_{cq}$ and $\tilde{i}_{cd}$ are the harmonic components of load current and the compensating currents in *d-q* frame, respectively. θ_a is obtained by v_a passing through the *PLL*. The harmonic reference voltage $v_{conxh}{}^*$ can be expressed as:

$$\begin{bmatrix} v^*_{conah} \\ v^*_{conbh} \\ v^*_{conch} \end{bmatrix} = \sqrt{\frac{2}{3}} \begin{bmatrix} \cos\theta_a & -\sin\theta_a \\ \cos\left(\theta_a - \frac{2\pi}{3}\right) & -\sin\left(\theta_a - \frac{2\pi}{3}\right) \\ \cos\left(\theta_a - \frac{4\pi}{3}\right) & -\sin\left(\theta_a - \frac{4\pi}{3}\right) \end{bmatrix} \begin{bmatrix} v^*_{conhd} \\ v^*_{conhq} \end{bmatrix} \tag{4.22}$$

The final reference converter voltage can be given as:

$$v^*_{conx} = v^*_{conxf} + v^*_{conxh} \tag{4.23}$$

where $v_{conxf}{}^*$ and $v_{conxh}{}^*$ are seen in (4.10) and (4.22), respectively.

C. *Summary of Overall Control Strategy of the Proposed HCIC*

In Fig. 4.1, SVC part impedance X_{SVCx} can be obtained from (4.11) by proposed *P-Q-α* control. Then, by comparing the impedance X_{SVCx} in (4.11) with (4.9), the firing angle α_x can be obtained. Since (4.9) does not have a closed-form solution, a look-up table has been created to obtain the firing angle α_x with known X_{SVCx}

Table 4.4 Parameters for case studies

	Parameters	Physical values
System parameters	V_x, f, L_s	110 V, 50 Hz,0.2 mH
HCIC parameters	L_c, C_c	5 mH, 80 μF
HCIC power	$S_{HIC}, Q_{LC},$	1250 W, 930 var
System power	$P_l, Q_{L1}, S_{Lh},$	600 W, 2000 var, 600 W

directly. By comparing the pre-calculated α_x of the phase θ_x, the trigger signals for the thyristor of the SVC part can be generated.

For the fundamental converter control, the fundamental voltage reference $v_{conxf}{}^*$ can be calculated through the proposed P-δ and Q-P-V droops from (4.5) and (4.6), and the reference converter voltage $v_{conxh}{}^*$ is obtained through the harmonic loop control in (4.19)–(4.22). Then, the final reference voltage is obtained by summing up $v_{conxf}{}^*$ and $v_{conxh}{}^*$. Finally, the trigger signals for the converter is generated by voltage pulse width modulation.

4.2.3 A Selective Power Droop Control for HCIC

Based on power flow analysis, the detailed selective power droop control scheme for the proposed HCIC is developed in the following subsections.

A. *Deduction of Power Range of the HCIC*

The required power transfer between the AC grid and LVDC grid can be expressed as:

$$S_{HCIC\_req}^2 = Q_{L1}^2 + P_I^2 + S_{Lh}^2 \tag{4.24}$$

For the $S_{HCIC\_req} > S_{HCIC}$ situation, the power ratings of the HCIC and its converter can be expressed as:

$$S_{HCIC}^2 = (k_P \cdot P_I)^2 + \left[Q_{LC} + k_Q \cdot (Q_{L1} - Q_{LC})\right]^2 + (k_H \cdot S_{Lh})^2 \tag{4.25}$$

$$S_{conv}^2 = (k_P \cdot P_I)^2 + \left[k_Q \cdot (Q_{L1} - Q_{LC})\right]^2 + (k_H \cdot S_{Lh})^2 \tag{4.26}$$

The Q_{LC} is expressed as:

$$Q_{LC} = \frac{3 \cdot V_x^2}{X_{Cc} - X_{Lc}} \tag{4.27}$$

Assuming that the S_{HCIC_req} in (4.24) is larger than S_{HIC} in (4.25), the k_P in terms of k_Q and k_H can be expressed based on (4.25) as:

$$k_P = \frac{\sqrt{\left[Q_{LC} + k_Q \cdot (Q_{L1} - Q_{LC})\right]^2 + (k_H \cdot S_{Lh})^2 - S_{HCIC}^2}}{P_I} \tag{4.28}$$

Taking the parameters in Table 4.4 as a case study, the relationship among S_{HIC}, Q_{L1}, and P_I (or S_{Lh}) can be plotted in Fig. 4.5. In Fig. 4.5, the required S_{HCIC_req} is calculated based on the full compensation situation by using (4.28) and Table 4.4, while the S_{HIC} is pre-set due to hardware limitation. Therefore, selective droop control is required for $S_{HCIC\_req} > S_{HCIC}$. The selective factors of a case study in Fig. 4.5 (redpoint) is provided in Fig. 4.6. In Fig. 4.6, since $S_{HCIC\_req} > S_{HCIC}$, the k_P, k_Q or k_H cannot equal to one at the same time.

B. *Proposed Selective Factor Calculation*

The priority mechanism of controlling selective factors is described as follows:

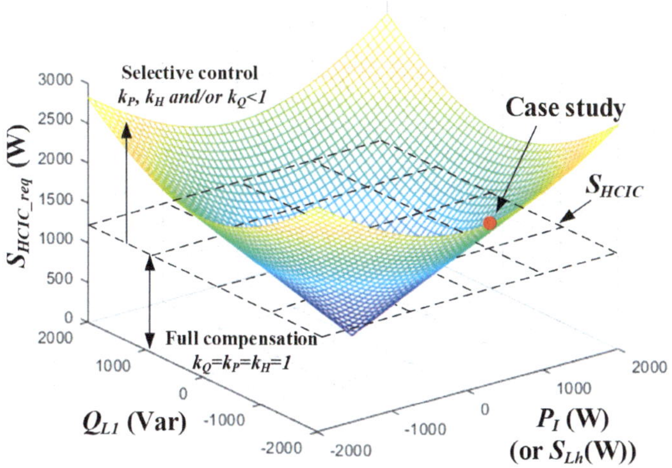

Fig. 4.5 The required S_{HCIC} in terms of Q_{L1} and P_I (or S_{Lh})

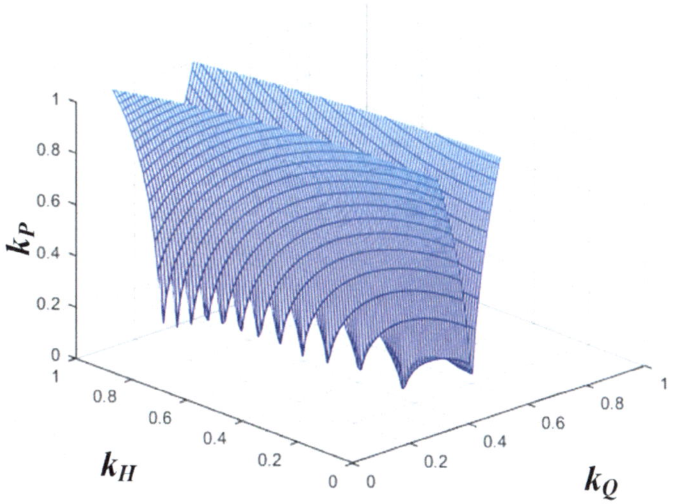

Fig. 4.6 The possible selection of k_P, k_Q or k_H

(1) Firstly, to make sure the primary function of the HCIC can be achieved, the active power is required to be transferred between AC and DC grids, which in turn keeps the AC and DC grids operating at acceptable variation levels;
(2) Under the fulfillment of the first requirement, as stated above, the HCIC needs to prevent harmonic pollution or harmonic resonance opportunities from occurring to the AC system to damp out the unstable problem. The harmonic performance is strictly limited by national or international standards;
(3) Finally, the reactive power is compensated when the above two criteria are fulfilled. The reactive power is highly related to system efficiency compared to the above two issues. Efficiency is important but is of lower priority compared to the basic function, stability, and operational standards.

Based on the above discussions, the priority has been assigned to the P_I first, then to the S_{Lh}, and finally to the Q_{L1}. If $S_{HCIC}{}^2 \geq P_I{}^2 + Q_{L1}{}^2 + S_{Lh}{}^2$, the HIC can perform full compensation with $k_P = k_H = k_Q = 1$. Otherwise, if $P_I{}^2 + Q_{L1}{}^2 + S_{Lh}{}^2 \geq S_{HCIC}{}^2 \geq P_I{}^2 + S_{Lh}{}^2 + Q_{LC}{}^2$, by substituting the $k_P = k_H = 1$ into (4.23), the k_Q can be calculated as:

$$k_Q = \left| \frac{\sqrt{S_{HCIC}^2 - P_I^2 - S_{Lh}^2} - Q_{LC}}{Q_{L1} - Q_{LC}} \right| \tag{4.29}$$

If $P_I{}^2 + S_{Lh}{}^2 + Q_{LC}{}^2 \geq S_{HCIC}{}^2 \geq P_I{}^2 + Q_{LC}{}^2$, the $k_P = 1$, $k_Q = 0$, and k_H can be calculated as:

$$k_H = \left| \frac{\sqrt{S_{HCIC}^2 - P_I^2 - Q_{LC}}}{S_{Lh}^2} \right| \tag{4.30}$$

If $S_{HCIC}{}^2 < P_I{}^2 + Q_{LC}{}^2$, the $k_Q = k_H = 0$ and k_P can be calculated as:

$$k_P = \left| \frac{\sqrt{S_{HCIC}^2 - Q_{LC}^2}}{P_I} \right| \tag{4.31}$$

Figure 4.7 shows the determination of selective factors. The instantaneous calculated Q_{L1}, S_{Lh}, P_I, and pre-designed S_{HCIC} as input signals are used to calculate the selective factors. If $S_{HCIC}{}^2 \geq P_I{}^2 + Q_{L1}{}^2 + S_{Lh}{}^2$, the S_{HCIC} is sufficient for active, reactive, and harmonic power; thus $k_P = k_H = k_Q = 1$. Otherwise, if $P_I{}^2 + Q_{L1}{}^2 + S_{Lh}{}^2 \geq S_{HCIC}{}^2 \geq P_I{}^2 + S_{Lh}{}^2 + Q_{LC}{}^2$, the S_{HCIC} can only support Q_{LC}, P_I, and S_{Lh}, while the capacity left is not enough for the rest of Q_{L1}. Thus $k_P = k_H = 1$ and k_Q is obtained from (4.29). If $P_I{}^2 + S_{Lh}{}^2 + Q_{LC}{}^2 \geq S_{HCIC}{}^2 \geq P_I{}^2 + Q_{LC}{}^2$, S_{HCIC} is just enough for P_I, Q_{LC}, and partial S_{Lh}, so that $k_P = 1$, $k_Q = 0$, and k_H can be calculated from (4.30). If $S_{HCIC}{}^2 < P_I{}^2 + Q_{LC}{}^2$, the S_{HIC} is used to support Q_{LC} and partial P_I, thus $k_Q = k_H = 0$ and k_P can be calculated as (4.31).

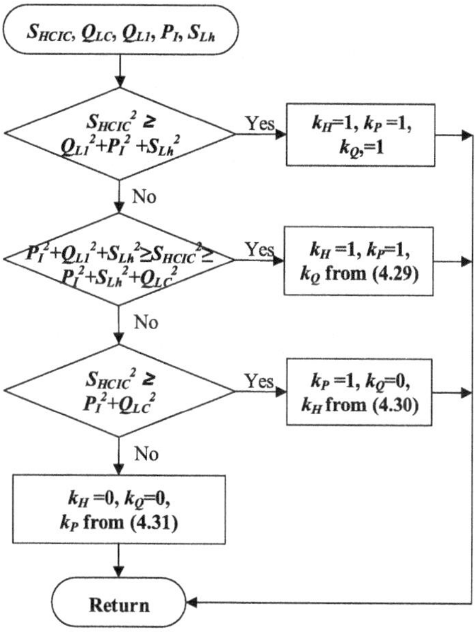

Fig. 4.7 Flowchart for determining k_Q, k_H and k_P

C. *Selective P-δ* and *Q-P-V droops for Converter Part*

The power flow between the converter and the grid can be calculated as follows:

$$P_{cxf}^* = \left(\frac{V_x V_{conxf}}{X}\cos\delta_x - \frac{V_x^2}{X}\right)\cos\varphi + \frac{V_x V_{conxf}}{X}\sin\delta_x\sin\varphi \tag{4.32}$$

$$Q_{cxf}^* = \left(\frac{V_x V_{conxf}}{X}\cos\delta_x - \frac{V_x^2}{X}\right)\sin\varphi + \frac{V_x V_{conxf}}{X}\sin\delta_x\cos\varphi \tag{4.33}$$

The phase angle δ_x operation region is a small region around $\delta_x = 0$. Thus, the P-δ and Q-P-V droop control theory in (4.32) and (4.33) for HCIC can be simplified as:

$$P_{cxf}^* = -\frac{V_x V_{conxf}}{X_{LC}}\sin\delta_x \approx -\frac{V_x V_{conxf}}{X_{LC}}\delta_x \tag{4.34}$$

$$Q_{cxf}^* = -\left(\frac{V_x V_{conxf}}{X_{LC}}\cos\delta_x - \frac{V_x^2}{X_{LC}}\right) \tag{4.35}$$

By substituting $P_{cxf}{}^* = k_P \cdot P_I/3$ and $Q_{cxf}{}^* = k_Q \cdot Q_{L1}/3$ into (4.34) and (4.35), it yields:

$$\delta_x = -\frac{X_{LC}}{V_x \cdot V_{conxf}}P_{cxf}^* = -\frac{m_x k_P \cdot P_I}{3} \tag{4.36}$$

$$\begin{aligned} V_{convf}^* &= \sqrt{\left(\frac{X_{LC}}{V_x}k_P \cdot \frac{P_I}{3}\right)^2 + \left(V_x - \frac{X_{LC}}{V_x}k_Q \cdot \frac{Q_{L1}}{3}\right)^2} \\ &= \sqrt{\left(\frac{n_x k_P \cdot P_I}{3}\right)^2 + \left(V_x - \frac{n_x k_Q \cdot Q_{L1}}{3}\right)^2} \end{aligned} \tag{4.37}$$

where m_x and n_x can be expressed as:

$$m_x = n_x / V_{conxf} \tag{4.38}$$

$$n_x = X_{LC} / V_x \tag{4.39}$$

In (4.34)–(4.39), the X_{LC} is given as:

$$X_{LC} = \omega L_c - \frac{1}{\omega C_c} \tag{4.40}$$

Three-phase reactive power of Q_{L1} from common loads can be instantaneously calculated as:

$$\begin{aligned} Q_{L1} &= \overline{q}_L \\ q_L &= -v_q i_{Ld} + v_d i_{Lq} \end{aligned} \tag{4.41}$$

where $\overline{q}_{Lq}$ is DC component of load reactive power. i_{Ld}, i_{Lq}, v_d, and v_q are expressed as:

$$\begin{bmatrix} i_{Ld} \\ i_{Lq} \end{bmatrix} = \sqrt{\frac{2}{3}} \begin{bmatrix} \cos\theta_a & \cos\left(\theta_a - \frac{2\pi}{3}\right) & \cos\left(\theta_a - \frac{4\pi}{3}\right) \\ -\sin\theta_a & -\sin\left(\theta_a - \frac{2\pi}{3}\right) & -\sin\left(\theta_a - \frac{4\pi}{3}\right) \end{bmatrix} \begin{bmatrix} i_{La} \\ i_{Lb} \\ i_{Lc} \end{bmatrix} \tag{4.42}$$

$$\begin{bmatrix} v_d \\ v_q \end{bmatrix} = \sqrt{\frac{2}{3}} \begin{bmatrix} \cos\theta_a & \cos\left(\theta_a - \frac{2\pi}{3}\right) & \cos\left(\theta_a - \frac{4\pi}{3}\right) \\ -\sin\theta_a & -\sin\left(\theta_a - \frac{2\pi}{3}\right) & -\sin\left(\theta_a - \frac{4\pi}{3}\right) \end{bmatrix} \begin{bmatrix} v_a \\ v_b \\ v_c \end{bmatrix} \tag{4.43}$$

Finally, the fundamental component ${v_{conxf}}^*$ is expressed as:

$$v_{conxf}^* = \sqrt{2} V_{conxf} \sin(\omega t + \vartheta_x + \delta_x) \tag{4.44}$$

where δ_x and V_{conxf} are from (4.26) and (4.27), respectively, the phase $\theta_x = \omega t + \vartheta_x$ can be directedly measured by the phase lock loop of the grid voltage.

D. *Proposed Selective Harmonic Power Control*

In the proposed selective control method, the harmonic power can be expressed as:

$$S_{Lh} + \tilde{s}_{Lh} = \sqrt{v_d^2 + v_q^2} \cdot \sqrt{\tilde{i}_{Ld}^2 + \tilde{i}_{Lq}^2} \tag{4.45}$$

where $\tilde{i}_{Lq}$ and $\tilde{i}_{Ld}$ are harmonic components of load current.

From Fig. 4.1, using Kirchhoff's voltage law, one can write:

$$v_{conx} = v_x - L_c \frac{di_{cx}}{dt} - \frac{1}{C_c} \int i_{cx} dt \tag{4.46}$$

The system is transformed into the synchronous orthogonal frame (d-q method) with the simplifications in:

$$v_{conhd}^* = 2\omega L_c i_{Lq} + \omega v_d - \omega \int v_q dt - u_d \tag{4.47}$$

$$v_{conhq}^* = -2\omega L_c i_{Lq} + \omega v_q + \omega \int v_d dt - u_q \tag{4.48}$$

In (4.47) and (4.48), i_{cq}, i_{cd}, u_d, and u_q can be expressed as:

$$\begin{bmatrix} i_{cd} \\ i_{cq} \end{bmatrix} = \sqrt{\frac{2}{3}} \begin{bmatrix} \cos\theta_a & \cos\left(\theta_a - \frac{2\pi}{3}\right) & \cos\left(\theta_a - \frac{4\pi}{3}\right) \\ -\sin\theta_a & -\sin\left(\theta_a - \frac{2\pi}{3}\right) & -\sin\left(\theta_a - \frac{4\pi}{3}\right) \end{bmatrix} \begin{bmatrix} i_{ca} \\ i_{cb} \\ i_{cc} \end{bmatrix} \tag{4.49}$$

$$\begin{bmatrix} u_d \\ u_q \end{bmatrix} = \begin{bmatrix} L_c \cdot \frac{di_{cd}}{dt} + \left(-\omega^2 L_c + \frac{1}{C_c}\right) i_{cd} \\ L_c \cdot \frac{di_{cq}}{dt} + \left(-\omega^2 L_c + \frac{1}{C_c}\right) i_{cq} \end{bmatrix} = \begin{bmatrix} K_p \cdot \left(k_H \cdot \tilde{i}_{Ld} - \tilde{i}_{cd}\right) \\ K_p \cdot \left(k_H \cdot \tilde{i}_{Lq} - \tilde{i}_{cq}\right) \end{bmatrix} \tag{4.50}$$

where $\tilde{i}_{Lq}$, $\tilde{i}_{Ld}$, $\tilde{i}_{cq}$ and $\tilde{i}_{cd}$ are harmonic components of load current and compensating currents in d-q frame, respectively. The harmonic reference voltage $v_{conxh}{}^*$ can be expressed as:

$$\begin{bmatrix} v_{conah}^* \\ v_{conbh}^* \\ v_{conch}^* \end{bmatrix} = \sqrt{\frac{2}{3}} \begin{bmatrix} \cos\theta_a & -\sin\theta_a \\ \cos\left(\theta_a - \frac{2\pi}{3}\right) & -\sin\left(\theta_a - \frac{2\pi}{3}\right) \\ \cos\left(\theta_a - \frac{4\pi}{3}\right) & -\sin\left(\theta_a - \frac{4\pi}{3}\right) \end{bmatrix} \begin{bmatrix} v_{conhd}^* \\ v_{conhq}^* \end{bmatrix} \tag{4.51}$$

The final reference converter voltage can be given as:

$$v_{conx}^* = v_{conxf}^* + v_{conxh}^* \tag{4.52}$$

where $v_{conxf}{}^*$ and $v_{conxh}{}^*$ are from (4.44) and (4.51), respectively.

E. *Overall Control Block of Proposed Method*

The overall control block of the proposed selective droop power management method for the HCIC is given in Fig. 4.8.

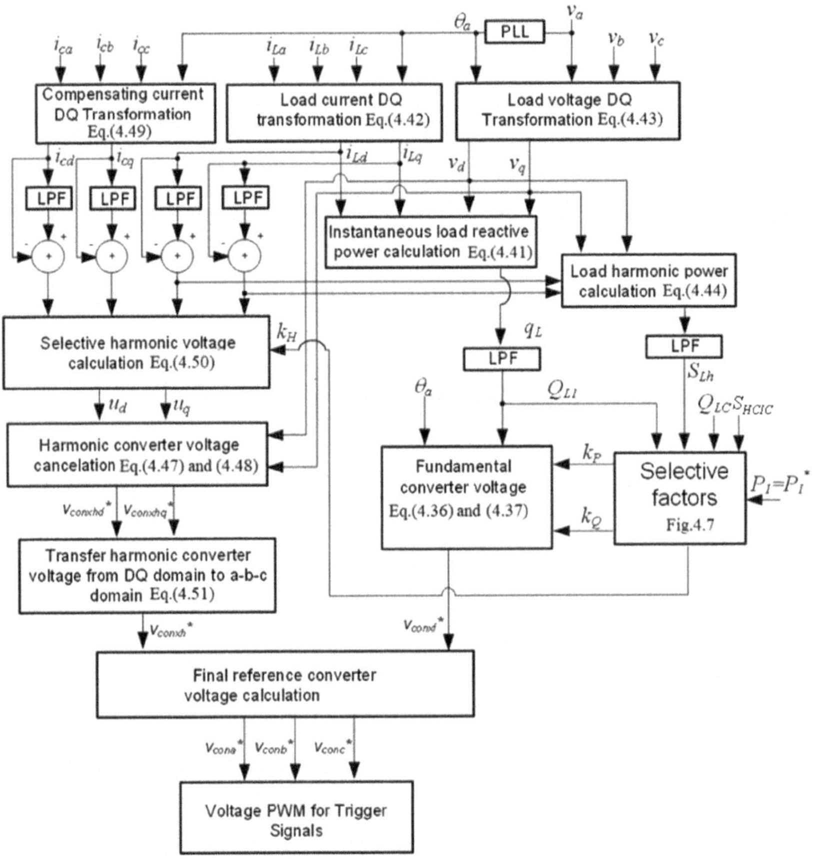

Fig. 4.8 The overall control of the proposed selective droop power management method

In practical applications, the voltages and currents (v_x, i_{Lx}, i_{cx}, and V_{DC}) are instantaneously measured by transducers and then input to the controller. As shown in Fig. 4.8, the reactive and harmonic powers (Q_{L1} and S_{Lh}) are calculated in the controller by using the measured voltage and current. With the Q_{L1}, S_{Lh}, P_I^* (from DC microgrid), pre-designed S_{HIC} and Q_{LC}, the selective factors are automatically obtained. Thus, the proposed selective droop has an automatic coefficient adjustment. After that, the reference voltages are calculated with these input signals, following the PWM control method to output the trigger signals for the converter.

4.2.4 Simulation Results

(a) *Case study of Droop Control for HCIC*

Table 4.5 Parameters of HCIC and microgrids

	Parameters	Physical values
AC microgrid	V_{LL}, f, L_s	200 V, 50 Hz, 1 mH
DC microgrid	V_{DC}	200 V
HCIC	L_c, L_{PF}, C_{PF},	5 mH, 30 mH, 160 μF
Controller	$K_{p1}, K_{p2}, K_{p3}, K_p, K_I$	19.45, 1.00, 0.06, 500, 20

In this part, the comparative simulation results are provided in *Part A* among the conventional ICIC, the recent ICIC + DC transformer, and the proposed HCIC in terms of the response time, THD, tracking accuracy, and power loss. In *Part B*, the simulation results are provided to verify the feasibility and validation of the proposed structure and the proposed droop control of HCIC. The parameters of the proposed HCIC and AC/DC microgrids for simulation are given in Table 4.5.

A. *Comparison of the Conventional ICIC, ICIC + DC Transformer, and the Proposed HCIC*

The switching loss of the conventional ICIC or ICIC + DC transformer is from the switching components (IGBTs) of the converter part. In contrast, the switching loss of the HCIC consists of two parts: (1) switching loss of the IGBTs in the converter part and (2) switching loss of the thyristors in the SVC part.

For the IGBT, the switching loss is obtained by measuring the collector-emitter voltage and current of each IGBT, the total converter switching losses can be found as:

$$P_{sw(IGBT)} = \sum_{n=0}^{n} P_{sw(IGBT),n} = \sum_{n=0}^{n} \frac{1}{T_f} \int_0^{T_f} v_{ce,n}(t) \cdot i_{ce,n}(t) dt \quad (4.53)$$

where $v_{ce,n}(t)$ and $i_{ce,n}(t)$ are the collector-emitter voltage and current of IGBT, respectively. T_f is the fundamental period, and n is the number of IGBT. For the thyristor, each thyristor of the SVC part only turns on one time in a fundamental period, which is different from that the IGBTs of the converter part do. In the HCIC applications, the on-state switching loss is the dominant loss of the thyristor that can be easily calculated as:

$$P_{sw(thyristor)} = \sum_{n=0}^{n=6} P_{thy,n} = \sum_{n=0}^{n=6} \left(U_{TO,n} \times I_{TAV,n} + r_T \times I_{TRMS,n}^2 \right) \quad (4.54)$$

where $I_{TRMS,n}$ is the RMS value of current passing through the thyristor, $I_{TAV,n}$ is the average value of current passing through the thyristor, U_{TO} is the thyristor threshold voltage, and r_T is the thyristor slope resistance. According to the datasheet of thyristor modules (PK110FG160), r_t = ~0.44 mΩ and $U_{TO} = 0.25$ V.

Seen from Fig. 4.9, the conventional ICIC has less response time than the proposed HCIC for the injected current and VAR demand, and the conventional ICIC + DC transformer has the same response time for the injected current and VAR demand as the conventional ICIC does, which is ignored to show in Fig. 4.9b. Specifically, the SVC part of the proposed HCIC has two back-to-back connected thyristors T_{1x} and T_{2x}, for each phase, and they are triggered alternately in every half cycle. Thus, the response time of SVC part is about one fundamental cycle (0.02 s). Thankfully, the converter part can improve the response time of the SVC part, so that the overall response time of HCIC can be less than 0.02 s. Moreover, the proposed HCIC has better steady-state tracking accuracy and less power losses than the conventional ICIC and ICIC + DC transformer do due to its high DC voltage. In addition, all three converters have satisfactory *THD* performances (<5%). Moreover, the detailed performances of these features are summarized in Table 4.6.

Compared with the conventional ICIC + DC transformer in Table 4.6, the proposed HCIC has a lower converter rating and switching loss. However, the transient time of HCIC is slightly longer than that of ICIC + DC transformer. However the transient time of HCIC is still less than one cycle.

B. *Dynamic Performance of the HCIC*

Figures 4.10, and 4.11 show the dynamic performance of the HCIC for absorbing active power (600 W) and injecting active power (−600 W) from the DC microgrid to the AC microgrid. In 50 ms, the AC local load current is suddenly increasing by about 100%. The total harmonic distortions of the load current (THD_{iLx}) are 12.8% and 5.8% for two sets of loads, respectively. The local load reactive power of the AC source current is compensated to about zero. Moreover, the AC source current THDisx is lower than 5%. Figure 4.12 presents the performance of the HCIC for unbalanced loads compensation. Seen from Fig. 4.12, the AC source reactive powers are compensated to about zero from the original 510, 290, and 508 var. Besides, the average value of source active power is 600 W ($P_{saf} = 460$, $P_{sbf} = 620$, and $P_{scf} = 710$ W) before compensation. After HCIC operation, the AC source active power is balanced to around 400 W with $P_{cxf} = 200$ W injection for each phase. Moreover, the AC source current *THD* is compensated to lower than 5% from 27.1% (showing the worst phase).

(b) *Case study of Droop Selective Control for HCIC*

This part provides comparative simulation studies of the proposed HCIC with droop selective control and other state-of-the-art converters. Also, this part verifies that the proposed HCIC with selective power control can selectively compensate active, harmonic, and reactive power components with satisfactory results. The parameters in Table 4.4 are also chosen for the simulation and experimental results.

A. *The Comparative Simulation Study*

For a fair comparison of the proposed HCIC and other converters, the selective factors in this situation are $k_P = k_Q = k_H = 1$. The switching loss is proportional to the connected DC link voltage and switch numbers. The conventional ICIC is connected

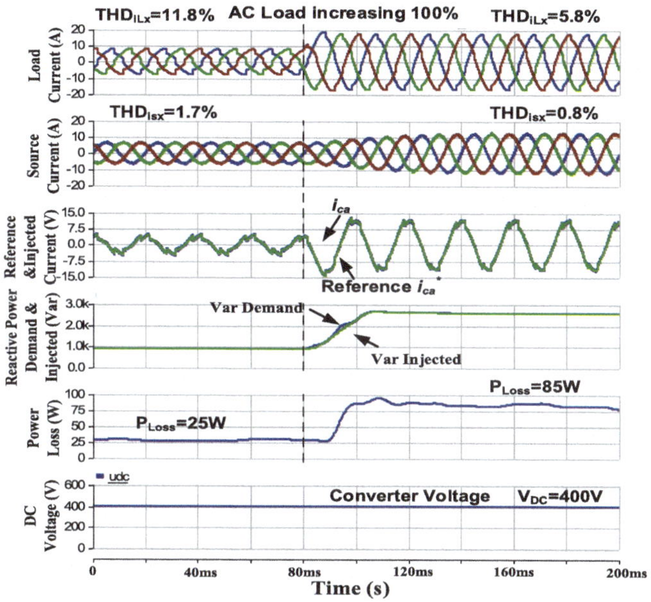

(a)

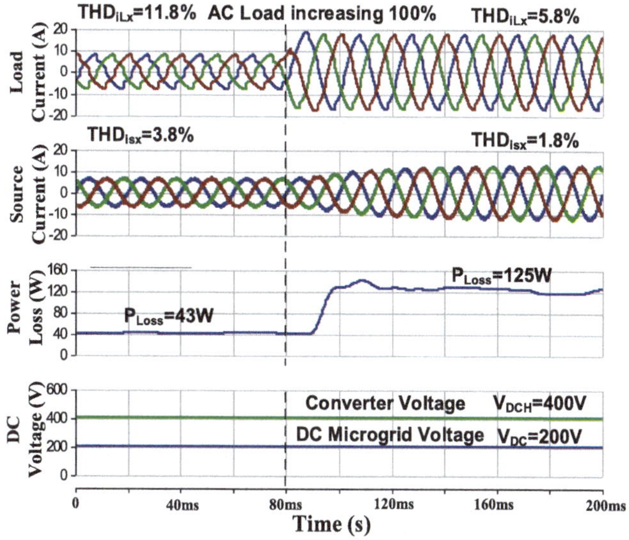

(b)

Fig. 4.9 Performances of interlinking converters in terms of THD, tracking accuracy, total power losses, and operation voltage: **a** ICIC, **b** ICIC + DC transformer, and **c** proposed HCIC

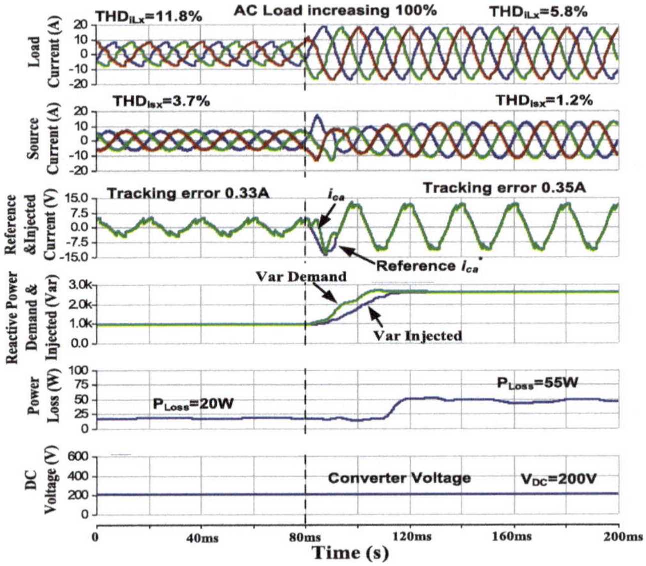

(c)

Fig. 4.9 (continued)

Table 4.6 Comparison of ICIC, ICIC + DC transformer and HCIC

Converter type		ICIC	ICIC + DC transformer	Proposed HCIC
Response time (ms)		<1	<1	10
Converter ratings (W)	1st load	1322	1322	611
	2nd load	3870	3870	1935
Tracking accuracy (%)	1st load	93.5	93.5	96.1
	2nd load	96.7	96.7	98.2
THD_{isx} (%)	1st load	1.7	3.8	3.7
	2nd load	0.8	1.8	1.2
Power loss (W)	1st load	25	43	20
	2nd load	85	125	55

to a 400 V DC bus so that the switching loss is high. To directly connect to the 200 V DC bus, the two-stage CLLC-DCT, conventional controlled HCIC and the proposed HCIC can be applied. Since both CLLC-DCT, and conventional controlled HCIC have more power switches, the power losses are higher than the proposed HCIC, as expected. Moreover. In Fig. 4.13a–d, all different structures reduce the source current total harmonic distortion (THD_{isx}) to less than 5% from the original 12.5% and 6.3%, respectively.

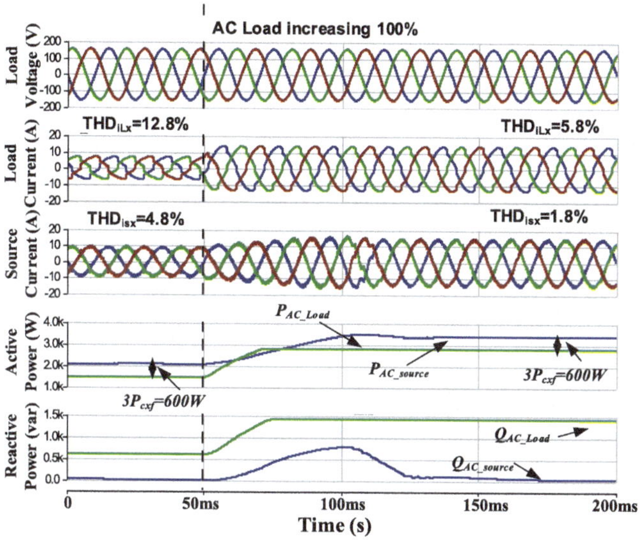

Fig. 4.10 Dynamic performance of HCIC for DC load charging

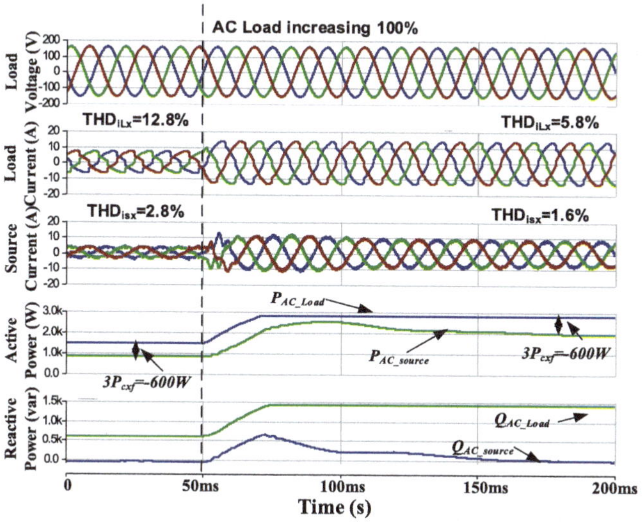

Fig. 4.11 Dynamic performance of HCIC for active power injection

B. *Proposed Selective droop control*

As shown in Fig. 4.14, with the same loads, the proposed method for the HIC can selectively compensate active, harmonic, and reactive power components with satisfactory results. In Fig. 4.14, for fixed Q_{LC} compensation by LC part ($k_P = k_H = k_Q$

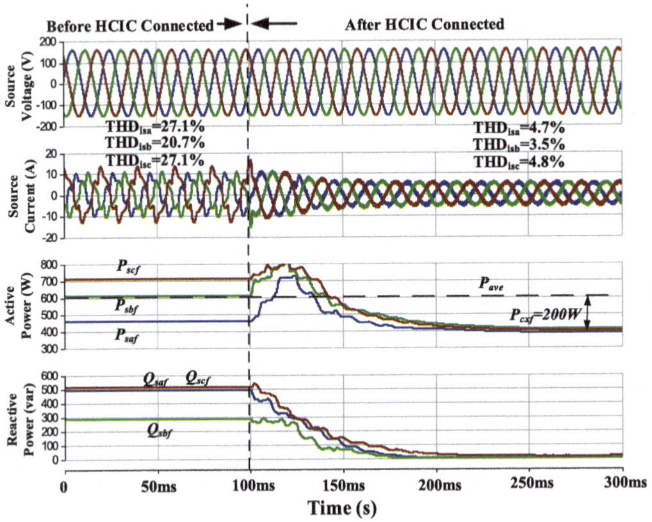

Fig. 4.12 Simulation results for unbalanced AC loads compensation before and after HCIC operation

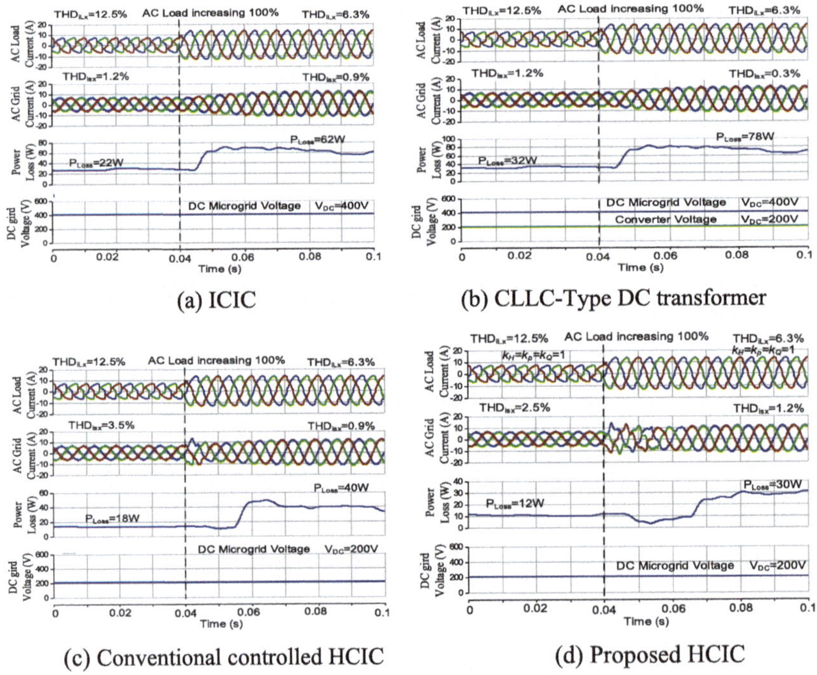

Fig. 4.13 The simulated performance of Interlinking Converters in terms of total power loss and operation voltage

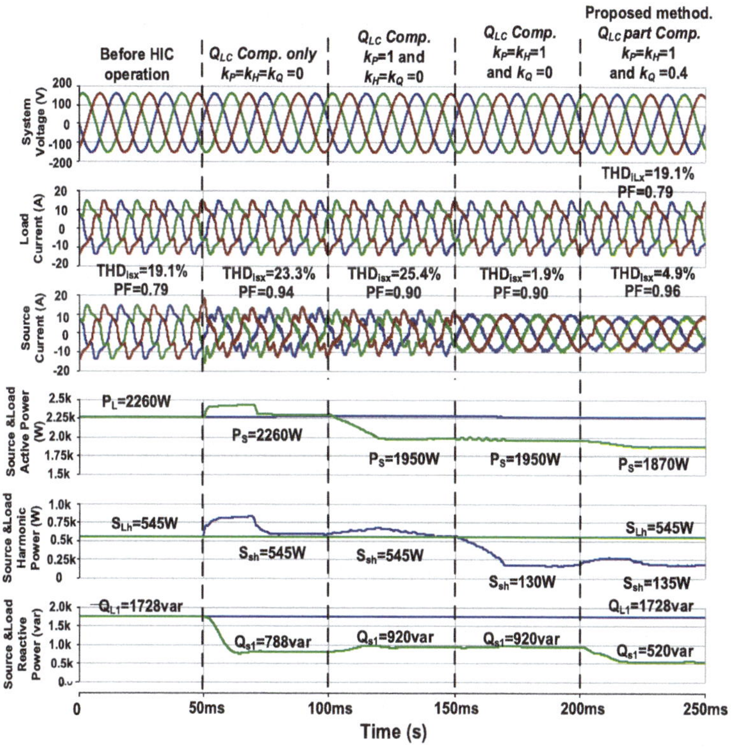

Fig. 4.14 Waveforms of system load voltage, load current, source current, source, and load reactive power, and source and load harmonic powers using the proposed selective droop control of HCIC

$= 0$), the source reactive power before and after the HIC compensation are 1728 var and 788 var ($Q_{LC} \approx 930$ var), respectively, while the harmonic and active power is kept the same. With the active power control ($k_P = 1, k_H = k_Q = 0$), the source active power is reduced from 2260 to 1950 W ($P_I = 310$ W). With the active and harmonic power control ($k_P = k_H = 1, k_Q = 0$), the source harmonic power is reduced from 545 to 130 W, and the THD_{isx} is improved from 25.4 to 1.9%. For the proposed selective droop control ($Q_{LC} \approx 930$ var, $k_P = k_H = 1$ and $k_Q = 0.4$), the source current THD_{isx} and PF are improved to 4.9% and 0.96%, respectively. The proposed control can selectively manage the active, reactive, and harmonic powers while satisfying IEEE standard 519-2014.

4.2.5 Experimental Results

The experimental prototype of the proposed HCIC (110 V–3.3 kW) is constructed in the laboratory, shown in Fig. 2.11. The digital control system of the HCIC consists of two paralleled digital signal processors (DSPs) TMS320F2812s, and the basic settings of both DSPs are the same. The sampling frequency of the control system is 25 kHz in the experiment (for both DSPs). For every 1/(25 kHz) period, the Timer will provide a signal to process analog to digital (A/D) conversion and the corresponding interrupt. After processing the proposed control strategy, the output hysteresis PWMs signal will be generated and the maximum PWM switching frequency is 12.5 kHz. Meanwhile, the SanRex PK110FG160 thyristors are used for the SVC part of the HCIC, and the Mitsubishi IGBT intelligent power modules PM300DSA60 are employed as the switching devices of the converter part of the HCIC. To capture the experimental results, the voltage and current waveforms are measured by an oscilloscope "Yokogawa DL750" and the current harmonic spectrums for three-phase are measured by a power quality analyzer named "Fluke 435". The parameters for experiments are the same as those for the simulation study. With reference to the IEEE standard 519-2014, the current THD_{isx} is required to be lower than 12% under the conditions of $I_{SC}/I_L \in [50, 100]$ based on the laboratory-scaled setups.

(a) *Case Study of Droop Control for HCIC*

Figs. 4.15, 4.16, 4.17 show the dynamic performances of the HCIC before and after operation for the unbalanced loads. In the unbalanced loads situations, the firing angles of the SVC part are different for each phase, and the compensating currents of each phase are also different. As shown in Figs. 4.15 and 4.17, the HCIC improved the source current THD_{isx} from 18.7 to 10.3% and source voltage THD_{Vx} from 4.7 to 3.0%, showing the worst phase. Moreover, the worst phase PF is improved from 0.59 to 0.99. In Fig. 4.16, the source currents are reduced from 5.7, 5.6, and 7.6 A to 4.3, 4.7, and 4.4 A, which are only slightly different, and after HCIC operation, the source voltage and current are in phase with each other as expected.

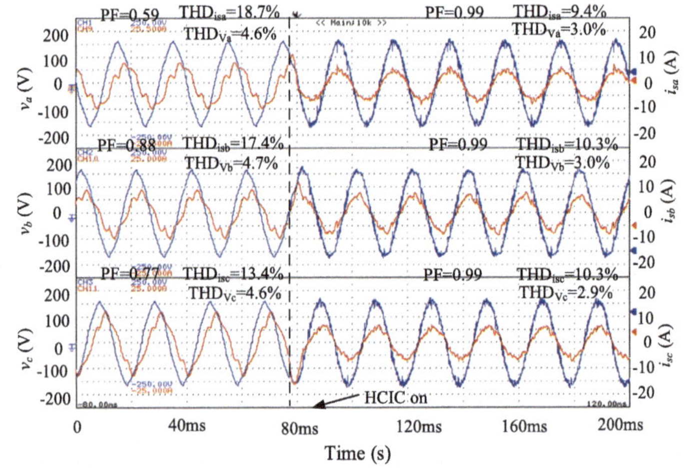

Fig. 4.15 Waveforms of v_x and i_{sx} for unbalanced reactive power and harmonic current compensation and active power injection at AC microgrid by using the HCIC

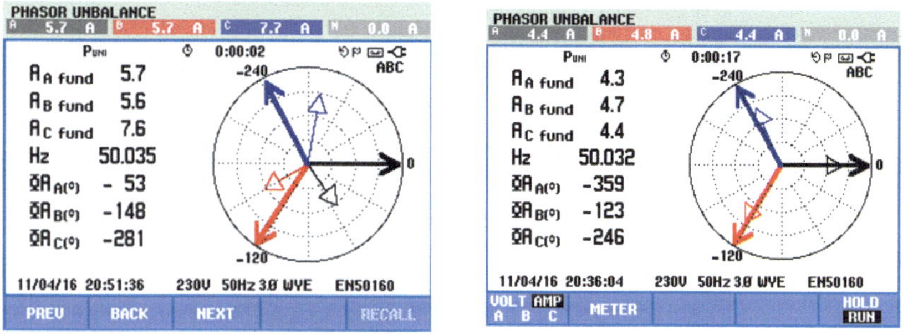

(a) (b)

Fig. 4.16 Experimental phasor diagrams of v_x and i_{sxf} for unbalanced inductive loads: **a** before HCIC operation and **b** after HCIC operation

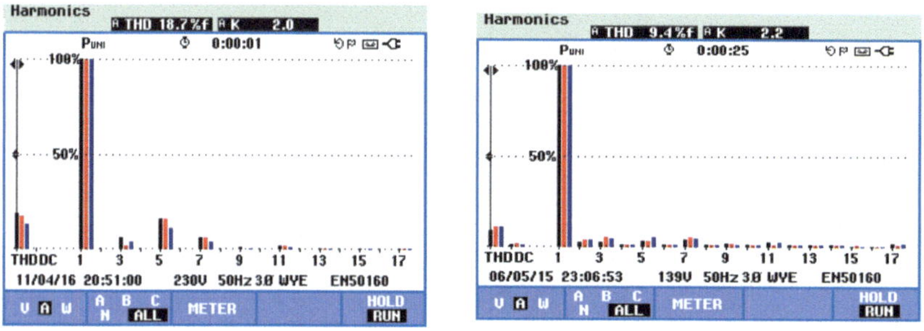

(a) (b)

Fig. 4.17 Experimental harmonic spectrum i_{sx} for unbalanced inductive loads: **a** before HCIC operation and **b** after HCIC operation

(b) *Case Study of Droop Selective Control for HCIC*

In Figs. 4.14, 4.15, 4.16, 4.17, 4.18, 4.19, 4.20, 4.21, 4.22, 4.23, 4.24, 4.25 the conventional ICIC can improve the power factor (PFs) from 0.75 and 0.84 to 0.99 and 0.98 for the 1st loading and 2nd loading, respectively. However, the THD_{isx} is increasing from 2.0 to 12.5% for the 1st linear load compensation of the worst phase. However, the THD_{isx} is just reduced from 16.6 to 13.2% for the 2nd nonlinear loading compensation of the worst phase among phases *a*, *b*, and *c*. In contrast, the proposed HIC with the proposed power management control can improve the *PF* to 0.99 and 0.97 for the 1st loading and 2nd loading, respectively, for the worst phase. Moreover, the THD_{isx} is reduced to lower than 8.0% for the 1st load and 2nd load of the worst phase.

From Figs. 4.21 and 4.22, the HCIC with the conventional methods can improve the *PF* from 0.85 to 0.97. The total active and reactive powers are reduced to 2400 W and 450 var from the original 2690 W and 1580 var. THD_{isx} is just reduced from 16.6 to 13.1% for the worst phase among phases *a*, *b*, and *c*. On the other hand,

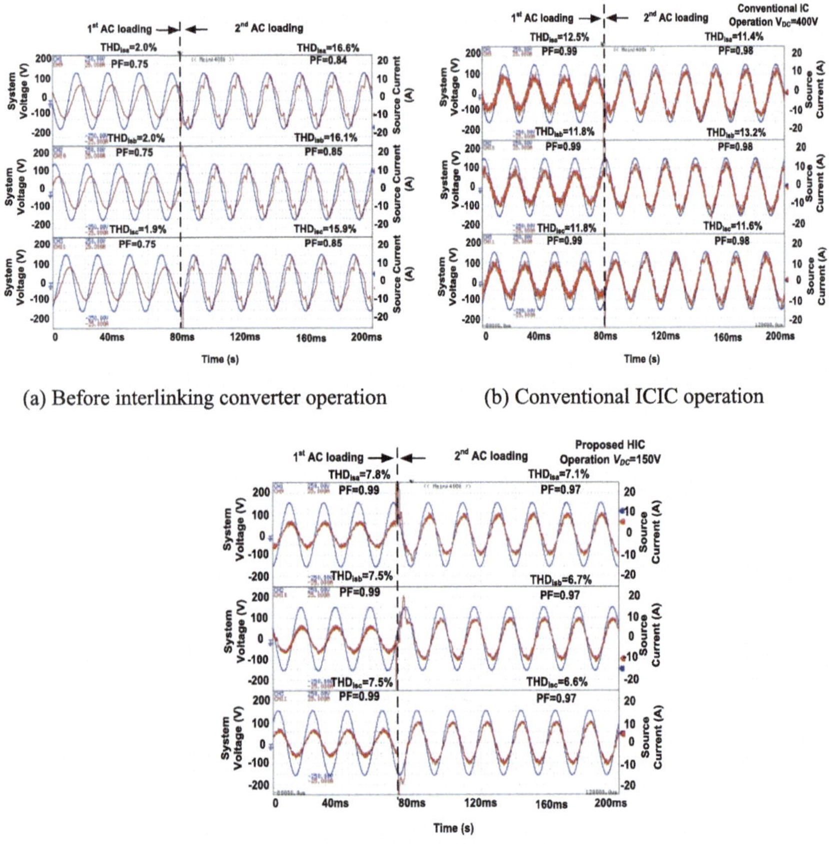

(a) Before interlinking converter operation

(b) Conventional ICIC operation

(c) Proposed HCIC operation

Fig. 4.18 Waveforms of v_x and i_{sx}

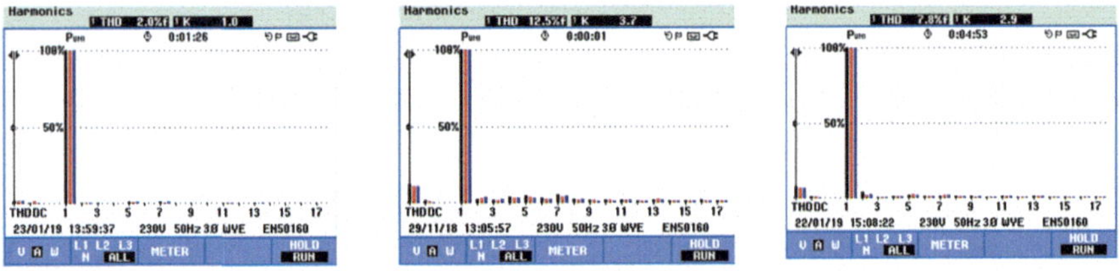

(a) Before ICIC operation

(b) Conventional ICIC operation

(c) Proposed HCIC operation

Fig. 4.19 The three-phase harmonic spectrums of the source current for the 1st loading (linear loading) are connected

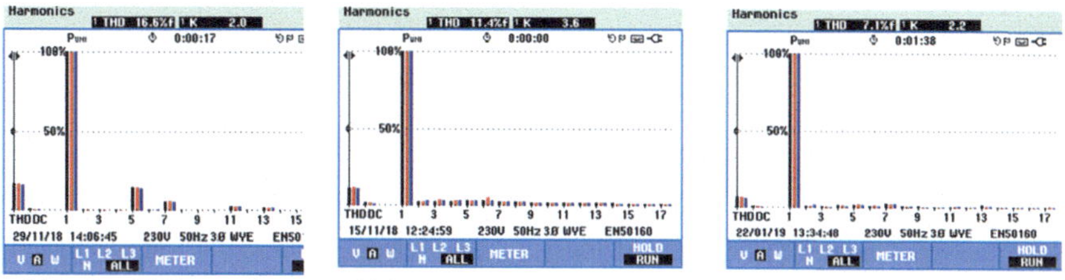

(a) Before ICIC operation (b) Conventional ICIC operation (c) Proposed HCIC operation

Fig. 4.20 The three-phase harmonic spectrums of source current for the 2nd loading connected (nonlinear loading)

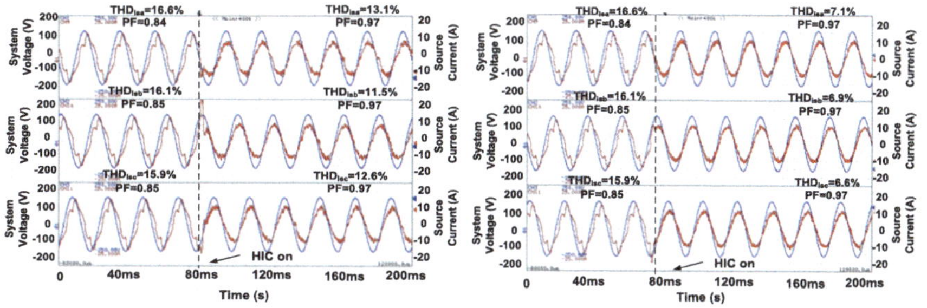

(a) Conventional control method (b) Proposed selective power management control

Fig. 4.21 Waveforms of v_x and i_{sx} after HCIC operation

the HIC with the proposed method (selective active, reactive, and harmonic power management control method) can improve the PF to 0.97 for the worst phase. The total active and reactive powers are reduced to 2370 W and 340 var from the original 2690 W and 1580 var. The THD_{isx} is reduced to 7.1% for the worst phase.

In Fig. 4.22, both the conventional and the proposed control methods cannot compensate the voltage and current in phase with each other since the load required reactive and harmonic power is beyond the designed capacity of the HCIC.

4.2.6 Section Summary

A new structure of interlinking converter, namely HCIC, composed of an SVC in series with a converter, is proposed for the hybrid AC/LVDC microgrid system. Moreover, a new droop control method has been correspondingly developed for the proposed HCIC system, and its fundamental active and reactive power flow in microgrids has been controlled by the proposed *P*-δ and *Q*-*P*-*V droop* controls in the converter part of HCIC, whereas *P*-*Q*-α droop has been proposed for the SVC part of HCIC. Furthermore, a fast harmonic control has been given based on a

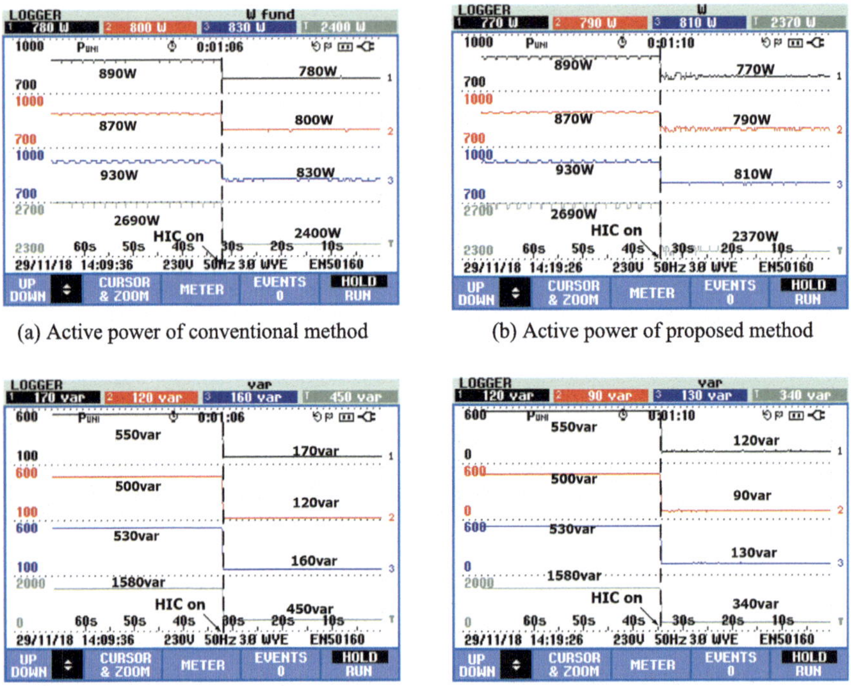

(a) Active power of conventional method

(b) Active power of proposed method

(c) Reactive power of conventional method

(d) Reactive power of proposed method

Fig. 4.22 Active and reactive power curves before and after HCIC operation, Active power curves: **a** conventional method and **b** proposed method; reactive curves: **c** conventional method and **d** proposed method

decoupled control strategy to improve the filtering characteristics of the SVC part of HCIC, suppress the harmonic power, and/or inject active power. Simulation results and experimental verification demonstrate that the proposed HCIC in the hybrid AC/LVDC microgrid has the ability of good power flow control and good power quality compensation. To avoid the over-capacity situation of HCIC, a selective power droop control was proposed to regulate active/reactive powers as well as improve power quality. Meanwhile, to avoid the potential over-capacity problem, the automatic selective factors adjustment control was proposed based on the tradeoff among active, reactive, and harmonic power. Simulation and experimental results convincingly verified the feasibility of the proposed selective power droop control.

4.3 Three-Port Power Hub Converter

4.3.1 System Configuration of Proposed Power Hub

The circuit configuration of the proposed power hub is depicted in Fig. 4.23. In the following analyses, the phases in the primary are expressed with capital letters subscripts *A*, *B*, and *C*, and the phases in the secondary side are expressed with lowercase letters subscripts *a*, *b*, *c*. In the structure, three-phase source power is transformed into two single phases, namely v_{ac}, and v_{bc}, via a V/V substation transformer. The full bridge-based inverter connecting with a static VAR compensator in series with adaptive impedance for power regulation is connected across v_{ac}. Similarly, the full bridge-based active inverter with an impedance is connected across v_{bc}. Single-phase loads are connected to the v_{ac} phase while the v_{bc} phase is unloaded. The power quality of the three-phase AC grid can be improved by providing power quality compensation from the secondary side to the primary side via the v_{ac} and v_{bc} phases.

In Fig. 4.23, i_X and v_X are the source current and voltage in the primary side, where the subscript "*X*" denotes phase $X = A, B$, and *C* in the primary side. i_L, i_{ac}, and i_{bc} are the load current, compensating current in the v_{ac} phase and in the v_{bc} phase. v_{invac} and v_{invbc} are the inverter voltage of the v_{ac} and v_{bc} phases, respectively.

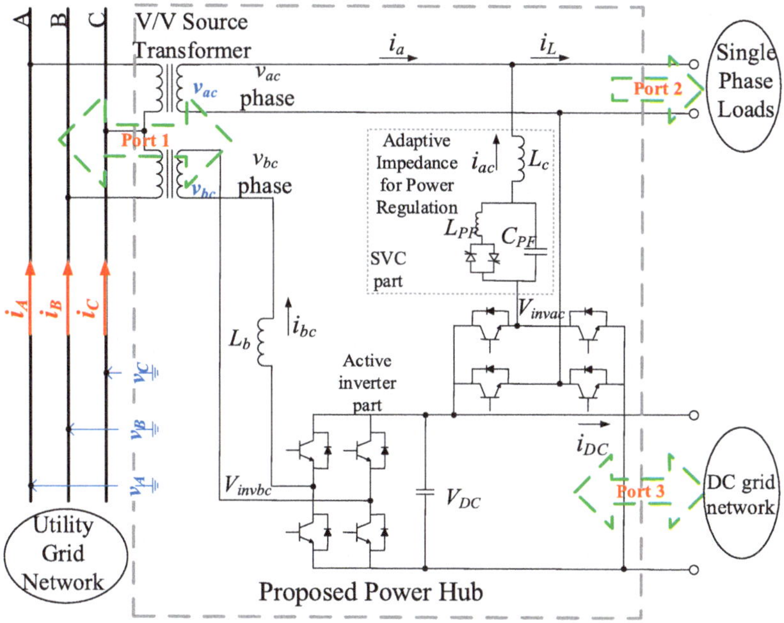

Fig. 4.23 Proposed structure of power hub interconnecting utility grid, AC load and DC grid in smart grid

The SVC part is composed of a coupling inductor L_c, a parallel capacitor C_{PF}, and a thyristor-controlled reactor with L_{PF}. The roles of the SVC part can be summarized as follows.

(1) The SVC part provides the optimum impedance for active and reactive power transfer. Thus, the operating range of the power hub can be adjusted.
(2) The SVC part provides a voltage drop. Thus, the power hub can operate at lower DC-link voltage. With the lower DC-link voltage, the converter power loss can be reduced.

These are achieved by controlling the firing angle α of the thyristors. The active inverter part is composed of a back-to-back connected single-phase inverter with a shared DC capacitor C_{DC}.

4.3.2 Operation Principle and Power Flow Analysis

The general power flow of the whole system is summarized as shown in Fig. 4.24. The proposed three-port converter interfaces two bidirectional power ports connecting the three-phase AC grid and DC grid, respectively, and one unidirectional power port connecting AC loads. At the same time, the mathematical fundamental power analysis is provided as well for the power hub interconnecting utility grid, AC load, and DC grid networks, respectively. Figure 4.24 summarizes the power flow analysis based on the mathematical calculations.

It is assumed that $\overline{V}_A = V_1 \angle 0°$, $\overline{V}_B = V_1 \angle -120°$, and $\overline{V}_C = V_1 \angle 120°$, where $1 \angle -120° = -1/2 - j\sqrt{3}/2$, $1 \angle 120° = -1/2 + j\sqrt{3}/2$. The single-phase loads are connected to the v_{ac} phase, and there is no direct connection between phase A and phase C, so the relationship $\vec{I}_{AC} = \vec{I}_A$ can be obtained. According to the above descriptions, the power from the transformers' primary sides flowing in the v_{ac} phase and the v_{bc} phase can be expressed as:

$$\begin{aligned}
P_{AC} + jQ_{AC} &= \vec{V}_{ac} \cdot \vec{I}_{ac} = \left(\vec{V}_A - \vec{V}_C\right) \cdot \vec{I}_A \\
&= \left(1\angle 0^{\circ} - 1\angle 120^{\circ}\right) \cdot V_1 \cdot \vec{I}_A = \left(\frac{3}{2} - j\frac{\sqrt{3}}{2}\right) \cdot (P_A + jQ_A) \\
&= \left(\frac{3}{2}P_A + \frac{\sqrt{3}}{2}Q_A\right) + j\left(-\frac{\sqrt{3}}{2}P_A + \frac{3}{2}Q_A\right)
\end{aligned} \tag{4.55}$$

$$\begin{aligned}
P_{BC} + jQ_{BC} &= \vec{V}_{bc} \cdot \vec{I}_{bc} = \left(\vec{V}_B - \vec{V}_C\right) \cdot \vec{I}_B \\
&= \left(\frac{3}{2}P_B - \frac{\sqrt{3}}{2}Q_B\right) + j\left(\frac{\sqrt{3}}{2}P_B + \frac{3}{2}Q_B\right)
\end{aligned} \tag{4.56}$$

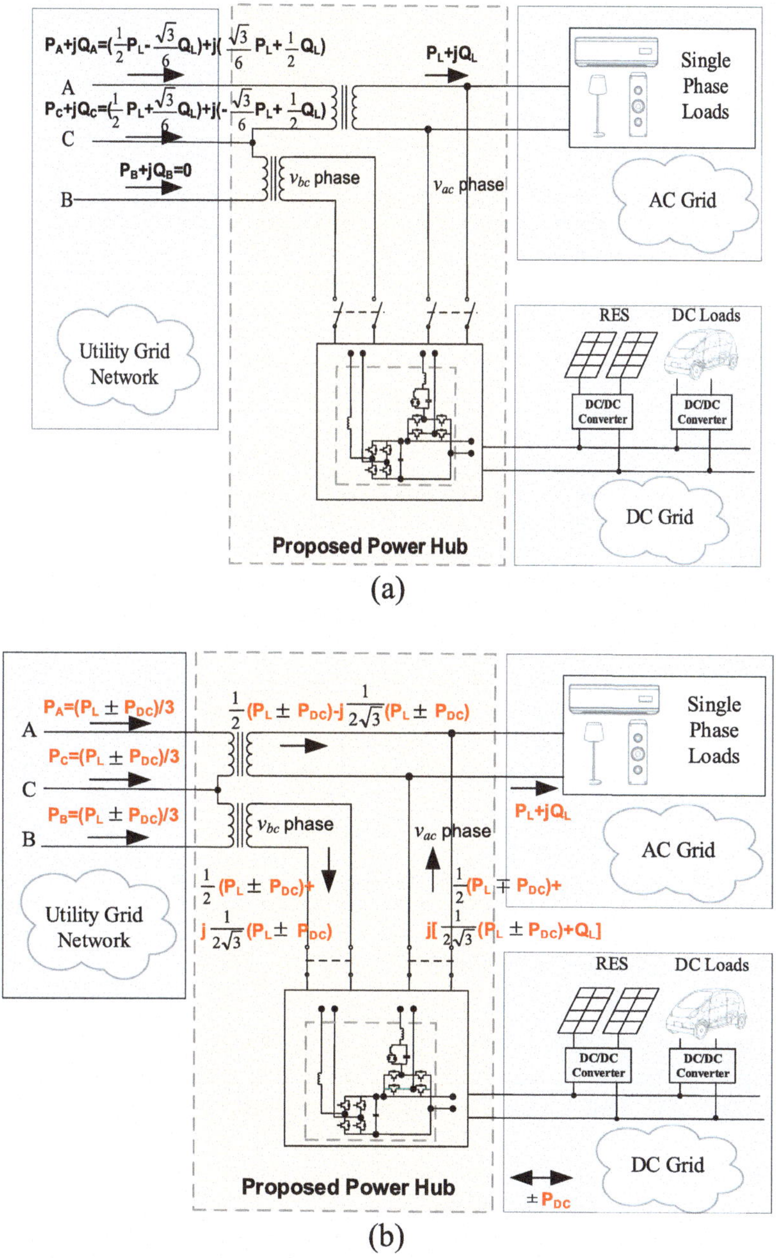

Fig. 4.24 Power flow analysis of power hub interconnecting utility grid, AC load, and DC grid **a** before power hub operation, **b** after power hub operation

Before compensation, the v_{ac} phase is connected to the single-phase loads while the v_{bc} phase is unloaded. Therefore, according to (4.55) and (4.56), the phase power can be expressed as (4.57)–(4.59). As shown in (4.57)–(4.59) and Fig. 4.29a, before the power hub operation, phase B is unloaded. Therefore, the unbalance factor (UF_i) is as large as 100%.

$$\begin{aligned} P_A + jQ_A &= \frac{P_{AC} + jQ_{AC}}{\frac{3}{2} - j\frac{\sqrt{3}}{2}} = \frac{P_L + jQ_L}{\frac{3}{2} - j\frac{\sqrt{3}}{2}} \\ &= \left(\frac{1}{2}P_L - \frac{\sqrt{3}}{6}Q_L\right) + j\left(\frac{\sqrt{3}}{6}P_L + \frac{1}{2}Q_L\right) \end{aligned} \tag{4.57}$$

$$P_B + jQ_B = 0 \tag{4.58}$$

$$P_C + jQ_C = \frac{P_L + jQ_L}{\frac{3}{2} + j\frac{\sqrt{3}}{2}} = \left(\frac{1}{2}P_L + \frac{\sqrt{3}}{6}Q_L\right) + j\left(-\frac{\sqrt{3}}{6}P_L + \frac{1}{2}Q_L\right) \tag{4.59}$$

The purposes of the power hub are to balance the active power $P_L \pm P_{DC}$ in the utility grid and, at the same time, to compensate reactive power caused by single-phase load reactive power Q_L. After ideal compensation, the P_L for household load and P_{DC} from the DC grid are transferred to each phase in the utility grid $P_A = P_B = P_C = 1/3(P_L \pm P_{DC})$, and the reactive power Q_L is compensated to zero ($Q_A = Q_B = Q_C = 0$). Substituting the $P_A = P_B = P_C = 1/3(P_L \pm P_{DC})$ and $Q_A = Q_B = Q_C = 0$ into (4.55) and (4.56), we obtain the active and reactive power flowing in the v_{ac} phase and the v_{bc} phase after compensation, given as below:

$$P_{AC'} + jQ_{AC'} = \frac{1}{2}(P_L \pm P_{DC}) - j\frac{1}{2\sqrt{3}}(P_L \pm P_{DC}) \tag{4.60}$$

$$P_{BC'} + jQ_{BC'} = \frac{1}{2}(P_L \pm P_{DC}) + j\frac{1}{2\sqrt{3}}(P_L \pm P_{DC}) \tag{4.61}$$

In (4.60), (4.61), and Fig. 4.29b, $\pm P_{DC}$ is active power from DC grid. the sign "+" stands for the active power received by DC grid (e.g. charging for the electric vehicle, etc.). The sign "-" stands for the active power sent by DC grid (e.g. injecting renewable energy to the utility grid via the power hub).

After obtaining the power flow in the v_{ac} and v_{bc} phases, the power hub compensating power can be calculated as:

$$\begin{aligned} P_{ac} + jQ_{ac} &= (P_L + jQ_L) - (P_{AC} + jQ_{AC}) \\ &= \frac{1}{2}(P_L \mp P_{DC}) + j\left[Q_L + \frac{1}{2\sqrt{3}}(P_L \pm P_{DC})\right] \end{aligned} \tag{4.62}$$

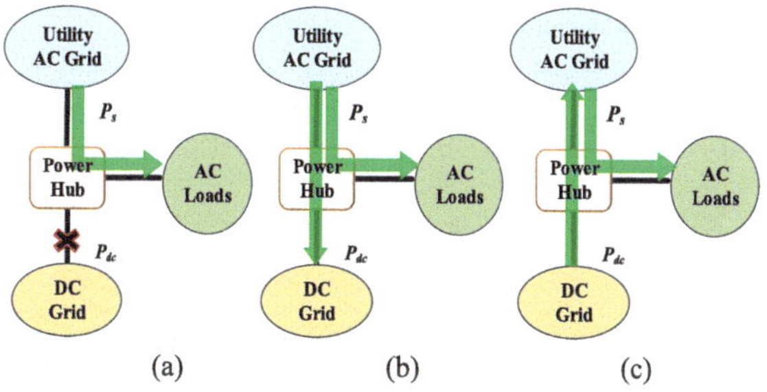

Fig. 4.25 Three cases of power flow for the proposed three-port converter. **a** DC grid disconnecting to the power hub; **b** DC grid connecting to the power hub and receiving power from the utility grid; **c** DC grid connecting to the power hub and sending power to the utility grid

$$\begin{aligned} P_{bc} + jQ_{bc} &= -(P_{BC} + jQ_{BC}) \\ &= -\frac{1}{2}(P_L \pm P_{DC}) - j\frac{1}{2\sqrt{3}}(P_L \pm P_{DC}) \end{aligned} \tag{4.63}$$

where P_{ac}, P_{bc}, Q_{ac}, and Q_{bc} are the active and reactive power provided by the power hub for v_{ac} and v_{bc}, respectively.

According to the above-detailed analysis and calculations, there are three cases of power flow for the proposed three-port converter, shown in Fig. 4.25. From Fig. 4.25a we can see that when the DC grid is disconnecting to the power hub, the three-phase AC grid only feeds the power to AC load. When the DC grid is connecting to the power hub shown in Fig. 4.25b, not only the AC load but also the three-phase AC grid feeds power to the DC grid. Meanwhile, in this case, the DC grid can feed power to three-phase AC grid as well, as seen in Fig. 4.25c. As previously mentioned, the power flow control between the three-phase AC grid and DC grid can be bidirectional via the three-port converter. A more detailed control strategy is developed in the following section.

4.3.3 Control Strategy of the Proposed Power Hub

A control strategy for the proposed power hub is proposed by coordinating control between the SVC part and back-to-back the inverter. The control strategy of the proposed power hub is divided into four parts for discussion: A. DC-link voltage control, B. inverter part control, C. SVC part control, and D. overall control strategy. The control block diagram of the power hub is summarized in Fig. 4.26.

A. *DC-Link Voltage Control*

The DC-link voltage reference $V_{DC}{}^*$ is designed to fulfill the active, reactive, and harmonic power transfer compensation requirements. The detailed DC-link voltage

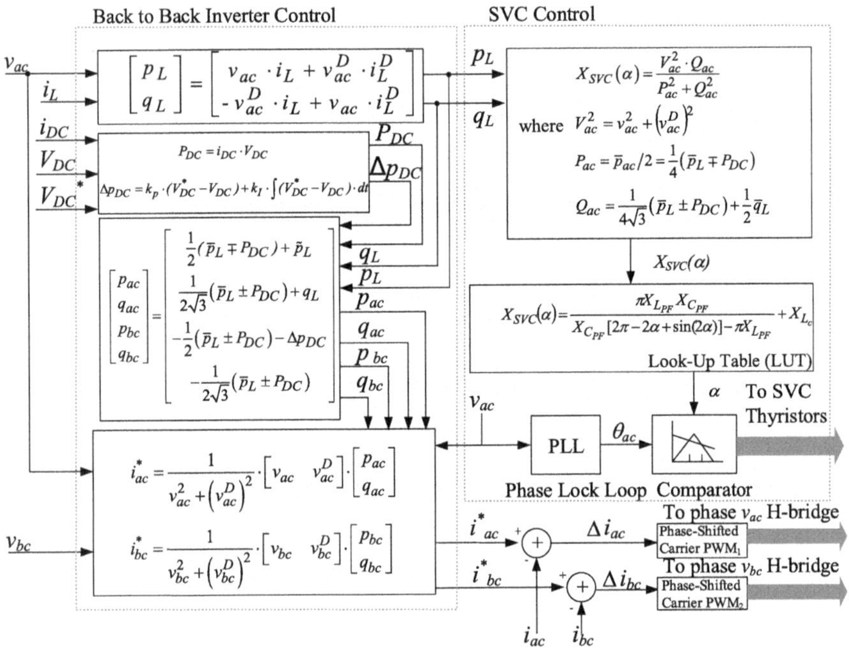

Fig. 4.26 Control strategy of the proposed power hub

design is provided as follows. The subscripts "*f*", "*h*", "*p*", "*q*", and "*n*" represent the fundamental, total harmonics, active component, reactive component, and the corresponding *n*th-order harmonic in the following, respectively. The phase v_{ac} inverter voltage considering the fundamental active and reactive components can be expressed as:

$$V_{invacf} = \sqrt{\left(V_{ac} - I_{acqf} X_{SVC}\right)^2 + \left(I_{acpf} X_{SVC}\right)^2} \tag{4.64}$$

Considering the fundament component and harmonic component together, the function of V_{invac} can be expressed as:

$$\begin{aligned} V_{invac} &= \sqrt{V_{invacf}^2 + V_{invach}^2} \\ &= \sqrt{\left(V_{ac} - I_{acqf} X_{SVC}\right)^2 + \left(I_{acpf} X_{SVC}\right)^2 + \sum_{n=2}^{\infty} I_{acn}^2 X_{SVCn}^2} \end{aligned} \tag{4.65}$$

In the proposed power hub system, the SVC part provides the optimum impedance for active and reactive power transfer with the minimum required DC-link voltage. In the condition of (4.76), the V_{invac} in (4.65) is the minimum, due to the derivative being equal to zero. The minimum DC-link voltage $V_{DC(Min)}$ is based on the minimum $V_{invac(Min)}$. Based on (4.65), (4.75), and (4.76), the minimum inverter voltage $V_{invac(Min)}$ can be expressed as:

$$V_{invac(Min)}=\sqrt{V_{invacf}^2+V_{invach}^2}$$
$$=\sqrt{V_{ac}^2\cdot\left[\left(1+\frac{Q_{ac}^2}{P_{ac}^2+Q_{ac}^2}\right)^2+\left(\frac{P_{ac}\cdot Q_{ac}}{P_{ac}^2+Q_{ac}^2}\right)^2\right]+\sum_{n=2}^{\infty}I_{acn}^2X_{SVCn}^2} \tag{4.66}$$

Considering the relationship between DC-link voltage and the inverter voltage, the minimum DC-link voltage $V_{DC(Min)}$ and the system $V_{DC}{}^*$ can be expressed as:

$$V_{DC}^* \geq V_{DC(Min)}=\sqrt{2}\cdot V_{invac(Min)}$$
$$=\sqrt{2V_{ac}^2\cdot\left[\left(1+\frac{Q_{ac}^2}{P_{ac}^2+Q_{ac}^2}\right)^2+\left(\frac{P_{ac}\cdot Q_{ac}}{P_{ac}^2+Q_{ac}^2}\right)^2\right]+2\sum_{n=2}^{\infty}I_{acn}^2X_{SVCn}^2} \tag{4.67}$$

The final designed DC-link voltage is required to be larger than this $V_{DC(Min)}$ as (4.67). The proposed power hub is composed of two single-phase converters with a back-to-back connection. The DC-link voltage is based on the maximum DC voltage required value between back-to-back converters, which are phase v_{bc} converter in this topology, due to the larger voltage drop on SVC than on L_b. Also, its control depends on the active component Δp_{DC} instead of the reactive component to recover the active power loss from the power hub itself since the reactive power cannot be consumed.

Therefore, the DC-link voltage control component can be obtained from (4.68), which should be calculated with the p_{bc} control together, as shown in (4.70).

$$\Delta p_{DC}=k_p\cdot\left(V_{DC}^*-V_{DC}\right)+k_I\cdot\int\left(V_{DC}^*-V_{DC}\right)\cdot dt \tag{4.68}$$

where V_{DC} and $V_{DC}{}^*$ are the actual and reference DC grid voltage, respectively. i_{DC} is the DC current, k_p is the proportional gain, and k_I is the integral gain.

B. *Inverter Part Control*

The AC loads are connected in phase v_{ac} only and left the phase v_{bc} unloaded. Before power hub compensation, the i_B is zero, and the unbalanced factor is 100%. Thus, one of the main functions of the power hub is to transfer the active power from the utility grid phase v_A, v_B, v_C to AC and DC loads, while keeping utility grid phase currents to be balanced. This function is achieved by taking active power from phase v_{bc} to phase v_{ac}. The other function of the power hub is to compensate the reactive and harmonic power for the single-phase nonlinear AC loads. This function is realized by including the power quality control function in the phase v_{ac} converter. Specifically, the reference compensating current $i_{ac}{}^*$ in the phase v_{ac} converter has all active, reactive, and harmonic components.

All the above-mentation power flow control and power quality compensation functions can be realized by limiting the compensating current i_{ac} and i_{bc} to their reference value $i_{ac}{}^{*}$ and $i_{bc}{}^{*}$. Thus, the back-to-back inverter also can solve the mistuning problem, the resonance problem, and the harmonic of the SVC part. The $i_{ac}{}^{*}$ and $i_{bc}{}^{*}$ are calculated based on the modified single-phase instantaneous pq theory. The instantaneous active and reactive power of the single-phase loads can be calculated as:

$$\begin{bmatrix} p_L \\ q_L \end{bmatrix} = \begin{bmatrix} v_{ac} \cdot i_L + v_{ac}^D \cdot i_L^D \\ -v_{ac}^D \cdot i_L + v_{ac} \cdot i_L^D \end{bmatrix} \tag{4.69}$$

where v_{ac} and i_L are the phase voltage and load current, $v_{ac}{}^D$, and $i_L{}^D$ are 90° delay of the phase voltage and load current. p_L and q_L are the instantaneous active and reactive power, which include DC components $\overline{p_L}$ and $\overline{q_L}$, and AC components. $\tilde{p}_L$ and $\tilde{q}_L$. $\overline{p_L}$ and $\overline{q_L}$ are obtained by passing p_L and q_L through low-pass filters. The load power obtained in (4.69) is used for the determination of the proposed power hub converter compensating both the active (p_{ac}) and reactive (q_{ac}) powers in v_{ac} phase, and the active (p_{bc}) and reactive (q_{bc}) powers in v_{bc} phase. Based on the power flow analysis in the previous section, the compensating power provided by the proposed power hub can be expressed as:

$$\begin{bmatrix} p_{ac} \\ q_{ac} \\ p_{bc} \\ q_{bc} \end{bmatrix} = \begin{bmatrix} \frac{1}{2}(\overline{p}_L \mp P_{DC}) + \tilde{p}_L \\ \frac{1}{2\sqrt{3}}\left(\overline{p}_L \pm P_{DC}\right) + q_L \\ -\frac{1}{2}\left(\overline{p}_L \pm P_{DC}\right) - \Delta p_{DC} \\ -\frac{1}{2\sqrt{3}}\left(\overline{p}_L \pm P_{DC}\right) \end{bmatrix} \tag{4.70}$$

In (4.70), the p_L and q_L are obtained from (4.69), and the P_{DC} can be obtained from (4.71).

$$P_{DC} = i_{DC} \cdot V_{DC} \tag{4.71}$$

After obtaining the compensating power, the reference compensating current of the power hub can be expressed as:

$$i_{ac}^{*} = \frac{1}{v_{ac}^2 + \left(v_{ac}^D\right)^2} \cdot \begin{bmatrix} v_{ac} & v_{ac}^D \end{bmatrix} \cdot \begin{bmatrix} p_{ac} \\ q_{ac} \end{bmatrix} \tag{4.72}$$

$$i_{bc}^{*} = \frac{1}{v_{bc}^2 + \left(v_{bc}^D\right)^2} \cdot \begin{bmatrix} v_{bc} & v_{bc}^D \end{bmatrix} \cdot \begin{bmatrix} p_{bc} \\ q_{bc} \end{bmatrix} \tag{4.73}$$

By comparing the actual compensating current (i_{ac} and i_{bc}) with the obtained reference ($i_{ac}{}^{*}$ and $i_{bc}{}^{*}$), the trigger signals for the back-to-back inverter can be generated through the current phase-shifted carrier modulation control.

C. *SVC Part Control*

The SVC part provides the adjustable impedance for a wide range of active and reactive power transfer with the minimum required DC-link voltage. The single-phase loads are directly connected to the v_{ac} phase. Based on the circuit analysis, the vector of the required inverter voltage of v_{ac} phase can be expressed as:

$$\begin{aligned}\vec{V}_{invac} &= \vec{V}_{ac} - jX_{SVC}(\alpha)\cdot\vec{I}_{ac}\\ &= V_{ac} - jX_{SVC}(\alpha)\cdot(I_{acp} + jI_{acq})\\ &= V_{ac} + X_{SVC}(\alpha)\cdot I_{acq} - jX_{SVC}(\alpha)\cdot I_{acp}\end{aligned} \tag{4.74}$$

where v_{ac}, I_{acq}, and I_{acp} are phase voltage, reactive, and active power components of the compensating current in the secondary side of phase v_{ac}. $X_{SVC}(\alpha)$ is the impedance of the SVC part, which is controlled by the firing angle α. According to (4.74), the ratio $\vec{V}_{invac}/\vec{V}_{ac}$ can be expressed in terms of active power and reactive power as:

$$\begin{aligned}\frac{\vec{V}_{invac}}{\vec{V}_{ac}} &= \sqrt{\left(1 + \frac{V_{ac}\cdot|I_{acq}|}{V_{ac}^2/X_{SVC}(\alpha)}\right)^2 + \left(\frac{V_{ac}\cdot|I_{acp}|}{V_{ac}^2/X_{SVC}(\alpha)}\right)^2}\\ &= \sqrt{\left(1 + \frac{Q_{ac}}{S_{SVC}(\alpha)}\right)^2 + \left(\frac{P_{ac}}{S_{SVC}(\alpha)}\right)^2}\end{aligned} \tag{4.75}$$

where $S_{SVC}(\alpha)$ is the apparent power of the SVC part, P_{ac} and Q_{ac} are the active and reactive power provided by the proposed power hub for the v_{ac} phase.

In (4.75), the V_{invac} is directly proportional to the voltage in the DC grid. To guarantee the inverter voltage is sufficient enough for reactive and active power transfer, the SVC impedance is to keep the V_{invac}/V_{ac} at its minimum value, and the minimum value of V_{invac}/V_{ac} can be obtained by differentiating (4.71) with respect to $S_{SVC}(\alpha)$, as given below:

$$\frac{\partial(V_{invac}/V_{ac})}{\partial(S_{SVC}(\alpha))} = 0,\ \ S_{SVC}(\alpha) = \frac{P_{ac}^2 + Q_{ac}^2}{Q_{ac}} \tag{4.76}$$

The ratio of $\vec{V}_{invac}/\vec{V}_{ac}$ can be calculated by substituting $S_{SVC}(\alpha)$ in (4.76) into (4.75).

$$\begin{aligned}X_{SVC}(\alpha) &= \frac{V_{ac}^2\cdot Q_{ac}}{P_{ac}^2 + Q_{ac}^2}\\ &= \frac{\pi X_{L_{PF}} X_{C_{PF}}}{X_{C_{PF}}[2\pi - 2\alpha + \sin(2\alpha)] - \pi X_{L_{PF}}} + X_{L_c}\end{aligned} \tag{4.77}$$

where V_{ac}, Q_{ac}, and P_{ac} are the RMS voltage of the v_{ac} phase, fundamental compensating reactive and active powers. When only the fundamental part is considered, the instantaneous expressions of V_{ac}, Q_{ac}, and P_{ac} can be expressed as:

$$V_{ac}^2 = v_{ac}^2 + \left(v_{ac}^D\right)^2 \tag{4.78}$$

$$P_{ac} = \overline{p}_{ac}/2 = \frac{1}{4}\left(\overline{p}_L \mp P_{DC}\right) \tag{4.79}$$

$$Q_{ac} = \overline{q}_{ac}/2 = \frac{1}{4\sqrt{3}}\left(\overline{p}_L \pm P_{DC}\right) + \frac{1}{2}\overline{q}_L \tag{4.80}$$

In (4.79) and (4.80), the $\overline{p}_L$ and $\overline{q}_L$ can be obtained from (4.69).

From (4.77), the firing angle α can be easily determined. Considering the complexity of (4.77), a look-up table between the firing angle α and the X_{SVC} is developed for the controller. Then, the trigger signals to the control of the SVC can be generated by comparing the firing angle α with θ_{ac}, which is the phase angle of the v_{ac} and θ_{ac} is obtained by using a phase lock loop.

D. *Overall Control Strategy*

The overall control block of the proposed power hub is provided in Fig. 4.26. The active back-to-back inverter in phase v_{ac} side is acting as a harmonic and reactive compensator while transferring active power from both the DC grid and the other AC grid (phase v_{bc}) simultaneously by limiting the compensating current (i_{ac} and i_{bc}) with the obtained references ($i_{ac}^*$and i_{bc}^*).

According to the modified single-phase instantaneous *pq* theory, the active and reactive powers of load are calculated through (4.70). Then, based on the aforementioned power analysis and (4.71), the compensating reactive and active powers for the power hub can be obtained. After that, the compensating reactive power and active power can be further transformed to $i_{ac}^*$and i_{bc}^* by (4.73) and (4.74). With the phase-shifted carrier modulation control method, the trigger signals of the back-to-back inverter can be generated by comparing the i_{ac} and i_{bc} with their references i_{ac}^* and i_{bc}^*. Moreover, the current phase-shifted carrier modulation control is applied in this study due to its advantages of simplicity of implementation, high robustness, etc.

For the SVC part control, the expressions in (4.78)–(4.80) are used to calculate the voltage of the vac phase, fundamental compensating active and reactive powers of the vac phase. Then, the required impedance of the SVC part $X_{SVC}(\alpha)$ can be obtained from (4.77). After that, the corresponding firing angle α of the SVC part can be obtained by comparing the left- and right-hand side terms of (4.77) and looking up the table. Finally, compared to the firing angle α with the phase angle θ_{ac}, the trigger signals for the thyristor of the SVC can be obtained.

4.3.4 Simulation Results

In this section, the simulation case studies are provided to show the feasibility of the proposed power hub. The below functions of the proposed power hub are verified in simulations: (1) bidirectionally active power transfer among three-phase AC grid and DC grid; (2) power quality compensation (mainly reactive, unbalanced, and harmonic power compensation for utility grid).

The dynamic performances of the proposed power hub are illustrated under three different situations, namely: (1) DC grid/load charging, (2) DC grid unloaded, and (3) DC grid active power injection. The parameters of different power grids are listed in Table 4.7.

From Table 4.8 and Fig. 4.27a, before the power hub operates, it can be seen that the single-phase AC grid loads can cause a large unbalance power in the three-phase utility grid. Moreover, the single-phase nonlinear loads can cause harmonic pollution in the three-phase utility grid ($THD_{iX} = 32.4\%$ for phases A and C).

During the DC grid/load charging, as shown in Table 4.8 and Fig. 4.27b, the DC grid absorbs the 2 kW from the three-phase utility grid through the power hub. In

Table 4.7 Parameters of different power grids

	Parameters	Physical values
Utility grid voltage	V_X, f_s	10 kV, 50 Hz
AC grid	V_{ac}, f	220 V, 50 Hz
DC grid	V_{DC}	200 V
Power hub	L_b, L_c, L_{PF}, C_{PF}	2.5 mH, 2.5 mH, 20 mH, 240 μF

Table 4.8 Simulation results of power hub for DC grid/load charging, DC grid unload and DC grid active power injection

Situation	Loads		PF	THD_{iX}(%)	P_{sx}(W)	Q_{sx}(var)	I_X(A)	P_{DC}(W)
Before power hub connected	Linear loads	A	0.54	0.1	1692	2600	0.54	–
		B	–	–	0	0	0	
		C	0.99	0.1	3100	170	0.54	
	Nonlinear loads	A	0.50	32.4	853	1472	0.31	–
		B	–	–	0	0	0	
		C	0.99	32.5	1703	5	0.31	
DC grid charging	Linear loads		1.00	1.3	2290	10	0.39	2000
	Nonlinear loads		1.00	1.7	1530	9	0.27	
DC grid unload	Linear loads		1.00	2.1	1600	9	0.27	0
	Nonlinear loads		1.00	2.3	880	9	0.15	
DC grid P_{DC} injection	Linear loads		1.00	2.1	1335	10	0.23	−800
	Nonlinear loads		1.00	4.1	600	10	0.11	

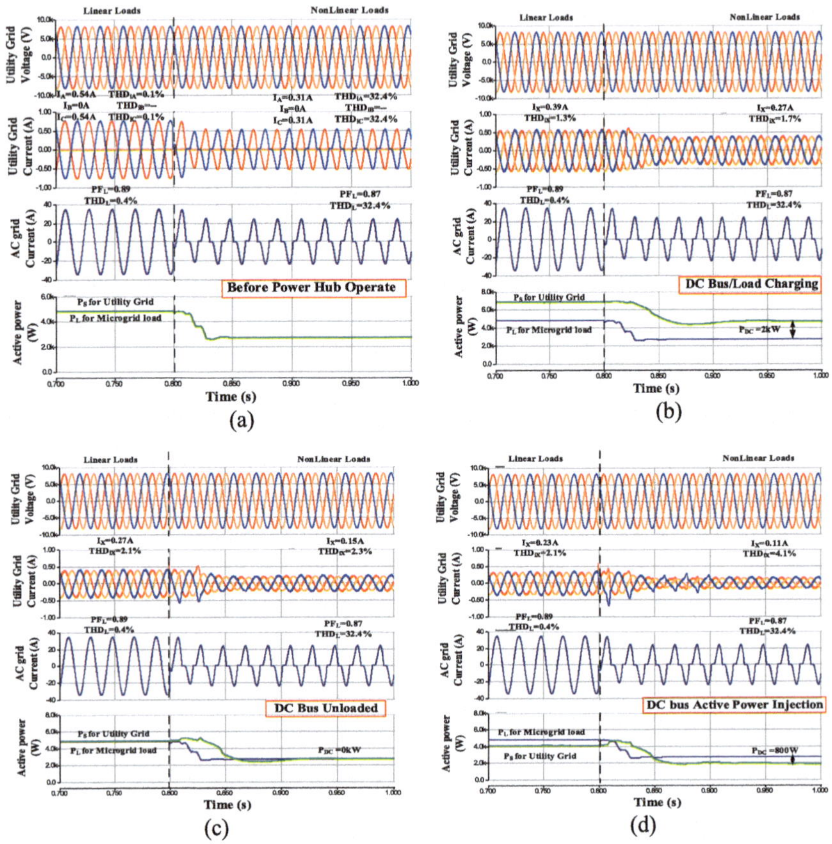

Fig. 4.27 Waveforms of utility grid voltage, utility grid current, AC grid current, and grid active power, **a** before the power hub operates, **b** DC grid/load charging, **c** DC grid unload, **d** DC grid active power injection

this case, the current of the utility grid is balanced to 0.39 and 0.27 A for linear load and nonlinear load. At the same time, the current THD_{iX} is correspondingly reduced to 1.3 and 1.7%.

During the DC grid unloading shown in Table 4.8 and Fig. 4.27c, the DC grid does not absorb active power. Even in this case, the current of the utility grid is still balanced to 0.27 and 0.15 A for linear load and nonlinear load. The current THD_{iX} is 2.1 and 2.3% after linear load and nonlinear load compensation.

During the DC grid unload, shown in Table 4.8 and Fig. 4.27d, the DC grid injects the active power to the three-phase utility grid through the power hub. The current of the utility grid is balanced to 0.23 and 0.11 A for linear load and nonlinear load. The current THD_{iX} is 2.1 and 4.1% after linear load and nonlinear load compensation.

4.3.5 Experimental Results

A scaled-down experimental prototype of the proposed power hub is constructed in the laboratory, as shown in Fig. 2.11. Due to the limitation of the laboratory, the utility grid line-to-line voltage is reduced from 10 kV to 50 V. The digital control system of the proposed power hub adopts a digital signal processor (DSP) TMS320F2812, which contains 2 event managers (EV). The Timer 2 of EVA is used to define the analog-to-digital conversion (ADC) module and limits the sampling frequency as 25 kHz. Every 1/(25 kHz), Timer 2 will provide a signal to the process and the corresponding interrupt. Timer 1 in EVA is used to generate phase v_{ac} H-bridge and thyristors PWM, and Timer 3 in EVB is used to generate phase v_{bc} H-bridge PWM. Thus, the same DSP can be used to control the system simultaneously. The switching devices for the active inverter part use Mitsubishi IGBT intelligent power modules PM300DSA060 (20 kHz maximum switching frequency). The thyristors for the SVC part use SanRex PK110FG160. Considering the 25 kHz sample frequency and hardware limitations, the average switching frequency in this paper is selected as 10 kHz.

In Fig. 4.28, the performances of the proposed power hub are illustrated under two representative situations: DC grid load charging and DC grid unloaded. The corresponding utility grid power and power factor and harmonic current spectrums are provided in Figs. 4.29 and 4.30, respectively. In Fig. 4.31, the dynamic performance of the voltages and currents are illustrated for the proposed power hub.

From Figs. 4.28a, 4.29a, and 4.30a, before the power hub working, the single-phase AC grid loads can cause the large unbalance power and harmonic pollution in three-phase utility grid. The unbalanced factor is 100%, and the worst phase THD_{iX} is 25.5%. From Figs. 4.28b, 4.29b, and 4.30b, the utility grid current becomes balanced, and the worst phase THD_{iX} is reduced to 6.1% after the power hub operating with the DC grid load connecting. Comparing Fig. 4.29a with Fig. 4.29b, we can know that the proposed power hub can transfer the 160 W from the utility grid to the DC grid. As shown in Figs. 4.28c, 4.29c, and 4.30c, the utility grid current also becomes balanced. The maximum phase current was reduced from 3.0 A (DC load charging) to 2.2 A (DC grid unloaded). Besides, the worst phase THD_{iX} is reduced to 9.6% from the original 25.5%. The total active powers in the utility grid are close in the cases before the power hub operates and during the DC grid unloaded. From Fig. 4.31, the transient time for the proposed power hub is above 1.5 cycles (30 ms) if suddenly double the grid load current. Moreover, it can be seen that the proposed power hub can compensate for the $THD_{iX} < 10\%$ results with different loads.

In this Section, the experiments are carried out in a low-capacity weak grid (150 VA). Thus, the harmonic distortions can easily affect the system current through the transformer. Therefore, the IEEE Standard 519-2014 relaxes the harmonic restrictions to be less than 15% for the weak grid. The simulated results are carried out in relatively ideal voltage and current conditions, thus, the unbalanced factor (UF_{is}) is reduced from 100 to 2%. Thus, the proposed topology and control method are verified.

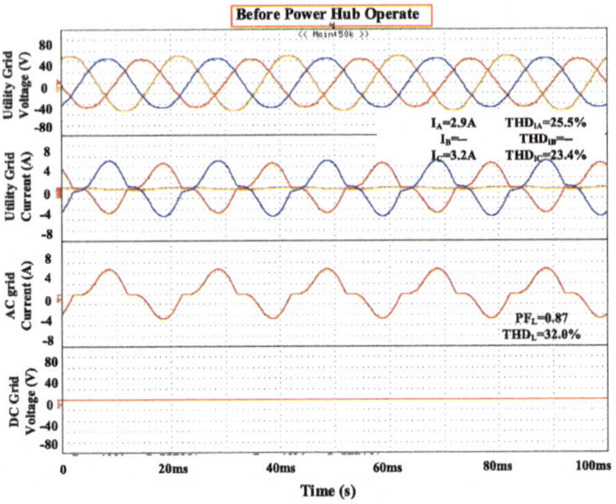

(a)

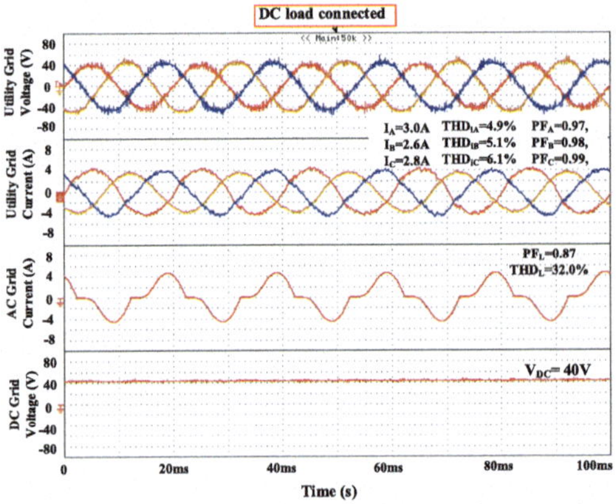

(b)

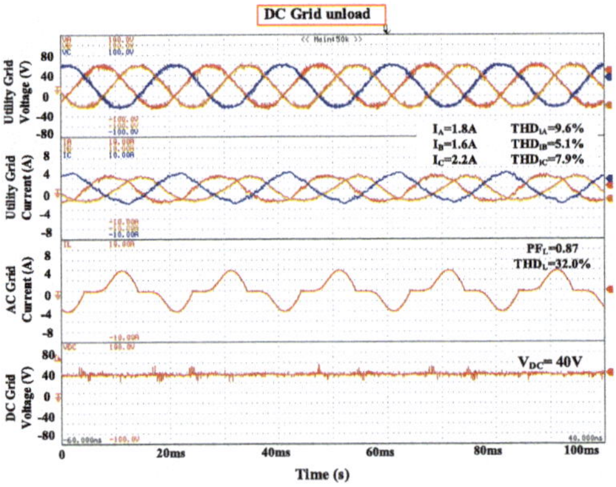

(c)

Fig. 4.28 Waveforms of utility grid voltage, utility grid current, grid current and DC grid voltage, **a** before the power hub operate, **b** DC load charging, **c** DC grid unload

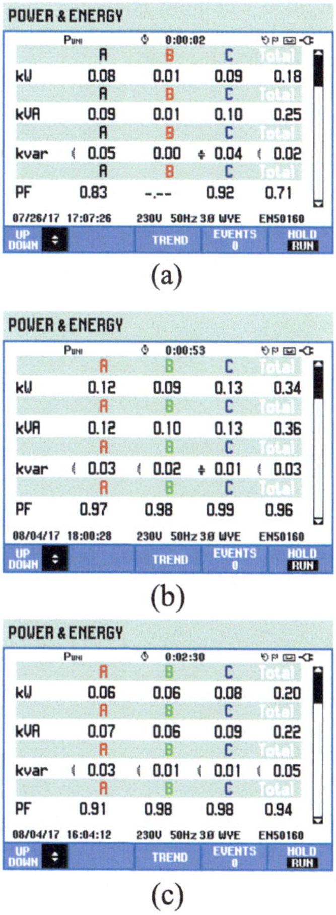

Fig. 4.29 The utility grid power and power factor **a** before the power hub operate, **b** DC load connected, **c** DC grid unload

4.3.6 *Section Summary*

A new topology of power hub, which consists of a V/V transformer, a back-back inverter, and a static VAR compensator with adaptive impedance for power regulation, is proposed in this Section for remote area residential power supply and generalized to smart city applications. The detailed system configuration and control algorithm were proposed and discussed. The proposed power hub offers an interface with different network ports: AC ports and DC ports. In this case, one AC port is for the three-phase AC grid, and the other AC port is for single-phase AC loads, and one DC port is for the DC grid. Simulation and experimental results verified that the proposed three-port converter not only achieves the bidirectional active power flow control between the three-phase AC grid and DC grid but also enables the reactive, unbalanced, and harmonic power compensation for the utility grid. It provides one of the promising multifunctional topological options for remote area residential power supply and

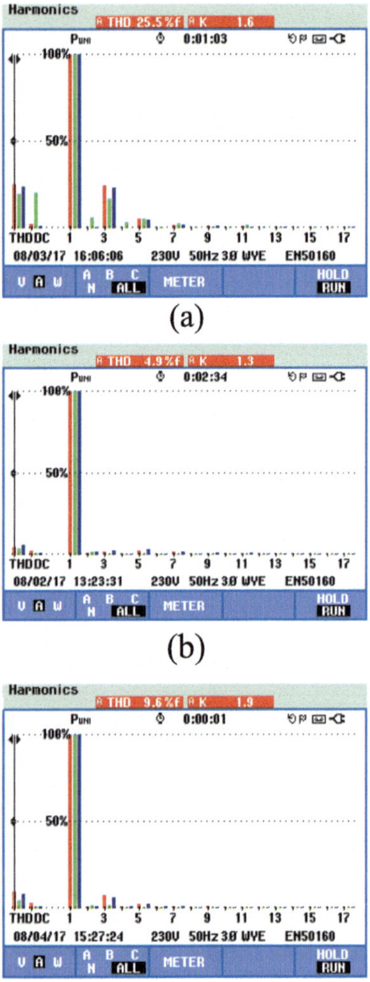

(c)

Fig. 4.30 The utility grid current harmonic spectrum **a** before the power hub operate, **b** DC load connected, **c** DC grid unload

for modern smart city applications to interface DC grid (different renewable energy sources, storage devices, electric vehicle charging stations, and so on) and different AC loads.

4.4 Chapter Summary

This chapter summarized the applications of adaptive impedance converters in interlinking the AC and DC microgrids. More importantly, this chapter proposed a hybrid coupled interlinking converter and its new droop control and selective droop control

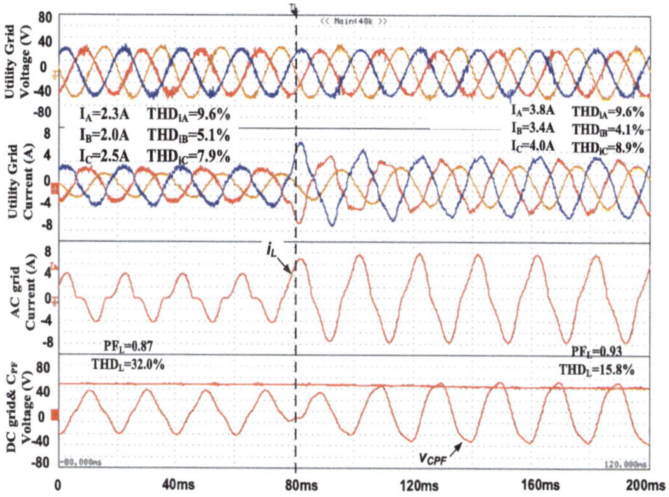

Fig. 4.31 Waveforms of utility grid voltage, utility grid current, AC grid current, DC grid voltage and capacitor voltage for AC load dynamic changing

for the hybrid AC/LVDC microgrid. Through the simulation and experimental results, it demonstrates that the proposed HCIC with the two controls has the ability to provide good power flow control and good power quality compensation. Moreover, the selective power droop control can effectively avoid the over-capacity situation of HCIC with proposed selective factors. Furthermore, a centralized adaptive impedance converter-based three-port power hub and its control are proposed to the dedicated loads in bidirectional or unidirectional, and at the same time isolate the power quality problems generated by the three-phase grid side and single-phase AC grid side, which is proposed for remote area power supply. It verified that the proposed three-port converter not only achieves the bidirectional active power flow control between the three-phase AC grid and DC grid, but also enables the reactive, unbalanced, and harmonic power compensation for the utility grid.

References

1. H. Xiao, A. Luo, Z. Shuai, G. Jin, Y. Huang, An improved control method for multiple bidirectional power converters in hybrid AC/DC microgrid. IEEE Trans. Smart Grid. **7**(1), 340–347 (2016)
2. G. Venkataramanan, M. Illindala, Microgrids and sensitive loads, in *Proceeding of the IEEE Power Engineering Society Winter Meeting, New York*, vol. 1 (2002), pp. 315–322
3. Z. Shuai, Y. Peng, J.M. Guerrero, Y. Li, J.Z. Shen, Transient response analysis of inverter-based microgrids under unbalanced conditions using dynamic phasor model. IEEE Trans. Ind. Electron. **66**(4), 2868–2879 (2019)
4. T.-P. Chen, Zero-sequence circulating current reduction method for parallel HEPWM inverters between AC bus and DC bus. IEEE Trans. Ind. Electron. **59**(1), 290–300 (2012)

5. H. Xiao, A. Luo, Z. Shuai, G. Jin, Y. Huang, An improved control method for multiple bidirectional power converters in hybrid AC/DC microgrid. IEEE Trans. Smart Grid **7**(1), 340–347 (2016)
6. Y. Xia, Y. Peng, P. Yang, M. Yu, W. Wei, Distributed coordination control for multiple bidirectional power converters in a hybrid AC/DC microgrid. IEEE Trans. Power Electron. **32**(6), 4949–4959 (2017)
7. C. Hou, Y. Huang, Diode rectifier with auxiliary converter for hybrid AC/DC microgrids. IEEE J. Emerg. Sel. Topics Power Electron. **2**(4), 1059–1069 (2014)
8. D. Salomonsson, L. Soder, A. Sannino, An adaptive control system for a DC microgrid for data centers. IEEE Trans. Ind. Appl. **44**(6), 1910–1917 (2008)
9. R. Majumder, A hybrid microgrid with DC connection at back to back converters. IEEE Trans. Smart Grid **5**(1), 251–259 (2014)
10. P.C. Loh, D. Li, Y.K. Chai, F. Blaabjerg, Hybrid AC–DC microgrids with energy storages and progressive energy flow tuning. IEEE Trans. Power Electron. **28**(4), 1533–1543 (2013)
11. P. Wang et al., Distributed control for autonomous operation of a three-port AC/DC/DS hybrid microgrid. IEEE Trans. Ind. Electron. **62**(2), 1279–1290 (2015)
12. J. Khajesalehi, K. Sheshyekani, M. Hamzeh, E. Afjei, Maximum constant boost approach for controlling quasi-Z-source based interlinking converters in hybrid AC–DC microgrids. IET Gener. Transmiss. Distrib. **10**(4), 938–948 (2016)
13. J. Huang, J. Xiao, C. Wen, P. Wang, A. Zhang, Implementation of bidirectional resonant DC transformer in hybrid AC/DC micro-grid. IEEE Trans. Smart Grid. **10**(2), 1532–1542 (2019)
14. J. Huang et al., Robust circuit parameters design for the CLLC-type DC transformer in the hybrid AC/DC microgrid. IEEE Trans. Ind. Electron **66**(3), 1906–1918 (2019)
15. H.C. Chiang, K.K. Jen, G.H. You, Improved droop control method with precise current sharing and voltage regulation. IET Power Electron. **9**(4), 789–800 (2016)
16. Y. Liu, T.C. Green, J. Wu, K. Rouzbehi, A. Raza, D. Xu, 'A new droop coefficient design method for accurate power-sharing in VSC-MTDC systems.' IEEE Access **7**, 47605–47614 (2019)
17. D. Pérez-Estévez, J. Doval-Gandoy, J.M. Guerrero, AC-voltage harmonic control for stand-alone and weak-grid-tied converter. IEEE Trans. on Ind. Appl. **56**(1), 403–421 (2020)
18. A. Micallef, M. Apap, C.S. Staines, J.M. Guerrero, J.C. Vasquez, Reactive power sharing and voltage harmonic distortion compensation of droop controlled single phase islanded microgrids. IEEE Trans. Smart Grid. **5**(3), 1149–1158 (2014)
19. G. Liu, T. Caldognetto, P. Mattavelli, P. Magnone, Suppression of second-order harmonic current for droop-controlled distributed energy resource converters in dc microgrids. IEEE Trans. Ind. Electron. **67**(1), 358–368 (2020)
20. F. Gao, S. Bozhko, A. Costabeber, G. Asher, P. Wheeler, Control design and voltage stability analysis of a droop-controlled electrical power system for more electric aircraft. IEEE Trans. Ind. Electron. **64**(12), 9271–9281 (2017)
21. W. Du et al., A comparative study of two widely used grid-forming droop controls on microgrid small signal stability. IEEE J. Emerg. Sel. Top. Power Electron. Early Access Article. https://doi.org/10.1109/JESTPE.2019.2942491
22. I.U. Nutkani, P.C. Loh, W. Peng, F. Blaabjerg, Cost-prioritized droop schemes for autonomous AC microgrids. IEEE Trans. Power Electron. **30**(2), 1109–1119 (2015)
23. I.U. Nutkani, P.C. Loh, F. Blaabjerg, Droop scheme with consideration of operating costs. IEEE Trans. Power Electron. **29**(3), 1047–1052 (2014)
24. L. Wang, C. Lam, M. Wong, The analysis of DC-link voltage, compensation range, cost, reliability and power loss for shunt (hybrid) active power filters, in *2018 IEEE PES Asia-Pacific Power and Energy Engineering Conference (APPEEC), Kota Kinabalu* (2018), pp. 640–645
25. L. Wang, C. Lam, M. Wong, Analysis, control, and design of a hybrid grid-connected inverter for renewable energy generation with power quality conditioning. IEEE Trans. Power Electron. **33**(8), 6755–6768 (2017)
26. D. Jovcic, W. Lin, Multiport high-power LCL DC hub for use in DC transmission grids. IEEE Trans. power Delivery. **29**(2), 760–768 (2014)

27. L. Wang, C.S. Lam, M.C. Wong, Design of a thyristor controlled LC compensator for dynamic reactive power compensation in smart grid. IEEE Trans. Smart. Grid. **8**(1), 409–417 (2017)
28. M. Rastegar, M. Fotuhi-Firuzabad, H. Zareipour, M. Moeini-Aghtaieh, A Probabilistic energy management scheme for renewable-based three-port converters. IEEE Trans. Smart Grid. **8**(5), 2217–2227 (2017)
29. V. Khadkikar, D. Xu, C. Cecati, Emerging power quality problems and state-of-the-art solutions. IEEE Trans. Ind. Electron. **64**(1), 761–763 (2017)

Chapter 5
Adaptive Impedance Converters in Co-phase Railway Traction Supply System

5.1 Introduction of Power Quality Conditioners in Railway Power Supply

Many single-phase 25-kV electrified railway systems today are loaded with conventional ac–dc thyristor-based locomotives, which can easily draw reactive, unbalanced, and harmonic currents [1]. The flow of reactive and harmonic currents through the feeder impedance can result in different power quality problems, such as low power factor (PF), harmonic pollution, and unbalanced problems. These power quality problems may lead to undesirable consequences in the railway system, such as large extra power loss and unexpected system power-off. Concerning the power quality problems introduced earlier, researchers proposed different power quality conditioners in the existing literature to solve the power quality issues in the railway system [2–16]. The single-phase railway power supply power is normally step-down from the three-phase high-voltage grid. As shown in Fig. 5.1a, three-phase STATCOM can be installed at the primary side of the electrical railway power system to relieve reactive power and negative-sequence current (NSC) problems [2, 3]. However, a costly, high-voltage, three-phase step-down transformer is required. The three-phase-to-single phase ac-to-ac inverter in Fig. 5.1b can directly address the railway NSC problem [4–6]. However, the failure of every component in the ac-to-ac inverter can lead to interruption of railway power supply, which can significantly decrease the reliability. Compared with STATCOMs in Fig. 5.1a and ac-to-ac inverters in Fig. 5.1b, the different railway power conditioners in Fig. 5.1c–e obtain lower power capacity than the STATCOMs and higher reliability than the ac-to-ac inverters. In the traditional traction power supply, as shown in Fig. 5.1c and d, the power in the three-phase grid is transformed into two single-phase outputs (phase Vac and Vbc). Since the two single-phase power grids (phase V_{ac} and V_{bc}) have different phases, neutral sections (NSs) are required to avoid the risk of phase mixing [7–13]. As locomotives run through NSs, they easily lose power and velocity. Therefore, the traditional traction power supply structure is not suitable for high-speed railways. The

L. Wang et al., *Design and Applications of Adaptive Impedance Converters*,
https://doi.org/10.1007/978-981-97-5319-2_5

locomotive loadings are connected across one single-phase output of the substation transformer only and leaving another phase unloaded [14–16] in the cophase power supply system, as shown in Fig. 5.1e, which is different from the traditional railway power supply, as shown in Fig. 5.1c and d. The purpose of using the "separator" is to isolate two adjacent traction substations on a long-distance railway traction power supply system. Thus, the traction substations can avoid circuiting currents, short circuits, andphase mixing problems between adjacent two substations [6]. Besides that, the separator allows the disconnection of the power grid section during maintenance. Otherwise, the whole traction power grid needs to be shut OFF, which causes significant economic loss. The cophase system is beneficial for the avoidance of NSs, which is more suitable for application in high-speed railways. The idea of a cophase traction power supply first appeared in 2009 [14]. The world's first cophase traction power supply device has already been implemented and put into trial operation at the MeiShan Substation of the Chengdu–Kunming Railway in 2012 in China. Based on the abovementioned discussions, the brief comparison is summarized in Table 5.1. As shown in Table 5.1, compared with the three-phase conditioners in Fig. 5.1a and c, the single-phase back-to-back-connected conditioners in Fig. 5.1d and e have better unbalanced compensation ability. Since the railway loading system can easily have an unbalance factor as large as 100%, the single-phase back-to-back-connected conditioners in Fig. 5.1d and e are more suitable for railway applications. For the high-speed railway, the structures shown in Fig. 5.1b and e are more suitable since they do not have the NSs.

Conventionally, the coupled impedances of the cophase RPCs and other different conditioners are inductance, which requires a high inverter operating voltage. As the cost of the passive part is lower than that of the active inverter part, the reduction of active inverter part capacity can lead to a decrease in the total cost of the whole system [1, 8, 17–28]. To reduce the inverter voltage rating, the large fixed capacitive impendence-coupled hybrid active power filters (HAPFs) have been proposed for reactive and harmonic power compensation in a distribution system [17–19]. Later on, the concept of capacitive-coupled impendence was applied to the railway power supply system [1, 8, 20–22], in which this capacitive-coupled power quality device is called HAPFs in the distribution system [17–19] and hybrid power quality conditioners (HPQCs) [1, 8, 20–22] in the cophase railway power supply system. However, the large capacitive impedances in both HAPFs and HPQCs limit their compensation range. Therefore, in order to enlarge the compensation range and keep at a low inverter operating voltage simultaneously, the adaptive impedance-coupled inverter topologies have been proposed, namely, thyristor-controlled LC-coupled (TCLC)-HAPFs for the distribution system and hybrid grid-connected inverters (HGCIs) for distribution generation (DG) systems [23–28]. From the point of view of realization methods, the adaptive impedance also can be realized by hardware topologies [23, 24] or control method, such as the virtual impedance method [23, 25, 27, 28], to change the coupled impedance and enlarge the compensation range. The comparisons of the conventional methods and the proposed method are provided above and summarized in Table 5.2. Based on the abovementioned analysis and discussions, a summary table is provided in Table 5.2. The cophase RPCs were first proposed in 2009 for traction

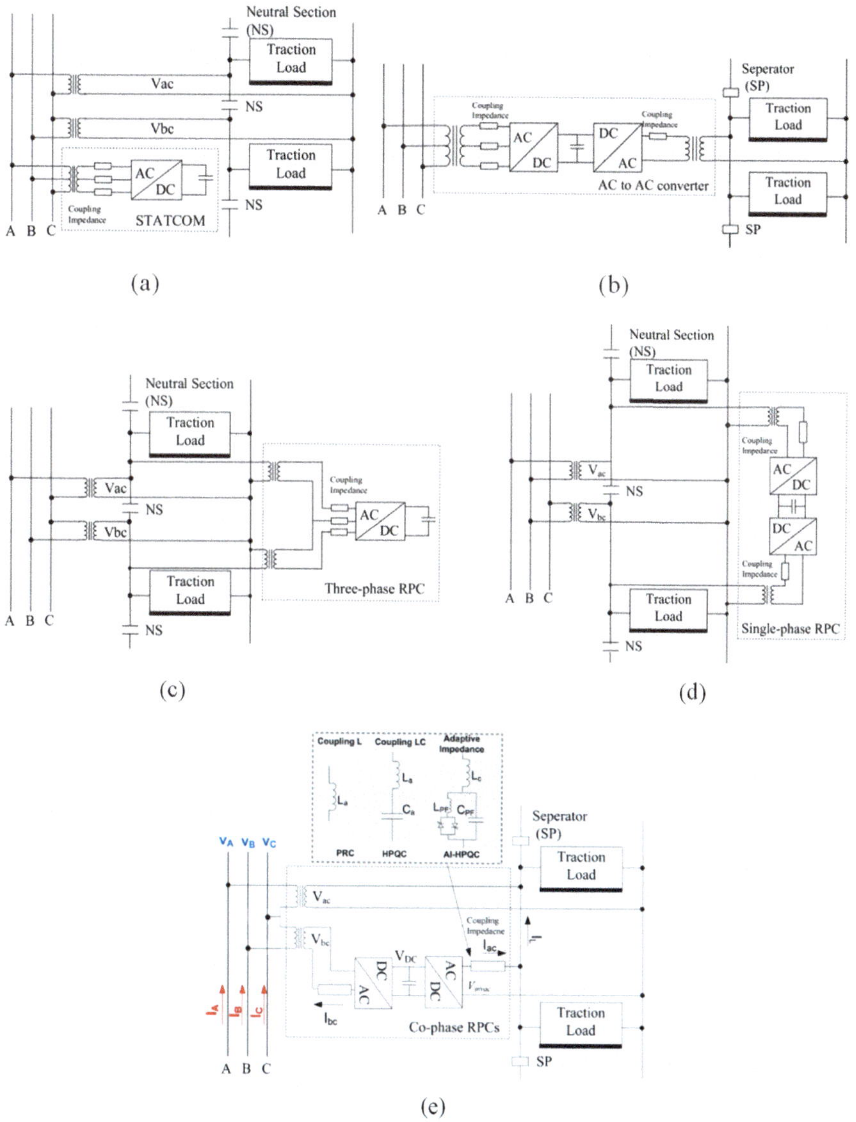

Fig. 5.1 Railway power supply with different power quality conditioners. **a** STATCOM. **b** AC-to-AC inverter. **c** Three-phase railway power quality compensator (RPC). **d** Single-phase RPC. **e** Cophase RPCs (RPC, HPQC, and the proposed AI-HPQC)

Table 5.1 The railway power supply with different power quality conditioners

	Developed year	Figure	Voltage level	Neutral sections (NSs)	Unbalanced comp. ability	Reliability	Cost
STATCOM	2001	Figure 5.1a	High	Required	Poor	Medium	Very high
AC to AC inverter	2004	Figure 5.1b	Medium	Not required	Good	Very low	Medium
3-phase RPC	2004	Figure 5.1c	Medium	Required	Medium	Low	Medium
1-phase RPC	2009	Figure 5.1d	Medium	Required	Good	High	Medium
Co-phase RPC	2009	Figure 5.1e	Medium	Not required	Good	High	Medium

Note that the shaded area means the undesirable characteristics

loads with a large compensation range. However, the cophase RPCs require high operation voltage and cost. To reduce the voltage and cost, the HPQCs have been proposed in 2013, but the compensation of the HPQCs is narrow. To overcome the disadvantages of both RPCs and HPQCs, the AI-HPQC, which has a wide compensation range and low operation voltage, is proposed in this section. In this chapter the adaptive impedance-coupled HPQC (AI-HPQC) is proposed in the cophase railway power supply system. The difference with the three-phase distribution system, the single-phase active, reactive, and harmonic powers are transferred from one single-phase converter of the AI-HPQC to the other. Therefore, the analysis, design, and control of the proposed AI-HPQC are different from the conventional structures [1, 8, 17–24]. Furthermore, the harmonic filter performance of the coupled impedance is guaranteed by operating the AI-HPQC in the quasi-linear region [29]. The main contributions of this article are to propose the structure, control method, and analysis of the AI-HPQC, which has desirable beneficial characteristics of low operation voltage and wide compensation range. The contributions and novelties of this article can be summarized as follows: (1) the AI-HPQC is proposed for co-phase railway traction power supply first to bring the advantages of adaptive impedance (reducing dc voltage, expanding application range, and improving the efficiency) into co-phase compensators; (2) the application differences of AI-HPQC between RPC and HPQC are compared and discussed in terms of the characteristic, compensation range, and operation voltage; (3) the power flow, control strategy, and parameters design of the AI-HPQC are proposed for the cophase railway traction power supply system; and 4) the comparative performances evaluations of the conventional RPC, HPQC, and the proposed AI-HPQC are given through simulations and experiments. The layout of this Chapter can be described as follows. In Sect. 5.2, the RPC, HPQC, and AI-HPQC are compared in terms of impedance, compensation range, and operation voltage based on cophase traction power loading conditions. After that, the control strategy is proposed for the AI-HPQC in Sect. 5.3. Following that, the comparative performance evaluations and verification are given for RPC, HPQC, and the AI-HPQC through simulations in Sect. 5.4 and experiments in Sect. 5.5. Finally, the Chapter summary and References is drawn in Sect. 5.5 and Sect. 5.6.

5.2 Active Impedance, Compensation Range, and Operation Voltage of RPC, HPQC, and AI-HPQC

In this Section, the adaptive impedance, compensation range, and operation voltage are discussed. The comparative studies are provided for the compensation range and operation voltage of RPC, HPQC, and AI-HPQC.

A. Adaptive Impedance of AI-HPQC

Referring to Fig. 5.1e, the vector of the required inverter voltage of V_{ac} phase can be expressed as:

$$\begin{aligned} \vec{V}_{invac} &= \vec{V}_{ac} - jX_{imp} \cdot \vec{I}_{ac} \\ &= V_{ac} - jX_{imp} \cdot \left(I_{acp} + jI_{acq}\right) \\ &= V_{ac} + X_{imp} \cdot I_{acq} - jX_{imp} \cdot I_{acp} \end{aligned} \tag{5.1}$$

where V_{ac}, I_{acq} and I_{acp} are phase voltage, reactive and active power components of the compensating current in secondary side of phase V_{ac}. The X_{imp} is the coupled impedance of conditioners.

Based on (5.1), the ratio of α can be expressed in terms of active and reactive power as:

$$\begin{aligned} \frac{\vec{V}_{invac}}{\vec{V}_{ac}} &= \frac{V_{ac} + X_{imp} \cdot I_{acq} - jX_{imp} \cdot I_{acp}}{V_{ac}} = 1 + \frac{X_{imp} \cdot I_{acq}}{V_{ac}} - j\frac{X_{imp} \cdot I_{acp}}{V_{ac}} \\ &= 1 + \frac{V_{ac} \cdot I_{acq}}{V_{ac}^2/X_{imp}} - -j\frac{V_{ac} \cdot I_{acp}}{V_{ac}^2/X_{imp}} \end{aligned} \tag{5.2}$$

The norm of the $\vec{V}_{invac}/\vec{V}_{ac}$ can be expressed as:

$$\begin{aligned} \left|\frac{\vec{V}_{invac}}{\vec{V}_{ac}}\right| &= \sqrt{\left(1 + \frac{V_{ac} \cdot I_{acq}}{V_{ac}^2/X_{imp}}\right)^2 + \left(\frac{V_{ac} \cdot I_{acp}}{V_{ac}^2/X_{imp}}\right)^2} \\ &= \sqrt{\left(1 + \frac{Q_{c\alpha}}{S_{imp}}\right)^2 + \left(\frac{P_{c\alpha}}{S_{imp}}\right)^2} \end{aligned} \tag{5.3}$$

In (5.3), the Q_{ca}, $P_{ca,}$ and S_{imp} are defined as:

Table 5.2 Comparisons of different structures, controls and operations

	Developed year	Applications	Structure	Control	Unbalanced comp. ability	Compensation range	Operation voltage
Co-phase RPC	2009	Railway system	Back to back converter	Q, P and H control	Strong	Wide	High
HAPFs	2012	Distribution system	LC part and three-phase inverter	Q and H control	Weak	Narrow	Low
HPQCs	2013	Railway system	LC part and back to back converter	Q, P and H control	Medium	Narrow	Low
TCLC–HAPFs	2014	Distribution system	Adaptive impedance part and three-phase inverter	1. Q and H control 2. Coordinate control of three-phase Adaptive impedance part and inverter part	Medium	Wide	Low
HGCIs	2016	DG system	Adaptive impedance part and three-phase inverter	Q, P and H control	Medium	Wide	Low
Proposed AI-HPQC	This chapter	Railway system	Adaptive impedance part and back to back converter	1.Q, P and H control 2. Coordinate control of single-phase adaptive impedance part and back to back inverter part	Strong	Wide	Low

$$
\begin{aligned}
Q_{c\alpha} &= V_{ac} \cdot I_{acq} \\
P_{c\alpha} &= V_{ac} \cdot I_{acp} \\
S_{imp} &= V_{ac}^2 / X_{imp}
\end{aligned}
\tag{5.4}
$$

where $P_{c\alpha}$ and $Q_{c\alpha}$ are the active and reactive power provided by conditioners for V_{ac} phase. The S_{imp} is the based power of conditioners related to the coupled impedance. In (5.4), the S_{imp} is the based power of conditioners related to the coupled impedance, which can be expressed as:

$$
S_{imp} = \frac{V_{ac}^2}{X_{imp}} = \begin{cases} S_{RPC} = \frac{V_{ac}^2}{X_{La}} \\ S_{HPQC} = \frac{V_{ac}^2}{X_{La} - X_{Ca}} \\ S_{AI-HPQC}(\alpha) = \frac{V_{ac}^2}{X_{AI}(\alpha)} = \frac{V_{ac}^2}{\frac{\pi X_{LPF} X_{CPF}}{X_{CPF}[2\pi - 2\alpha + \sin(2\alpha)] - \pi X_{LPF}} + X_{Lc}} \end{cases}
\tag{5.5}
$$

where the S_{imp} is the based power of conditioners related to the coupled impedance. X_{imp} can be the coupling impedance of RPC, HPQC, and AI-HPQC. For the RPC and HPQC, the impedances are small inductance $X_{La} > 0$ and large capacitance $X_{La} - X_{Ca} << 0$ ($X_{La} << X_{Ca}$). For the AI-HPQC, the coupling adaptive impedance can be expressed as [27–31]:

$$
X_{AI}(\alpha) = \frac{\pi X_{L_{PF}} X_{C_{PF}}}{X_{C_{PF}}[2\pi - 2\alpha + \sin(2\alpha)] - \pi X_{L_{PF}}} + X_{L_c}
\tag{5.6}
$$

where X_{Lc}, X_{CPF}, and X_{LPF} are the impedances of the coupling inductor L_c, the paralleled capacitor C_{PF} and inductor L_{PF}, respectively.

Based on the above (5.5) and (5.6), the based power and impedance versus with firing angle can be shown in Figs. 5.2 and 5.3. From Figs. 5.2 and 5.3, it can be seen that the based power and adaptive impedance can be dynamically controlled by controlling the firing angle α. Compared with S_{RPC} and S_{HPQC}, the variability of $S_{AI\text{-}HPQC}$ is based on the trigger angle α, which gives the proposed AI-HPQC a flexible and wider compensation range.

According to the railway traction load power factor and (5.3), the expression related to railway traction load power factor PF_L can be further expressed as:

$$
\begin{aligned}
\left| \frac{\vec{V}_{invac}}{\vec{V}_{ac}} \right| &= \sqrt{\left(1 + \frac{(0.2887 \cdot PF_L + \sqrt{1 - PF_L^2}) \cdot S_L}{S_{imp}}\right)^2 + \left(\frac{(0.5 \cdot PF_L) \cdot S_L}{S_{imp}}\right)^2} \\
&= \sqrt{\left(1 + \frac{k_q \cdot r \cdot S_{rated}}{S_{imp}}\right)^2 + \left(\frac{k_p \cdot r \cdot S_{rated}}{S_{imp}}\right)^2}
\end{aligned}
\tag{5.7}
$$

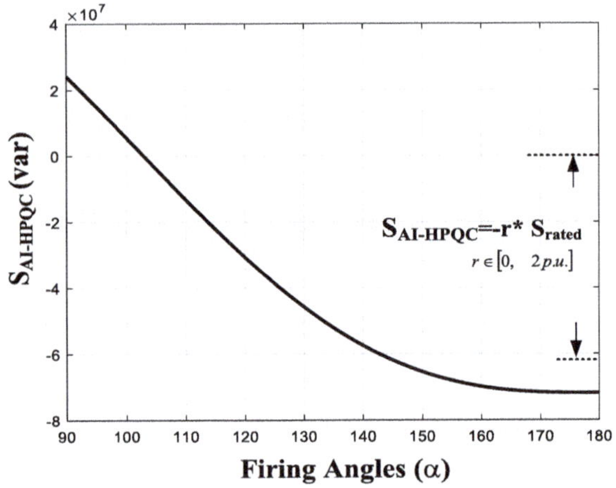

Fig. 5.2 Based power versus with the firing angle α

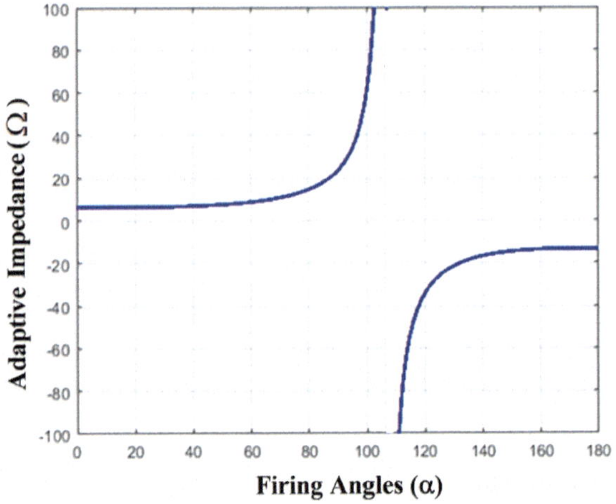

Fig. 5.3 Adaptive impedance versus with the firing angle α

where the k_p and k_q are the load active and reactive capacity ratios related to the power factor PF_L. S_L and S_{rated} are the load's apparent power and load-rated power. And the S_L can be expressed by a ratio and rated load power. The k_p, $k_{q,}$ and S_L can be given as:

$$\begin{aligned} k_p &= 0.5 \cdot PF_L \\ k_q &= 0.2887 \cdot PF_L + \sqrt{1 - PF_L^2} \\ S_L &= r \cdot S_{rated} \end{aligned} \tag{5.8}$$

In (5.7), the S_{imp} for the RPC and HPQC are fixed values. However, the S_{imp} for the AI-HPQC is controllable, which can be used to keep the ratio V_{invac} and V_{ac} at the minimum value. By letting the derivative of (5.2) and (5.7) equal to zero, the S_{imp} ($S_{AI\text{-}HPQC}$) can be deduced as:

$$\partial\left(\frac{\overrightarrow{V}_{invac}}{\overrightarrow{V}_{ac}}\right) = 0 \; S_{AI-HPQC}(\alpha) = \frac{P_{c\alpha}^2 + Q_{c\alpha}^2}{Q_{c\alpha}} = \frac{k_p^2 + k_q^2}{k_q} \cdot r \cdot S_{rated} \tag{5.9}$$

where r is the ratio between S_L and S_{rated}, k_p and k_q can be found from (5.8). Based on (5.5), (5.8), and (5.9), the adaptive impedance X_{AI} can be expressed as:

$$X_{AI}(\alpha) = \frac{\pi X_{L_{PF}} X_{C_{PF}}}{X_{C_{PF}}[2\pi - 2\alpha + \sin(2\alpha)] - \pi X_{L_{PF}}} + X_{L_c} = \frac{V_{ac}^2 \cdot Q_{c\alpha}}{P_{c\alpha}^2 + Q_{c\alpha}^2} \tag{5.10}$$

Based on (5.9) and (5.10), the adaptive impedance X_{AI} can be calculated through k_p and k_q. Assuming a load traction feeder voltage of 27.5 kV, the impedance that yields the lowest Railway HPQC operation voltage under different load power factors and capacity conditions is shown in Fig. 5.4. Furthermore, the variation of SVC adaptive impedance with different firing angles is plotted in Fig. 5.3. By comparing Figs. 5.3 and 5.4, the design allows the adaptive impedance railway HPQC to operate at minimum voltage by varying the impedance using different firing angles.

From Fig. 5.4, it can be seen that the required adaptive impedance X_{AI} varies with $P_{c\alpha}$ and $Q_{c\alpha}$. Moreover, by varying the firing angle α, the reactive power compensation capability of the AI-HPQC can be controlled. And the active part of AI-HPQC can compensate harmonics caused by the switching angle of the adaptive impedance.

B. Compensation Range and Operation Voltage of RPC, HPQC, and AI-HPQC

The purpose of the adaptive impedance X_{AI} is to keep the ratio of V_{invac}/V_{ac} at the minimum value. Compared with adaptive impedance X_{AI}, the inductive impedance

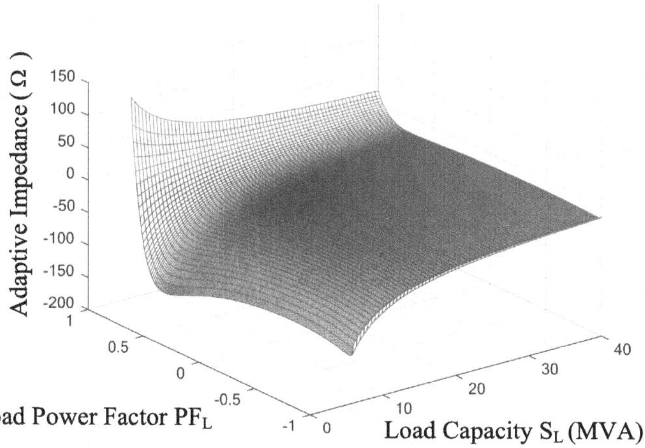

Fig. 5.4 Variation of adaptive impedance with load power factor and capacity

Table 5.3 Data extracted from China Wuqing substation

	Minimum	Maximum	Rated
Active power (MW)	0	36.5	–
Reactive power (MVar)	6.0	10.5	–
Total capacity (MVA)	6.5	38	31.5
Total capacity (p.u.)	0.2	1.2	1.0
Load power factor (PF_L)	0	1	–

in RPC and capacitive impedance in HPQC are fixed, so the ratio of V_{invac}/V_{ac} of RPC and HPQC is larger than AI-HPQC. The below analysis is based on practical data from China Wuqing substation [21–25] as Table 5.3. In Table 5.3, with a rated capacity of 31.5 MVA, the traction load capacity ranges from 6.5 to 36.5 MVA (0.2–1.16 p.u.). Since locomotive load *PF* usually ranges from 0.8 to 0.9, the rated load *PF* is around 0.85 [21–25]. In the analysis and simulation sections, the traction load capacity range is designed to be 0–2.0 p.u., which is wider than the total capacity. Therefore, the typical ratio r between S_L and S_{rated}, is also from 0 to 2.0.

Based above analysis in (5.2)–(5.9), the operation voltage ratio comparisons among the RPC, HPQC, and AI-HPQC can be plotted as Fig. 5.5. The compensation range is defined as the range of loading capacity (in per unit), in which the compensation device can provide satisfactory compensation performance under a certain railway traction load power factor (PF). In Fig. 5.5a, the HPQC is a V-shaped surface, while the RPC is a flat surface. However, the range is limited for V_{invac} of the HPQC and is smaller than the RPC, which implies its narrow compensation range characteristics with low dc-link operating voltage. From Fig. 5.5b, it can be seen that the AI-HPQC can keep at low V_{invac}/V_{ac} level for a wide *r* range.

In order to further compare the V_{invac} between HPQC and AI-HPQC, the inverter ratio of $V_{invHPQC}/V_{invAI\text{-}HAPC}$ can be plotted as in Fig. 5.6. Based on Fig. 5.6, it is seen that for most of the conditions, the operation inverter voltage requirement of the proposed AI-HPQC is lower than that of HPQC. According to the 24-h monitoring power quality data in the WuQing substation, the active power varies from 0 W to 39 MW, and the PF_L varies from 0 to 1. And the load power S_L varies from 6.5 MVA to 38 MVA. Due to the large fluctuation of PF_L and S_L, the AI-HPQC with a wider compensation range and lower inverter rating has a stronger practical value than the conventional HPQC, according to Fig. 5.6.

5.3 Control and Parameter Design Strategy of the Proposed AI-HPQC

In this chapter, the adaptive impedance, compensation range, and operation voltage are discussed. The comparative studies are provided for the compensation range and operation voltage of RPC, HPQC, and AI-HPQC.

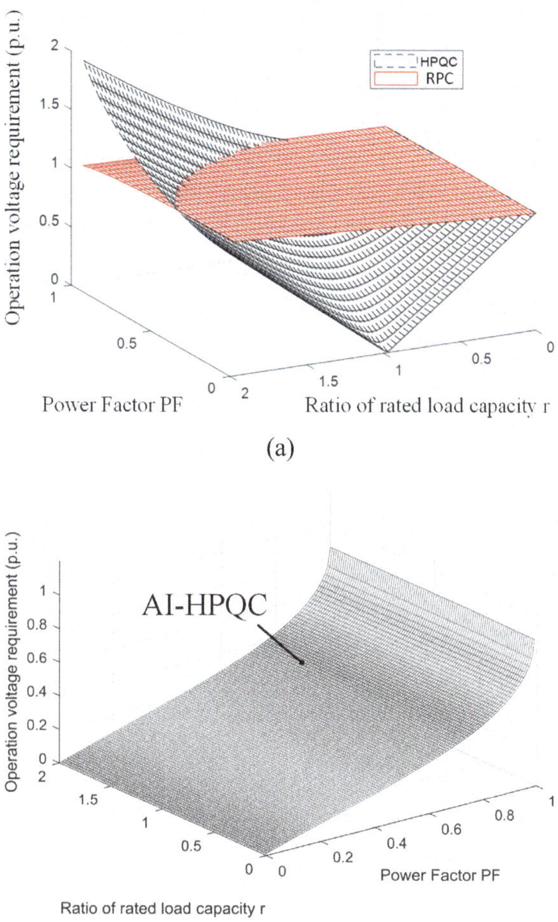

Fig. 5.5 Ratio of operation voltage requirement in term of *PF* and load capacity ratio *r* in **a** RPC and HPQC, **b** the proposed AI-HPQC

A. Back-to-Back Inverter Part Control

Back-to-back inverter part control is based on power follow analysis in Fig. 5.7. In Fig. 5.7 i_X and v_X are the source current and voltage on the primary side, where the subscript 'X' denotes phase X = A, B, C on the primary side. i_L, $i_{ac,}$ and i_{bc} are the load current, compensating current in V_{ac} phase and V_{bc} phase.

As shown in Fig. 5.7, it is assumed $\vec{V}_A = V_1 < o^{\circ}$, $\vec{V}_B = V_1 < -120^{\circ}$ and $\vec{V}_c = V_1 < 120^{\circ}$, where $1 < -120° = -{}^1\!/_2 - j\sqrt{{}^3\!/_2}$ and $1 < -120° = -{}^1\!/_2 + j\sqrt{{}^3\!/_2}$. The traction loads are connected to V_{ac} phase and V_{bc} phase is unloaded so that the relationship $\vec{I}_{AC} = \vec{I}_A$ and $\vec{I}_{BC} = \vec{I}_B$ can be obtained. Based on the above descriptions, the power flowing in the V_{ac} phase and V_{bc} phase can be expressed as:

$$P_{AC} + jQ_{AC} = \vec{V}_{AC} \cdot \vec{I}_{AC} = \left(\vec{V}_A - \vec{V}_C\right) \cdot \vec{I}_A$$

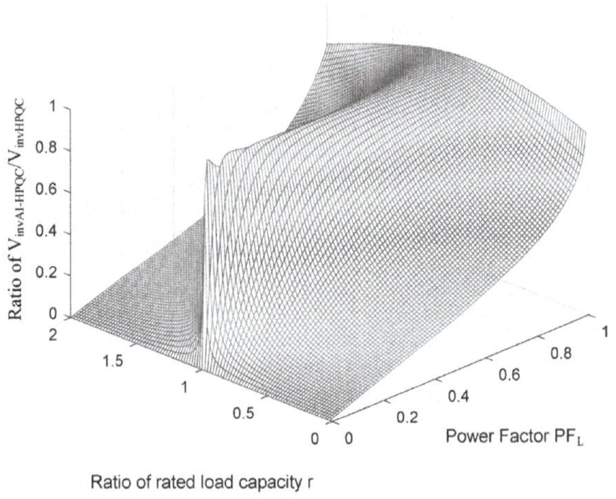

Fig. 5.6 Ratio of $V_{invAI\text{-}HPQC}/V_{invHPQC}$ in terms of *PF* and load capacity ratio *r*

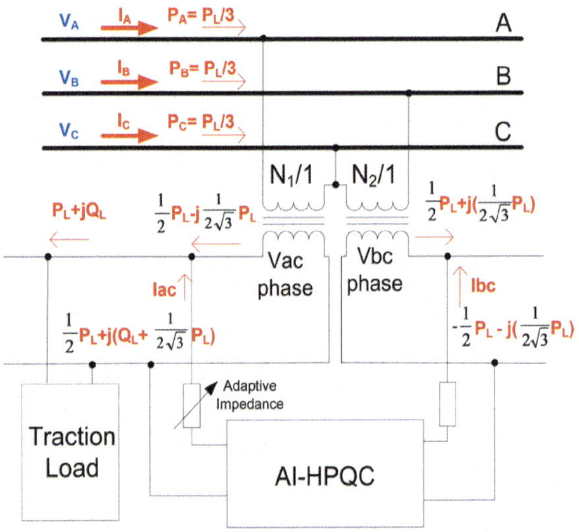

Fig. 5.7 Power flow analysis of co-phase AI-HPQC railway power supply

$$= \left(1\angle 0^o - 1\angle 120^o\right) \cdot V_1 \cdot \vec{I}_A = \left(\frac{3}{2} - j\frac{\sqrt{3}}{2}\right) \cdot (P_A + jQ_A)$$

$$= \left(\frac{3}{2}P_A + \frac{\sqrt{3}}{2}Q_A\right) + j\left(-\frac{\sqrt{3}}{2}P_A + \frac{3}{2}Q_A\right) \tag{5.11}$$

$$P_{BC} + jQ_{BC} = \vec{V}_{BC} \cdot \vec{I}_{BC} = \left(\vec{V}_B - \vec{V}_C\right) \cdot \vec{I}_B$$

$$= \left(\frac{3}{2}P_B - \frac{\sqrt{3}}{2}Q_B\right) + j\left(\frac{\sqrt{3}}{2}P_B + \frac{3}{2}Q_B\right) \tag{5.12}$$

The AI-HPQC's purpose is to compensate load reactive power QL and balance the active power P_L. After ideal compensation, the P_L is transferred to each phase ($P_A = P_B = P_C = 1/3P_L$), and the reactive power Q_L is compensated to zero ($Q_A = Q_B = Q_C = 0$). By substituting the $P_A = P_B = P_C = 1/3P_L$ and $Q_A = Q_B = Q_C = 0$ into (5.11) and (5.12), the active and reactive power flowing in V_{ac} phase and V_{bc} phase after compensation can be given as:

$$P_{AC} + jQ_{AC} = \frac{1}{2}P_L - j\frac{1}{2\sqrt{3}}P_L \tag{5.13}$$

$$P_{BC} + jQ_{BC} = \frac{1}{2}P_L + j\frac{1}{2\sqrt{3}}P_L \tag{5.14}$$

After obtaining the power flow in V_{ac} and V_{bc} phase, the compensating power provided by the AI-HPQC can be calculated as:

$$P_{c\alpha} + jQ_{c\alpha} = (P_L + jQ_L) - (P_{AC} + jQ_{AC}) = \frac{1}{2}P_L + j\left(Q_L + \frac{1}{2\sqrt{3}}P_L\right) \tag{5.15}$$

$$P_{c\beta} + jQ_{c\beta} = -(P_{BC} + jQ_{BC}) = -\frac{1}{2}P_L - j\frac{1}{2\sqrt{3}}P_L \tag{5.16}$$

where $P_{c\alpha}$, $P_{c\beta}$, $Q_{c\alpha}$, and $Q_{c\beta}$ are the active and reactive power provided by the AI-HPQC for V_{ac} and V_{bc} phase. Referring to (5.4) and (5.5), the instantaneous active and reactive compensating power can be expressed as:

$$\begin{bmatrix} p_{c\alpha} \\ q_{c\alpha} \\ p_{c\beta} \\ q_{c\beta} \end{bmatrix} = \begin{bmatrix} \frac{1}{2}\overline{p}_L + \tilde{p}_L \\ \frac{1}{2\sqrt{3}}\overline{p}_L + q_L \\ -\frac{1}{2}\overline{p}_L \\ -\frac{1}{2\sqrt{3}}\overline{p}_L \end{bmatrix} \tag{5.17}$$

where p_L and q_L are the instantaneous active and reactive power, which include DC components $\overline{p}_L$ and $\overline{q}_L$, and AC components $\tilde{p}_L$ and $\tilde{q}_L$. In (5.17), the instantaneous active and reactive power of the traction loads can be calculated by the modified instantaneous *pq* theory [32] as:

$$\begin{bmatrix} p_L \\ q_L \end{bmatrix} = \begin{bmatrix} v_{ac} \cdot i_L + v_{ac}^D \cdot i_L^D \\ -v_{ac}^D \cdot i_L + v_{ac} \cdot i_L^D \end{bmatrix} \tag{5.18}$$

where v_{ac} and i_L are the phase voltage and load current, $v_{ac}{}^D$ and $i_L{}^D$ are 90° delay of the phase voltage and load current. p_L and q_L are the instantaneous active and reactive power, which includes DC components $\overline{p}_L$ and $\overline{q}_L$, and AC components $\tilde{p}_L$ and $\tilde{q}_L$. $\overline{p}_L$ and $\overline{q}_L$ are obtained by passing p_L and q_L through low-pass filters. The load power obtained in (5.18) is used for the determination of the inverter compensating active ($p_{c\alpha}$) and reactive ($q_{c\alpha}$) power in the V_{ac} phase, and active ($p_{c\beta}$) and reactive ($q_{c\beta}$)

power in V_{bc} phase. The back-to-back inverter is used to improve the characteristic of the adaptive impedance part by limiting the compensating current $i_{c\alpha}$ and $i_{c\beta}$ to their reference value $i_{c\alpha}^*$ and $i_{c\beta}^*$, so that the mistuning problem, the resonance problem, and the harmonic injection problem can be avoided for the AI-HPQC. The $i_{c\alpha}^*$ and $i_{c\beta}^*$ is calculated based on the modified instantaneous pq theory [32] as:

$$i_{c\alpha}^* = \frac{1}{v_{ac}^2 + \left(v_{ac}^D\right)^2} \cdot \begin{bmatrix} v_{ac} & v_{ac}^D \end{bmatrix} \cdot \begin{bmatrix} p_{c\alpha} \\ q_{c\alpha} \end{bmatrix} \tag{5.19}$$

$$i_{c\beta}^* = \frac{1}{v_{bc}^2 + \left(v_{bc}^D\right)^2} \cdot \begin{bmatrix} v_{bc} & v_{bc}^D \end{bmatrix} \cdot \begin{bmatrix} p_{c\beta} \\ q_{c\beta} \end{bmatrix} \tag{5.20}$$

By comparing the actual compensating current ($i_{c\alpha}$ and $i_{c\beta}$) with the obtained reference ($i_{c\alpha}^*$ and $i_{c\beta}^*$), the trigger signals for the back-to-back inverter part can be generated through the current phase shifted carrier modulation control.

B. Adaptive Impedance Part Control

As discussed in (5.9), the adaptive impedance part is used to transfer the active and reactive current while considering the minimum dc-link voltage. Based on (5.10), the $X_{AI}(\alpha)$ can be expressed in terms of V_{ac}, $Q_{c\alpha}$, and $P_{c\alpha}$. And the instantaneous expression of V_{ac}, $Q_{c\alpha}$, and $P_{c\alpha}$ can be expressed as:

$$V_{ac}^2 = v_{ac}^2 + \left(v_{ac}^D\right)^2 \tag{5.21}$$

$$P_{c\alpha} = \overline{p}_{c\alpha}/2 = \frac{1}{4}\overline{p}_L \tag{5.22}$$

$$Q_{c\alpha} = \overline{q}_{c\alpha}/2 = \frac{1}{4\sqrt{3}}\overline{p}_L + \frac{1}{2}\overline{q}_L \tag{5.23}$$

In (5.22) and (5.23), the $\overline{p}_L$ and $\overline{q}_L$ can be obtained from (5.18) and low pass filters.

By comparing the calculated left-hand side term and the right-hand side term of X_{AI} expression in (5.10), the firing angle α can be determined. Because (5.10) is complicated, a look-up table between the firing angle α and the X_{AI} is installed inside the controller. The demo waveforms of v_{ac} and control signals of two thyristors are illustrated in Fig. 5.8.

C. Overall Control Strategy

The overall control block of the AI-HPQC is provided in Fig. 5.9. The purpose of the back-to-back inverter part is to balance active power and compensate reactive power and harmonic power by limiting the compensating current ($i_{c\alpha}$ and $i_{c\beta}$) with the obtained references ($i_{c\alpha}^*$ and $i_{c\beta}^*$).

By using the modified instantaneous pq theory, the load active and reactive power can be calculated through (5.18). Then, according to the power analysis, the compensating active and reactive power can be obtained for the AI-HPQC. After that, the

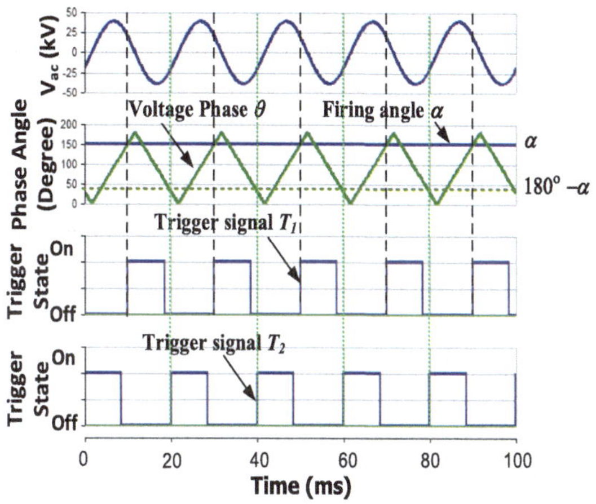

Fig. 5.8 Demo waveforms of v_{ac} and control signals of two thyristors

compensating active and reactive power can be transformed to $i_{c\alpha}{}^*$ and $i_{c\beta}{}^*$through (5.19) and (5.20). Through phase shifted carrier modulation control method, the trigger signals of the back to back inverter part can be generated by comparing the $i_{c\alpha}$ and $i_{c\beta}$ with their references $i_{c\alpha}{}^*$ and $i_{c\beta}{}^*$. In this paper, the current phase-shifted carrier modulation control is applied due to its advantages of simplicity of implementation, high robustness, etc.

For the adaptive impedance part control, the expressions in (5.21)–(5.23) are used to calculate the voltage of V_{ac} phase, fundamental compensating active and reactive powers of V_{ac} phase. Then, the required impedance of adaptive impedance part $X_{AI}(\alpha)$ can be obtained from (5.10). After that, the corresponding firing angle α of the adaptive impedance part can be obtained by comparing the left and right-hand side terms of (5.10). As can be seen from Fig. 5.8, θ_{ac} is the phase angle of the voltage, which can be instantaneously measured by the phase lock loop. And, α is the calculated firing angle for the thyristor. The adaptive impedance part has two back-to-back connected thyristors T_1 and $T_{2,}$ for each phase, and they are triggered alternately in every half cycle. When $\theta_{ac} > \alpha$, the gate pulse for thyristor T_1 can be generated. On the other hand, when $\theta_{ac} < 180°\text{-}\alpha$, the gate pulse for thyristor T_2 can be generated.

D. Parameter Design Strategy

(1) Selection of L_{PF}

Figure 5.10 shows the reactive power/current compensation range under different L_{PF}. As shown in Fig. 5.10, with smaller L_{PF}, the adaptive impedance part can provide a larger inductive reactive. However, with the smaller L_{PF}, if the firing angles change to the same degree, the compensating reactive power changes much larger than with the larger L_{PF}. This larger change of reactive power can bring a large negative influence on the small rating AI-HPQC system.

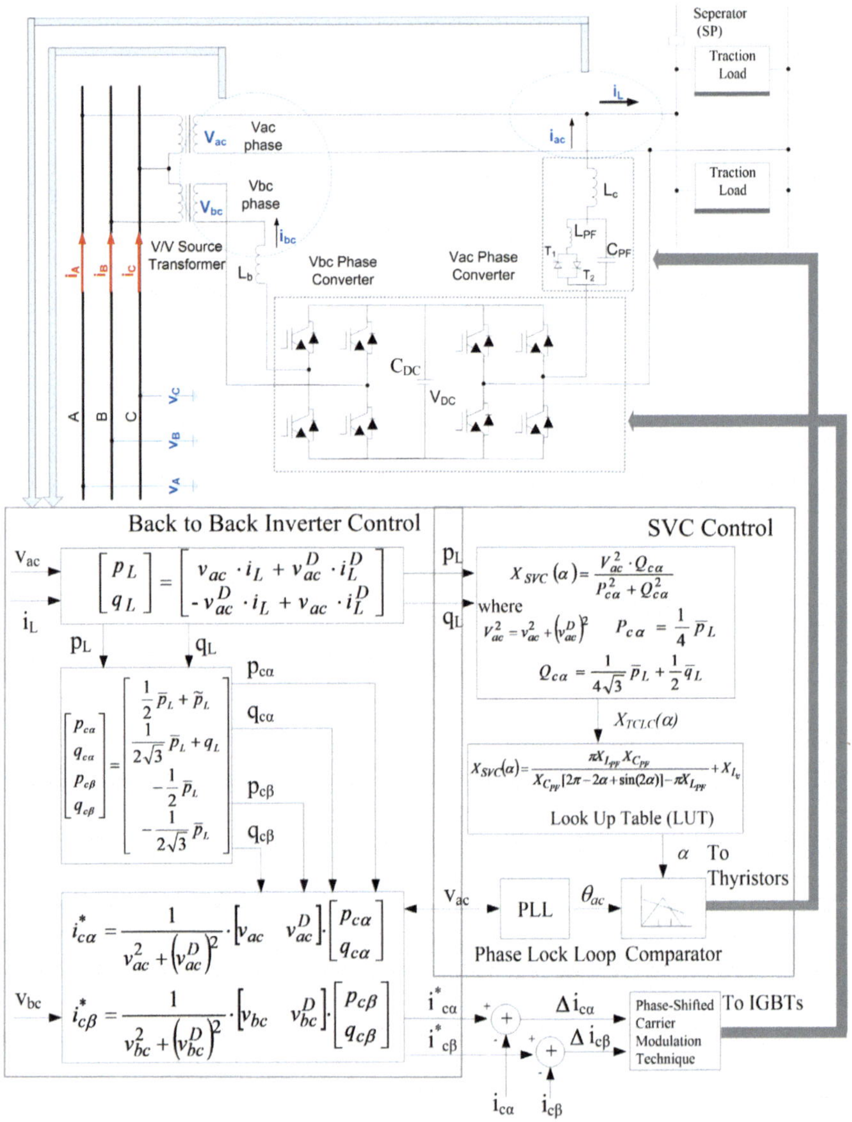

Fig. 5.9 Overall control strategy of AI-HPQC

The design of L_{PF} is inversely proportional to the reactive power/current compensation range. In this paper, due to the limitation of the laboratory facilities, a 1:100 k ratio power rating reduced 150 VA AI-HPQC (from 15 MVA to 150 VA) is designed and built in the laboratory for testing purposes, so that the value of L_{PF} is selected to be large to limit the compensation range in a small range based on the power rating. It is worth mentioning that the reactive power and current ratings of the designed

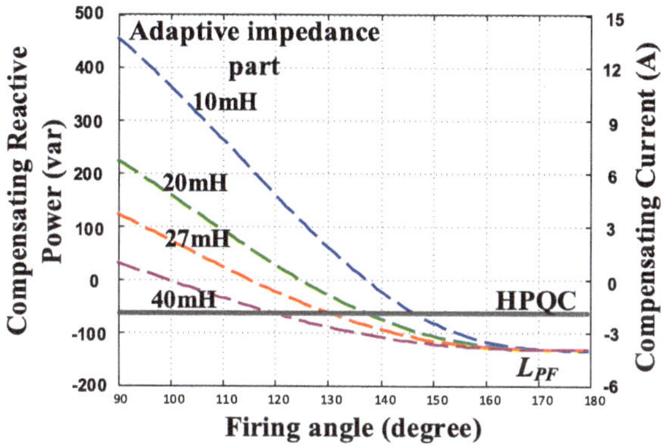

Fig. 5.10 Comparison of reactive power compensation range between adaptive impedance part and HPQC passive part with the different L_{PF}

hardware are 100 Var and 5 A. However, for industrial real applications, the reactive power compensation range is much larger, and the design value of L_{PF} can be significantly reduced. With the low value of the L_{PF}, the size, weight, and loss can be reduced.

In this section, load reactive power and current ratings vary from 0 to 100 Var and 0–5 A. Based on Fig. 5.10, the value of L_{PF} should be around 30 mH, and combined with our laboratory, 27 mH (≈30 mH) inductor is selected as L_{PF}.

(2) Selection of L_c and C_{PF}

The main purpose of L_c is to filter the harmonics ripple caused by power switches, and the purpose of L_{PF} is to adjust the equivalent capacitance values. The relationship between L_{PF} and L_c is given as (5.28). The parameter design of AI-HPQC is summarized in Table 5.4 and explained as follows.

If L_c can fulfill the (5.24), it can theoretically filter out the switching ripple as:

$$L_c \geq \frac{V_{DC}}{8 \cdot f_{sw} \cdot \Delta i_{ac\,\max}} \tag{5.24}$$

where $\Delta i_{Lc\max}$ is the maximum allowed output current ripple value.

The adaptive impedance X_{AI} can be expressed as (5.6), α is the firing angle of the two back-to-back connected thyristors, and they are triggered alternately in every half-cycle. When $\alpha = 180°$ (thyristors are opened for the whole cycle), the adaptive impedance part has the maximum capacitive impedance $X_{Cap(\text{Max})}$ (<0). On the other hand, when the firing angle $\alpha = 90°$ (one of the thyristors is closed for the whole cycle), the adaptive impedance part has the minimum inductive impedance $X_{Ind(\text{Min})}$ (>0). Therefore, the relationship can be expressed as:

Table 5.4 Parameters design criterion for the adaptive impedance part

	L_c design	L_{PF} design	C_{PF} design
Function	$L_c \geq \frac{V_{DC}}{8 \cdot f_{sw} \cdot \Delta i_{ac\,\max}}$	$L_{PF} > \frac{L_c}{\omega^2 L_c C_{PF} - 1}$	$\frac{1}{\omega^2 L_c} < C_{PF} < \frac{1}{\omega^2 L_c - \omega X_{AI}}$

$$X_{Cap(Max)} = X_{L_c} - X_{C_{PF}} = \omega L_c - 1/\omega C_{PF} < 0 \tag{5.25}$$

$$X_{Ind(Min)} = \frac{X_{L_{PF}} X_{C_{PF}}}{X_{C_{PF}} - X_{L_{PF}}} + X_{L_c} = \frac{\omega L_{PF}}{1 - \omega^2 L_{PF} C_{PF}} + \omega L_c > 0 \tag{5.26}$$

where ω (= $2\pi f$) is the angular frequency. (5.25) and (5.26) can be simplified as:

$$C_{PF} < 1/\omega^2 L_c \tag{5.27}$$

$$L_{PF} > L_c/\left(\omega^2 L_c C_{PF} - 1\right) \tag{5.28}$$

In the railway power supply system, the traction loads are inductive unbalanced loads, and required adaptive impedance X_{AI} obtained in (5.6) is capacitive ($X_{AI} < 0$). To compensate inductive loads, the design maximum capacitive impedance $X_{Cap(\mathrm{Max})}$ in (5.25) is required to be larger than the required X_{AI} in (5.6). Therefore, the X_{AI} design criteria are given as follows:

$$X_{AI} < X_{Cap(Max)} < 0 \tag{5.29}$$

Combining with (5.25), the tradeoff to select C_{PF} is obtained as:

$$\frac{1}{\omega^2 L_c - \omega X_{AI}} < C_{PF} < \frac{1}{\omega^2 L_c} \tag{5.30}$$

Finally, the parameters design criterion for the adaptive impedance part in AI-HPQC can be summarized in Table 5.4.

5.4 Simulation Results

The comparative simulation verifications are provided for the traditional RPC, HPQC, and the proposed AI-HPQC through the PSCAD simulation.

The system parameters of the RPC, HPQC, and AI-HPQC used in simulations are shown in Table 5.5. In the simulation results, the inductor L_c and L_{PF} are considered as their corresponding constant values. Once the L_c and L_{PF} are set in simulations, the adaptive impedance (combinations of L_c, $L_{PF,}$ and C_{PF}) is controlled by the firing angles of the thyristors. The adaptive impedance versus the firing angle α is provided as Fig. 5.3. Figure 5.11 shows the waveforms of source voltage and current, load current, and dc-link voltage before and after the RPC, HPQC, and AI-HPQC compensation. And, Fig. 5.12 shows the dynamic performance of RPC, HPQC, and AI-HPQC. Table 5.6 summarizes the simulation results.

A. Before Compensation

Table 5.5 Parameters used in the analysis and simulation of co-phase railway traction power supply with RPC, HPQC, and AI-HPQC

Descriptions	Parameters	Values
Source voltage	V_{ac}	27.5 kV
DC bus capacitance	C_{DC}	10,000 μF
RPC	L_a, L_b	4 mH, 5 mH
HPQC	L_a, C_a, L_b	4 mH, 140 μF, 5 mH
AI-HPQC	L_c, L_{PF}, C_{PF}, L_b	4 mH, 10 mH, 240 μF, 5 mH

As shown in Fig. 5.11a and Table 5.6, the system current I_B is equal to zero before compensation since the loads are connected to phases A and C while phase B is disconnected with loads. And, the source current total harmonic distortion (THD) of phases A and C are $THD_{IA} = THD_{IC} = 0.8\%$ for the linear loads and $THD_{IA} = THD_{IC} = 32.4\%$ for nonlinear loads. And the power factors are $PF_{\mathrm{A}} = 0.74$ and $PF_{\mathrm{C}} = 0.95$ for linear loads and $PF_{\mathrm{A}} = 0.50$, and $PF_{\mathrm{C}} = 1.0$ for nonlinear loads. In this paper, the unbalanced factor (UF) stands for the system's current UF_i on the primary side, defined as the negative sequence divided by the positive sequence. The source current UF_i on the primary side can be expressed as:

$$UF_i = \frac{I_-}{I_+} \times 100\% \tag{5.31}$$

where I_- and I_+ are the negative and positive sequences of the source current on the primary side. Since the phase B is disconnected with loads, the UF_i is 100%.

B. Using Traditional RPC Compensation

After traditional RPC compensation with $V_{DC} = 60$ kV as shown in Fig. 5.11b and Table 5.6, the PFs on the source side are improved to unity. And the source current THD_{iX} are improved to 2.6 and 4.8% for the worst phase of linear loads and nonlinear loads compensation. Moreover, the UF_i is reduced to 6.8 and 8.6% for linear loads and nonlinear loads compensation.

C. Using HPQC Compensation

After HPQC compensation with $V_{DC} = 35$ kV, as shown in Fig. 5.11c and Table 5.6, the PFs on the source side are also improved to unity. And the source current THD_{iX} is improved to 1.9 and 2.9% for the worst phase of linear loads and nonlinear loads after compensation. Moreover, the UF_i is reduced to 5.6 and 5.1% for linear loads and nonlinear loads compensation.

However, if the dc-link voltage of HPQC is reduced to $V_{DC} = 27$ kV as shown in Fig. 5.11d and Table 5.6, the HPQC cannot provide satisfactory source current THD_{iX} performance ($THD_{iX} = 26.7\%$) for the worst phase of linear loads after compensation. The UFi is just reduced to 18.8%, and the worst PF is improved to 0.94 for linear loads compensation. This result verifies the HPQC's narrow compensation range characteristics.

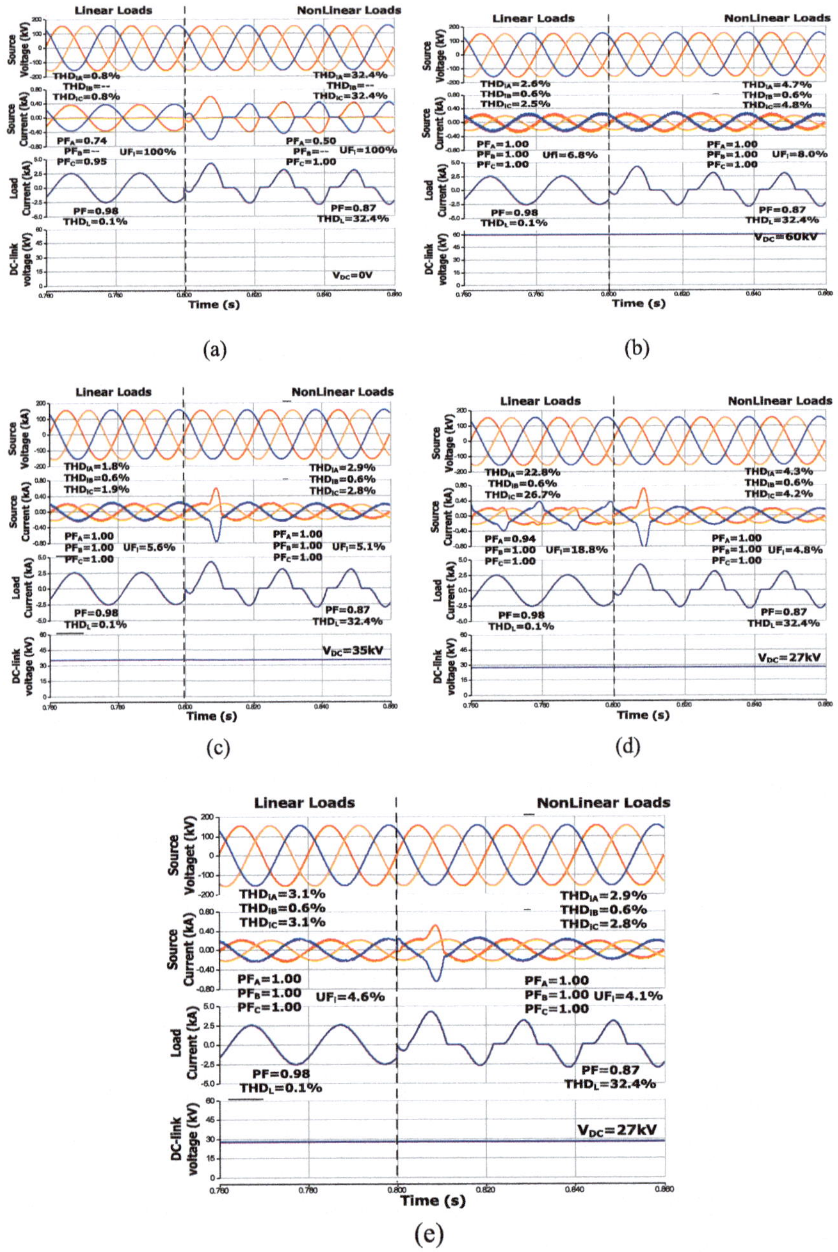

Fig. 5.11 Waveforms of source voltage and current, load current and dc-link voltage: **a** before compensation, **b** after RPC compensation with $V_{DC} = 60$ kV, **c** after HPQC compensation with $V_{DC} = 35$ kV, **d** after HPQC compensation with $V_{DC} = 27$ kV, **e** after proposed AI-HPQC compensation with $V_{DC} = 27$ kV

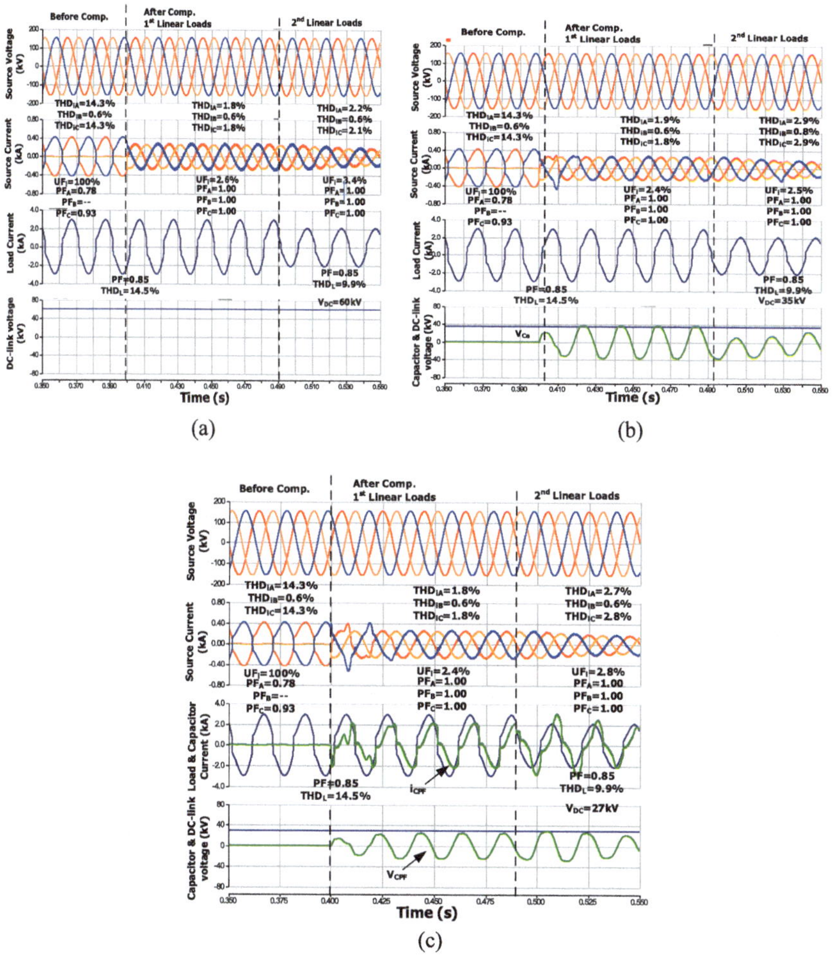

Fig. 5.12 Waveforms of source voltage and current, load current, and capacitor current, capacitor voltage and dc-link voltage: **a** after RPC compensation with $V_{DC} = 60$ kV, **b** after HPQC compensation with $V_{DC} = 35$ kV, **c** after proposed AI-HPQC compensation with $V_{DC} = 27$ kV

D. Using Proposed AI-HPQC

On the other hand, with the same dc-link voltage $V_{DC} = 27$ kV, the proposed AI-HPQC can provide the satisfactory source current THD_{iX} performance with the worst phase $THD_{iX} = 3.1\%$ for linear loads and 2.9% for nonlinear loads compensation as shown in Fig. 5.11e and Table 5.6. The PFs are improved to unity. In addition, the UF_i is reduced to 4.6% and 4.1% for linear and nonlinear load compensation, respectively. The simulation results as shown in Fig. 5.11 and Table 5.6 clearly verify that the proposed AI-HPQC can achieve the best compensation results with the lowest dc-link voltage.

Table 5.6 Summary of the simulation results for linear loads and nonlinear loads

		Before comp	RPC comp	HPQC comp		AI-HPQC comp
DC voltage (V)		–	60 k	35 k	27 k	27 k
Linear loads	THD_{isx} (worst phase) (%)	0.8	2.6	1.9	26.7	3.1
	PF (worst phase)	0.74	1.00	1.00	0.94	1.00
	Source current *UFi* (%)	100	6.8	5.6	18.8	4.6
Non-linear loads	THD_{isx} (worst phase) (%)	32.4	4.8	2.9	4.3	2.9
	PF (worst phase)	0.50	1.00	1.00	1.00	1.00
	Source current *UFi* (%)	100	8.0	5.1	4.8	4.1

E. Dynamic Performances of RPC, HPQC, and Proposed AI-HPQC

With sufficient V_{DC} of RPC, HPQC, and proposed AI-HPQC in the above *Parts A-D*, all compensators can obtain satisfactory compensation performances, as shown in Fig. 5.12

In Fig. 5.12a, the RPC can compensate *PF* to unity and $THD_{isx} < 2.2\%$ for both the 1st and 2nd loads compensation. The transient time for RPC is negligible for the start-up process and load change process. In Fig. 5.12b, the HPQC also provides satisfactory *PF* and THD_{isx} results. The transient time for HPQC is about 0.015 s during the start-up process. The transient time for the load change process is negligible.

For the proposed AI-HPQC, the adaptive impedance part has two back-to-back connected thyristors in each phase, and they are triggered alternately in every half cycle so that the control period of the adaptive impedance part is one cycle (0.02 s). However, the proposed adaptive impedance part in series with an instantaneous operated active inverter part can significantly improve the overall response time of AI-HPQC. According to Fig. 5.12c, the transient time for sudden start-up is about 1.5 cycles (0.03 s). The transient time for dynamic load change is negligible.

Figure 5.12b shows the performance of the HPQC. The current distortion takes place at current peak value where the capacitor voltage is still oscillation (transient-state). This indicates that the compensation current cannot be fully supported by the dc voltage during the transient-state. This distortion is another shortage of HPQC, but it does not appear in the proposed AI-HPQC.

5.5 Experimental Results

In this section, the scaled-down experimental prototype of the RPC, HPQC, and AI-HPQC is built in the laboratory. The appearance of the hardware prototype of the co-phase power with Railway AI-HPQC is provided in Fig. 2.11. Due to the laboratory power rating limitation, the rated load capacity is reduced from 15 MVA to 150 VA (ratio: 1:100 k), and the S_L varies from 0.2 p.u to 1.2 p.u. (30–180 VA). Moreover, the load voltage is reduced from 27.5 kV to 50 V (ratio: 1:550). The digital control system of the RPC, HPQC, and AI-HPQC is a digital signal processor (DSP) TMS320F2812, and the sampling frequency of the control system is 25 kHz. For every 1/25 kHz(s) period, the timer will provide a signal to process analog to digital (A/D) conversion and the corresponding interrupt. The switching devices for the active inverter part are Mitsubishi IGBT intelligent power modules PM300DSA060. The switching devices for the adaptive impedance part of AI-HPQC are thyristors SanRex PK110FG160. The parameters of the RPC, HPQC, and AI-HPQC are from Table 5.5.

Figures 5.13 and 5.14 show the waveforms of source voltage and current, load current, dc-link voltage and/or L_{PF} current for linear and nonlinear loads compensation before and after the RPC, HPQC, and AI-HPQC compensation, and Table 5.6 summarizes the experimental results.

A. Before Compensation

As shown in Figs. 5.13 and 5.14, and Table 5.6, the current unbalanced factor UF_i is 100% before compensation. The worst source current is $THD_{IX} = 2.4\%$ for the linear loads and $THD_{IX} = 25.4\%$ for nonlinear loads. The worst phase PFs are 0.74 and 0.45 for linear loads and nonlinear loads, respectively.

B. Using Traditional RPC Compensation

After traditional RPC compensation with $V_{DC} = 90$ V as shown in Figs. 5.13b and 5.14b, and Table 5.6, the worst phase PFs are improved to 0.97 and 0.91 for linear loads and nonlinear loads. The source current THD_{iX} is improved to 12.9% and 11.1% for the worst phase of linear loads and nonlinear loads compensation. Moreover, the UF_i is reduced to 22.1% and 21.6% for linear loads and nonlinear loads compensation.

C. Using HPQC Compensation

After traditional HPQC compensation with $V_{DC} = 60$ V as shown in Figs. 5.13c and 5.14c, and Table 5.6, the worst phase PFs are improved to 0.93 and 0.91 for linear loads and nonlinear loads. The source current THD_{iX} is compensated to 13.7 and 12.3% for the worst phase of linear loads and nonlinear loads compensation. Moreover, the UF_i is reduced to 25.8 and 24.3% for linear loads and nonlinear loads compensation.

D. Using Proposed AI-HPQC

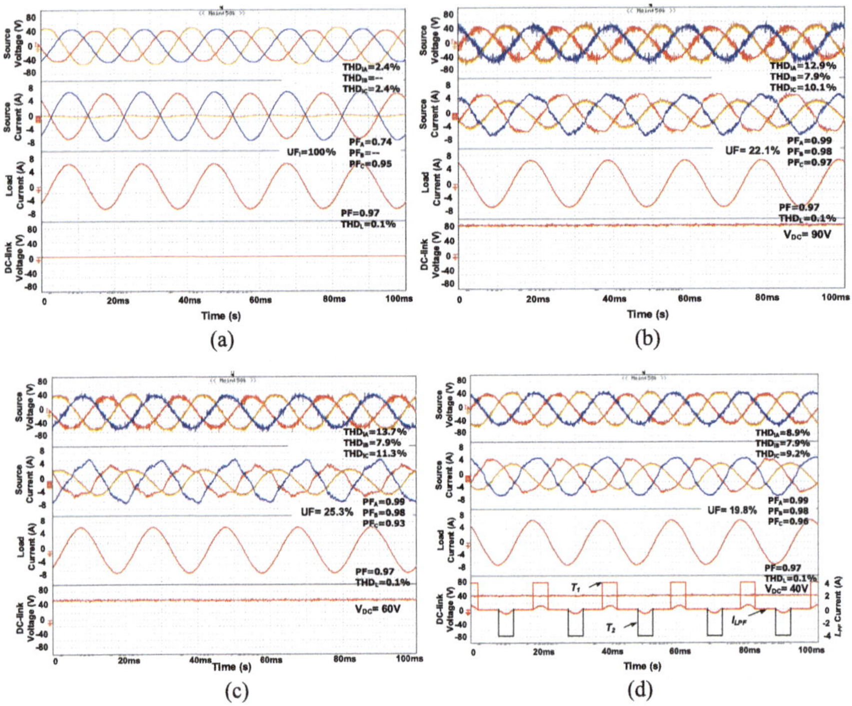

(a) (b) (c) (d)

Fig. 5.13 Waveforms of source voltage and current, load current, dc-link voltage and/or L_{PF} current for linear loads compensation: **a** before compensation, **b** after RPC compensation with $V_{DC} = 90$ V, **c** after HPQC compensation with $V_{DC} = 60$ V, **d** after proposed AI-HPQC compensation with $V_{DC} = 40$ V

On the other hand, with the lowest dc-link voltage $V_{DC} = 40$ V, the proposed AI-HPQC can provide the best source current THD_{iX} performance with the worst phase of 9.2% for linear loads and 9.6% for nonlinear loads compensation, as shown in Figs. 5.13d and 5.14d, and Table 5.6.

The *PFs* are improved to unity. In addition, the UF_i is reduced to 18.9% and 21.2% for linear and nonlinear load compensation, respectively.

The experimental results as shown in Figs. 5.13 and 5.14, and Table 5.7 clearly verify that the proposed AI-HPQC can achieve the best compensation results with the lowest dc-link voltage. The higher THD performance obtained in experiments compared to simulation ones is mainly caused by the noise in the experiment equipment, the digital computation error, inaccurate component parameters, and the resolution of the transducers.

E. Experimental Results Discussions

The simulated results are carried out in relatively ideal voltage and current conditions. Thus the unbalanced factor (UF_{is}) is reduced from 100 to 2%. Thus, the proposed topology and control method are verified. However, experiments are carried

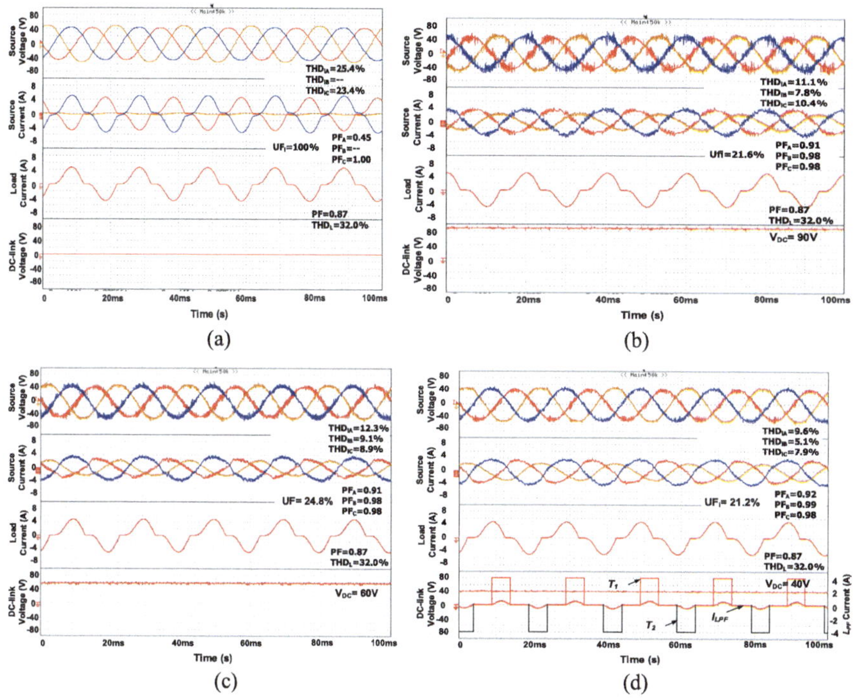

Fig. 5.14 Waveforms of source voltage and current, load current, dc-link voltage and/or L_{PF} current for nonlinear loads compensation: **a** before compensation, **b** after RPC compensation with $V_{DC} = 90$ V, **c** after HPQC compensation with $V_{DC} = 60$ V, **d** after proposed AI-HPQC compensation with $V_{DC} = 40$ V

Table 5.7 Summary of the experimental results for linear loads and nonlinear loads

		Before comp	RPC comp	HPQC comp	AI-HPQC comp
DC-link voltage (V)		–	90	60	40
Linear loads	THD_{isx} (worst phase) (%)	2.4	12.9	13.7	9.2
	PF (worst phase)	0.74	0.97	0.93	0.96
	Source current *UFi* (%)	100	22.1	25.3	19.8
Nonlinear loads	THD_{isx} (worst phase) (%)	25.4	11.1	12.3	9.6
	PF (worst phase)	0.45	0.98	0.91	0.92
	Source current *UFi* (%)	100	21.6	24.8	21.2

out in a low-capacity weak grid (downscaled from 15 MVA to 150 VA), and there are some significant differences between experiments and simulations. The reasons for some remain unbalance in the experiments can be summarized as: (1) The experimental results are significantly affected by *ESR* in component parameters. (2) The resolution of the transducers, the digital computation error, and the noise always exist in the experiment to affect the results. (3) The low-capacity weak grid experimental conditions are especially significantly increasing the system noise, which affects the system performance.

The following steps or methods can be done to improve the unbalanced performances and to maintain the balanced operation for future works or other researchers' reference as (1) For the experimental system, the voltage and power capacity can be increased. The original isolated transformer can be replaced to get a stronger experimental grid. (2) For the components, the resolution of the transducers can be improved. Higher switching IGBTs can be applied to get a higher accuracy current. (3) For the control system, the sampling frequency can be increased, and the control line can be replaced by optical fiber to improve the accuracy of the DSP calculation.

5.6 Summary

This chapter summarized the application of adaptive impedance converters in co-phase railway traction supply. The design and analysis of adaptive impedance structures were given for the co-phase railway traction supply power quality conditioner. The compensation characteristics of the AI-HPQC were proposed in comparison with the traditional co-phase RPC and HPQC. After comparisons, the proposed AI-HPQC has the characteristics of a wide compensation range and low dc-link voltage. Also, the control strategy for the AI-HPQC was proposed for the coordination control of the adaptive impedance part and the active inverter part. The simulation and experimental results were provided to verify the proposed AI-HPQC. The proposed AI-HPQC can obtain satisfactory results with the lowest dc-link voltage compared with the traditional RPC and HPQC.

References

1. S.M. Mousavi Gazafrudi, A. Tabakhpour Langerudy, E.F. Fuchs, K. Al-Haddad, Power quality issues in railway electrification: a comprehensive perspective. IEEE Trans. Ind. Electron. **62**(5), 3081–3090 (May 2015)
2. L. Battistelli, D. Lauria, P. Vernillo, Control strategy of advanced 25 kV–50 Hz electrified railway systems. Proc. Inst. Elect. Eng. **148**, 97–104 (2001)
3. A. Bueno, J.M. Aller, J.A. Restrepo, R. Harley, T.G. Habetler, Harmonic and unbalance compensation based on direct power control for electric railway systems. IEEE Trans. Power Electron. **28**(12), 5823–5831 (2013)

4. G.W. Chang, H.-W. Lin, S.-K. Chen, Modeling characteristics of harmonic currents generated by high-speed railway traction drive converters. IEEE Trans. Power Del. **19**(2), 766–773 (2004)
5. X. He, A. Guo, X. Peng, Y. Zhou, Z. Shi, Z. Shu, A traction three-phase to single-phase cascade converter substation in an advanced traction power supply system. Energies **8**(9), 9915–9929 (2015)
6. X. He et al., Advanced cophase traction power supply system based on three-phase to single-phase converter. IEEE Trans. Power Electron. **29**(10), 5323–5333 (2014)
7. Z. Sun, X. Jiang, D. Zhu, G. Zhang, A novel active power quality compensator topology for electrified railway. IEEE Trans. Power Electron. **19**(4), 1036–1042 (2004)
8. S. Hu et al., A new integrated hybrid power quality control system for electrical railway. IEEE Trans. Ind. Electron. **62**(10), 6222–6232 (2015)
9. L. Zhou, Q. Fu, X. Li, C. Liu, A novel multilevel power quality compensator for electrified railway, in *Proceeding of the IEEE 6th International Power Electronics and Motion Control Conference* (May 2009), pp. 1141–1147
10. D. Zhang, Z. Zhang, W. Wang, Y. Yang, Negative sequence current optimizing control based on railway static power conditioner in V/v traction power supply system. IEEE Trans. Power Electron. **31**(1), 200–212 (2016)
11. S. Hu et al., A power factor-oriented railway power flow controller for power quality improvement in electrical railway power system. IEEE Trans. Ind. Electron. **64**(2), 1167–1177 (2017)
12. M. Miranbeigi, H. Iman-Eini, M. Asoodar, A new switching strategy for transformer-less back-to-back cascaded H-bridge multi-level converter. IET Power Electron. **7**(7), 1868–1877 (2014)
13. F. Ma, Z. He, Q. Xu, A. Luo, L. Zhou, M. Li, Multilevel power conditioner and its model predictive control for railway traction system. IEEE Trans. Ind. Electron. **63**(11), 7275–7285 (2016)
14. M. Chen, Q. Li, and G. Wei, Optimized design and performance evaluation of new cophase traction power supply system, in *Proceeding of the Asia-Pacific Power and Energy Engineering Conference (APPEEC)* (Mar. 2009), pp. 1–6.
15. Z. Shu, S. Xie, Q. Li, Single-phase back-to-back converter for active power balancing, reactive power compensation, and harmonic filtering in traction power system. IEEE Trans. Power Electron. **26**(2), 334–343 (2011)
16. Z. Shu et al., Digital detection, control, and distribution system for cophase traction power supply application. IEEE Trans. Ind. Electron. **60**(5), 1831–1839 (2013)
17. Y. Wang, J. Xu, L. Feng, C. Wang, A novel hybrid modular three-level shunt active power filter. IEEE Trans. Power Electron. **33**(9), 7591–7600 (2018)
18. S.C. Ferreira, R.B. Gonzatti, R.R. Pereira, C.H. da Silva, L.E.B. da Silva, G. Lambert-Torres, Finite control set model predictive control for dynamic reactive power compensation with hybrid active power filters. IEEE Trans. Ind. Electron. **65**(3), 2608–2617 (2018)
19. S.D. Swain, P.K. Ray, K.B. Mohanty, Improvement of power quality using a robust hybrid series active power filter. IEEE Trans. Power Electron. **32**(5), 3490–3498 (2017)
20. B. Chen, C. Zhang, C. Tian, J. Wang, J. Yuan, A hybrid electrical magnetic power quality compensation system with minimum active compensation capacity for V/V cophase railway power supply system. IEEE Trans. Power Electron. **31**(6), 4159–4170 (2016)
21. K.-W. Lao, M.-C. Wong, N. Dai, C.-S. Lam, C.-K. Wong, L. Wang, Analysis in the effect of co-phase traction railway HPQC coupled impedance on its compensation capability and impedance-mapping design technique based on required compensation capability for reduction in operation voltage. IEEE Trans. Power Electron. **32**(4), 2631–2646 (2017)
22. K.-W. Lao, M.-C. Wong, N. Dai, C.-S. Lam, L. Wang, C.-K. Wong, Analysis of the effects of operation voltage range in flexible DC control on railway HPQC compensation capability in high-speed co-phase railway power. IEEE Trans. Power Electron. **33**(2), 1760–1774 (2018)
23. S. Rahmani, A. Hamadi, K. Al-Haddad, L.A. Dessaint, A combination of shunt hybrid power filter and thyristor-controlled reactor for power quality. IEEE Trans. Ind. Electron. **61**(5), 2152–2164 (2014)

24. L. Wang, C.-S. Lam, M.-C. Wong, Analysis, control, and design of a hybrid grid-connected inverter for renewable energy generation with power quality conditioning. IEEE Trans. Power Electron. **33**(8), 6755–6768 (2018)
25. L. Wang, C.-S. Lam, M.-C. Wong, Modeling and parameter design of thyristor-controlled LC-coupled hybrid active power filter (TCLC-HAPF) for unbalanced compensation. IEEE Trans. Ind. Electron. **64**(3), 1827–1840 (2017)
26. L. Wang, C.-S. Lam, M.-C. Wong, Design of a thyristor controlled LC compensator for dynamic reactive power compensation in smart grid. IEEE Trans. Smart Grid **8**(1), 409–417 (2017)
27. R.S.R. Chilipi, N. Al Sayari, K.H. Al Hosani, A.R. Beig, Adaptive notch filter-based multipurpose control scheme for grid-interfaced three-phase four-wire DG inverter. IEEE Trans. Ind. Appl. **53**(4), 4015–4027 (Jul. 2017)
28. X. Sun, Z. Chen, Y. Tian, Adaptive decoupled power control method for inverter connected DG. IET Renew. Power Gener. **8**(2), 171–182 (2014)
29. A. Tan, K.Ç. Bayindir, M.U. Cuma, M. Tümay, Multiple harmonic elimination-based feedback controller for shunt hybrid active power filter. IET Power Electron. **10**(8), 945–956 (2017)
30. E.H. Watanabe, M. Aredes, J.L. Afonso, J.G. Pinto, L.F.C. Monteiro, H. Akagi, Instantaneous p–q power theory for control of compensators in micro-grids, in *Proceeding of the International School on Nonsinusoidal Currents and Compensation* (2010), pp. 17–26
31. L. Wang, C.-S. Lam, M.-C. Wong, Minimizing inverter capacity design and comparative performance evaluation of SVC-coupling hybrid active power filters. IEEE Trans. Power Electron. **34**(2), 1227–1242 (2019)
32. A.K. Yadav, K. Gopakumar, K.R.R.L. Umanand, S. Bhattacharya, W. Jarzyna, A hybrid 7-Level inverter using low-voltage devices and operation with single DC-link. IEEE Trans. Power Electron. **34**(10), 9844–9853 (Oct. 2019)

Chapter 6
Adaptive Impedance Matching Networks for Power Amplifiers

6.1 Introduction of Impedance Matching Networks for Power Amplifiers

Power amplifiers can support the load with the adjustable frequency and amplitude voltage [1–3]. With stable broadband output characteristics, power amplifiers are widely applied as the voltage source of sonar, radio transmitters, etc. [4]. These systems usually require high power transfer utilization as their special or essential application scenarios, such as resource detection [5], communication [6], etc. However, many factors reduce the power utilization factor (PUF) between the power amplifiers and the loads, such as device power losses, line losses, low power factor, etc. [3, 4]. Among them, the low power factor is the most common and is usually caused by the load inductance or capacitance [7]. Theoretically, the resistive system transfers all of the active power from the source to the load [8].

Therefore, based on the impedance matching purpose, many methods are proposed to realize this resistive system. The impedance matching networks are generally developed from the uncontrollable to controllable [9–13]. Firstly, the passive fixed capacitor [9], as Fig. 6.1a was employed as the simplest impedance-matching network to match only a fixed load impedance. With the development of power electronics devices, varactor diodes were proposed as the first controllable impedance matching network in a specific range, as shown in Fig. 6.1b [10]. However the shortages are its small load impedance and frequency matching ranges. To improve the performance in some frequency bands of the special fixed capacitor, the *LLC* or *LCC* with the complex design was invented [11] as Fig. 6.1c. Furthermore, a group of capacitors is controlled with several switches [12] Fig. 6.1d was applied to match several fixed load conditions, but the switching processes may cause overvoltage problems. To overcome the above-mentioned problems, the variable capacitors group [13] was proposed as Fig. 6.1e. Unfortunately, although the variable capacitors group can solve the above problems, it has a high device cost. For one of the most common applications of power amplifiers, underwater sonar detection, and communication,

L. Wang et al., *Design and Applications of Adaptive Impedance Converters*,
https://doi.org/10.1007/978-981-97-5319-2_6

the optimal acoustics frequency range is 100 Hz–1000 Hz [14]. Too high frequency will cause the attenuation in the water to be too fast and spread distance too close, while too low frequency will cause the intensity of the detection target to be too low [15]. Therefore, one thyristor-controlled impedance matching network (TCIMN), as Fig. 6.1f is proposed in Sect. 6.2 for the underwater sonar power amplifiers to realize the adaptive broadband (300 Hz–1000 Hz) impedance matching and only with a medium-cost by referring to the concept of thyristor-controlled series capacitors (TCSCs) in the medium-voltage transmission grid. TCSCs are applied to increase the power transfer capabilities of the transmission grid [16] and mitigate the sub-synchronous resonance phenomenon [17] of the long transmission lines. Although the concept of TCSCs already exists and has been developed with some state-of-the-art technologies, e.g., stochastic [18] and data-driven [19] optimization. Especially nowadays, fast-switching thyristors (FSTs) (100 Hz–10 kHz) have been developed [20], which can be applied in wide frequency. But until now, it still cannot be directly utilized with varying frequencies and loads. There are still several significant differences and obstacles when this structure is applied with a power amplifier system for impedance matching. Compared to conventional TCSCs, the significantly different points for TCIMNs are: (1) Working frequency: TCSCs in transmission grid always work under one fundamental frequency, while TCIMN is proposed for wide frequency band application. (2) Impedance characteristics: TCSCs in the transmission grid are used to match the narrow-varying line impedance, while TCIMN is proposed for a wide range of load impedance (Table 6.1).

Besides the abovementioned passive adaptive impedance networks, the active adaptive impedance has also been widely studied. The AC-DC converter is widely used in the distribution grid, but this method also cannot achieve broadband impedance matching. For one of the most common power amplifier applications, long-range underwater sonar detection, the commonly-used acoustic frequency range is 100 Hz–1000 Hz. The high frequency causes significant attenuation in the water and shorter the transmission distance, while too low frequency will lead to the high diffraction of the detection target. To achieve long range detection, the system frequency operating range is usually 100 Hz–400 Hz. In a complete operation process of transducers, it operates at a few discrete frequencies, e.g., 100 Hz, 200 Hz, 300 Hz, and 400 Hz, or some frequency points with smaller intervals. Compared with the single-frequency continuous wave, the frequency-hopping (FH) wave can resist frequency-selective fading of underwater acoustic channels. FH signal has a wider bandwidth, and all of the above matching methods cannot achieve good matching performance or have other defects. To fill these gaps, this Chapter proposes a shunt adaptive broadband active impedance matching network (SAIMN) in Sect. 6.3. The proposed SAIMN is realized by injecting a capacitive reactive current into the system, while the power amplifier only outputs active power. Moreover, the proposed SAIMN can reduce the current stress of the power amplifier switching devices while supplying the same active power to the load.

According to the topology of the power amplifier, the design difficulty of high-voltage is much greater than that of high-current. Therefore, the series matching method to reduce the power amplifier and the matching network voltage is preferred

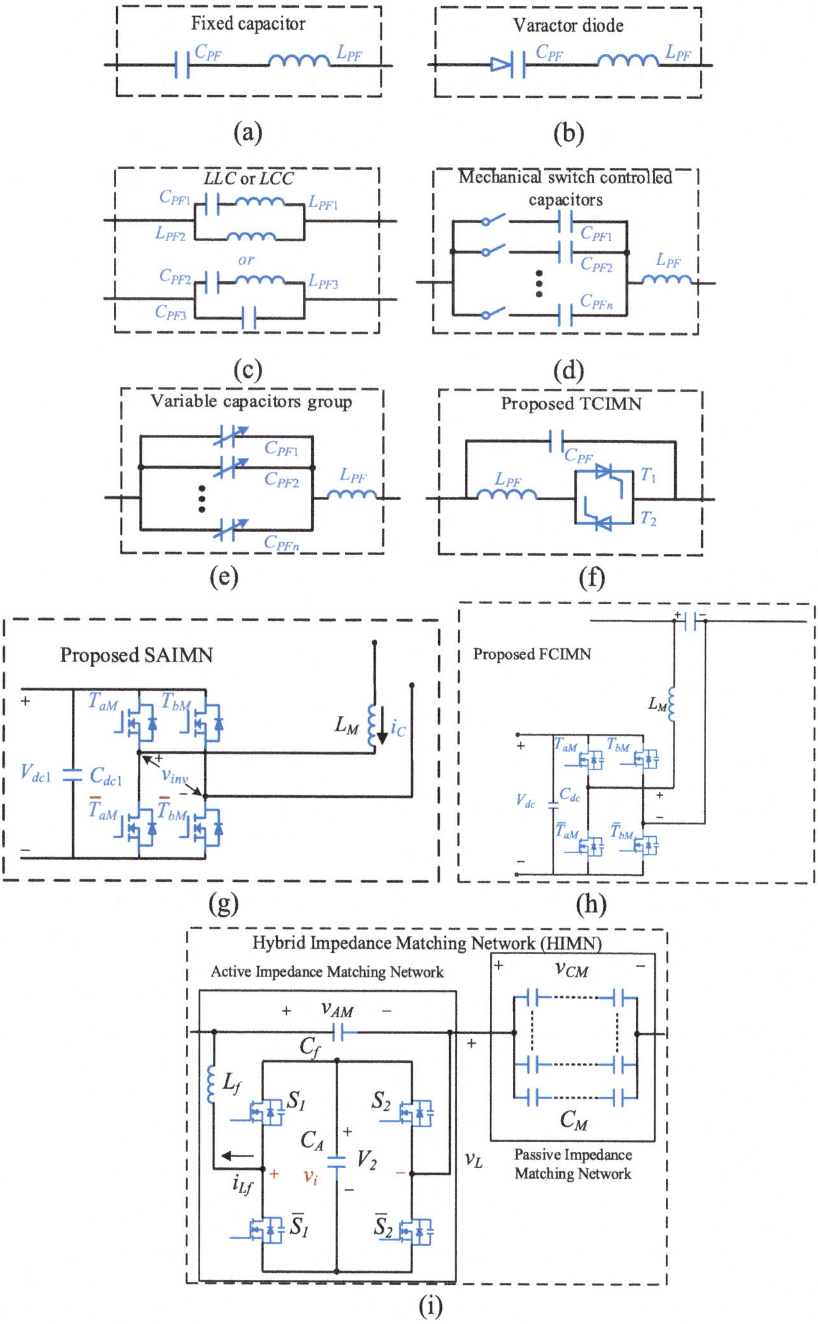

Fig. 6.1 Impedance matching network comparison of **a** fixed capacitor, **b** varactor diode, **c** *LLC* or *LCC*, **d** mechanical switch-controlled capacitors, **e** variable capacitor group, **f** proposed TCIMN, **g** proposed SAIMN, **h** proposed FCIMN, and **i** proposed HIMN

Table 6.1 Comparisons of different passive impedance matching methods for power amplifier

	Figure	Controllable	Broadband matching	Cost	Matching load range
Fix capacitor	Fig. 6.1(a)	No	No	Low	Fixed
Varactor diode	Fig. 6.1(b)	Yes	No	Low	Narrow
LLC or *LCC*	Fig. 6.1(c)	No	No	Low	Fixed
Switch contr	Fig. 6.1(d)	Yes	No	Medium	Wide
Variable cap	Fig. 6.1(e)	Yes	Yes	High	Wide
TCIMN	Fig. 6.1(f)	Yes	Yes	Medium	Wide
SAIMN	Fig. 6.1(g)	Yes	Yes	Medium	Wide
FCIMN	Fig. 6.1(h)	Yes	Yes	Low	Wide
HIMN	Fig. 6.1(i)	Yes	Yes	Medium	Wide

more than the shunt matching methods. Therefore, an active adaptive broadband full-bridge converter impedance matching network (FCIMN) is proposed in Section in 6.4 referring to the concept of series dynamic voltage restorer. The general dynamic voltage restorers all operate at the fundamental frequency 50/60 Hz. Differently, the proposed FCIMN operates at a wide frequency range to realize broadband impedance matching. Since the previous research can hardly extend to a wide frequency range, a segmented PR controller is accordingly proposed for FCIMN.

The active impedance matching method can achieve broadband impedance matching between the power amplifier and the UAT. However, the matching value of this method is calculated after delaying the output voltage of the power amplifier by a 3/4 period. Section 6.5 proposes a hybrid impedance matching network (HIMN), which consists of a passive impedance matching network and an active impedance matching network. The passive matching part is responsible for the fundamental matching and cannot be adjusted. On this basis, the active matching is added, which can be dynamically adjusted. Therefore, compared with the conventional impedance matching method, the proposed HIMN can further improve the impedance matching range. To realize the fast control of HIMN, a two-step model predictive control (MPC) method is also proposed. In addition, the power amplifier and the impedance matching network are controlled synchronously to reduce the delay of matching value calculation.

The layout of this chapter is given as follows. Section 6.1 presents the development of impedance-matching networks for power amplifiers. Then, the adaptive broadband thyristor-controlled impedance matching network is given in Sect. 6.2. After that, the shunt an series of adaptive broadband active impedance matching networks are provided in Sects. 6.3 and 6.4, respectively. The hybrid adaptive broadband active impedance matching network is provided in Sect. 6.5. Finally, the Chapter summary is given in Sect. 6.6.

6.2 Adaptive Broadband Thyristor-Controlled Impedance Matching Network

6.2.1 *Circuit Configurations and Impedance Functions of the Proposed TCIMN*

In this section, the circuit configurations of the power amplifier and proposed TCIMN are given as Fig. 6.2. The relationship between TCIMN impedance, firing angle, and its parameters are provided.

A. *Power Amplifier Circuit Configuration*

As shown in Fig. 6.2, the power amplifier is applied as a controllable voltage source with adjustable output voltage amplitude and frequency. It consists of n modular insulated gate bipolar transistors (IGBTs) H-bridges and its output inductor L_o, capacitor C_o. V_{dcn}, v_{Co}, v_{TC}, and v_L are the voltages of C_{dcn}, C_o, C_{PF}, and load network, respectively. i_s and i_L are the amplifier output and load current, respectively. The load equivalent impedance is expressed as:

$$Z_L = R + jX_L = R + j(2\pi fL) \tag{6.1}$$

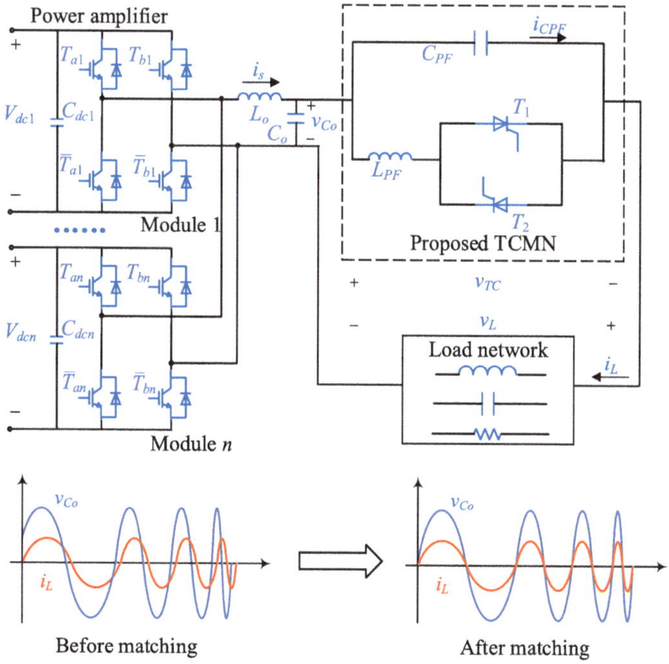

Fig. 6.2 Proposed TCIMN circuit configuration and matching performance

where R is the load equivalent resistance, j is the imaginary unit, X_L is the load equivalent inductive resistance, and L is the load equivalent inductance. In actual applications, since the loads applied with the power amplifier (e.g., sonar, radio transmitter, etc.) are normally inductive loads, this paper is mainly focused on the inductive load conditions.

B. *Proposed TCIMN Circuit Configuration*

As shown in Fig. 6.2, the proposed TCIMN consists of a pair of anti-parallel thyristors coupling with inductor L_{PF}, which is called the thyristor-controlled reactor (TCR). TCR is parallel with a capacitor C_{PF}. v_{TC} and v_L are the voltage of the TCIMN and the load network, respectively. The load network is in series with the proposed TCIMN and then parallel with the power amplifier output capacitor C_o. Thus, the power amplifier provides power to the load and its matching network through C_o with the adjustable amplitude and frequency voltage.

C. *Impedance Functions*

Normally, the loads connected to the power amplifier are all inductive. Thus, the matching network is needed to be capacitive to match the inductive load. The equivalent impedance of the TCIMN X_{TC} depends on the value of L_{PF}, C_{PF}, and the thyristor firing angle α ($0^\circ \leq \alpha \leq 90^\circ$), which determines the duration of thyristor conduction in every period.

The TCR part equivalent impedance X_{TCR} is:

$$X_{TCR} = \frac{\pi X_{L_{PF}}}{2\alpha - \sin 2\alpha} \tag{6.2}$$

where X_{LPF} is the fundamental impedance of the L_{PF}, therefore, the proposed TCIMN equivalent impedance X_{TC} is deduced as:

$$X_{TC}(\alpha) = \frac{X_{TCR}(\alpha) \cdot X_{C_{PF}}}{X_{C_{PF}} - X_{TCR}(\alpha)} = \frac{\pi X_{L_{PF}} X_{C_{PF}}}{X_{C_{PF}}(2\alpha - \sin 2\alpha) - \pi X_{L_{PF}}} \tag{6.3}$$

where X_{CPF} is the fundamental impedance of the C_{PF}.

6.2.2 Parameters Design and Matching Analysis

In this section, a parameter design strategy is discussed in two parts, namely, the selection of C_{PF} and the selection of L_{PF}. The matching frequency ranges and the firing angle operating ranges are analyzed based on the parameters. In addition, the harmonic currents are analyzed to prove TCIMN can bring less harmonic pollution to the load and amplifier.

A. *Selection of* C_{PF}

The purpose of the impedance matching network is to provide the capacitive impedance, that is, the reverse of the inductive impedance of the loads. In other words, the TCIMN should provide capacitive reactive power Q_{TC}, which has the same amplitude as the load inductive reactive power Q_L. Their relationship can be expressed as:

$$X_{TC} = \frac{V_{TC}^2}{Q_{TC}} = -\frac{V_L^2}{Q_L} \tag{6.4}$$

To realize (6.4) with the varying Q_L, α should be controlled to adjust (6.2) and (6.3). Thyristors T_1 and T_2 are triggered alternately in every half-cycle. The two opposite extreme situations ($\alpha = 0°$ and $\alpha = 90°$) represent the minimum and maximum capacitive reactive power provided by the proposed TCIMN. When the firing angle $\alpha = 0°$, the thyristors are turned off for the whole cycle, and the branch of L_{PF} can be treated as an open circuit. TCIMN produces the minimum capacitive impedance $X_{TC(\mathrm{Min})}$ and reactive power $Q_{TC(\mathrm{Min})}$. When the firing angle $\alpha = 90°$, each thyristor is turned-on alternately in every period, which is equivalent to the C_{PF} in parallel with the L_{PF} and producing the maximum capacitive impedance $X_{TC(\mathrm{Max})}$ and reactive power $Q_{TC(\mathrm{Max})}$. Based on the above analysis, the $X_{TC(\mathrm{Min})}$ and $X_{TC(\mathrm{Max})}$ can be expressed as follows:

$$X_{TC(\mathrm{Min})} = \left.\frac{X_{TCR}(\alpha)\cdot X_{C_{PF}}}{X_{C_{PF}} - X_{TCR}(\alpha)}\right|_{\alpha=0^{\circ}} = X_{C_{PF}} = -\frac{1}{2\pi f C_{PF}} \tag{6.5}$$

$$X_{TC(\mathrm{Max})} = \left.\frac{X_{TCR}(\alpha)\cdot X_{C_{PF}}}{X_{C_{PF}} - X_{TCR}(\alpha)}\right|_{\alpha=90^{\circ}} = \frac{X_{L_{PF}} X_{C_{PF}}}{X_{L_{PF}} + X_{C_{PF}}} = \frac{1}{\frac{1}{2\pi f L_{PF}} - 2\pi f C_{PF}} \tag{6.6}$$

where f is the power amplifier output frequency.

To ensure that $X_{TC(\mathrm{Min})}$ can provide enough $Q_{TC(\mathrm{Min})}$ under the power amplifier minimum output frequency f_{Min}. Thus, C_{PF} should satisfy (6.7), and C_{PF} range can be deduced as (6.8).

$$\frac{V_{TC}^2}{X_{TC(\mathrm{Min})}} \geq -\frac{V_L^2}{L_{\mathrm{Min}}} \tag{6.7}$$

$$C_{PF} \leq \frac{Q_{L(\mathrm{Min})}}{2\pi f_{Min}\cdot V_{TC}^2} \tag{6.8}$$

The C_{PF} determines the minimum matching frequency f_{Min} and the minimum matching load power $Q_{L(\mathrm{Min})}$.

B. *Selection of* L_{PF}

The TCR part of the proposed TCIMN can adjust the TCIMN equivalent impedance. When there is current passing through L_{PF}, TCIMN equivalent capacitance decreases, and the Q_{TC} increases as (6.4) illustration. Thus, with TCR, the TCIMN can match larger load impedance or under higher frequency. L_{PF} determines the maximum matching frequency f_{Max} and the maximum matching load power $Q_{L(\text{Max})}$.

$$\frac{V_{TC}^2}{X_{TC(\text{Max})}} \leq -\frac{V_L^2}{L_{\text{Max}}} \tag{6.9}$$

$$L_{PF} \leq \frac{1}{4\pi^2 f_{\text{Max}}^2 C_{PF} + 2\pi f_{Max} \frac{V_{TC}^2}{Q_{L(\text{Max})}}} \tag{6.10}$$

C. *Matching Range Analysis*

The proposed TCIMN can be considered as a broadband variable capacitor. Its matching frequency range is depended on the L_{PF}, C_{PF}, and load equivalent inductance L. The minimum matching frequency f_{Min} can be realized when the firing angle $\alpha = 0°$ and depends on the C_{PF}, which can be expressed as:

$$f_{\text{Min}} = \frac{1}{2\pi\sqrt{LC_{PF}}} = \sqrt{\frac{Q_L}{2\pi C_{PF}}} \cdot \frac{1}{V_L} \tag{6.11}$$

The maximum matching frequency f_{Max} can be realized when the firing angle $\alpha = 90°$ and depends on the L_{PF} and C_{PF}, which can be expressed as:

$$f_{\text{Max}} = \sqrt{\frac{L_{PF} + L}{4\pi^2 L_{PF} C_{PF} L}} = \sqrt{\frac{2\pi Q_L L_{PF} + V_L^2}{4\pi^2 L_{PF} C_{PF} V_L^2}} \tag{6.12}$$

The users can determine the L_{PF} and C_{PF} based on the load conditions and the amplifier output voltage frequency range (f_{Min} and f_{Max}). Based on the above-mentioned analysis, the matching frequency range can be drawn as Fig. 6.3 with the fixed L_{PF} and L. Moreover, with the variable L_{PF} and C_{PF} consideration, the relationship among the maximum matching frequency, L_{PF}, and C_{PF} can be drawn as Fig. 6.4.

D. *Firing Angle Operating Area Analysis*

It should be noted that the TCIMN should keep capacitive in the selected α conditions ($\alpha = 0°$–$90°$). Therefore, X_{TC} should also satisfy (6.13), and the α limitation can be deduced as (6.14).

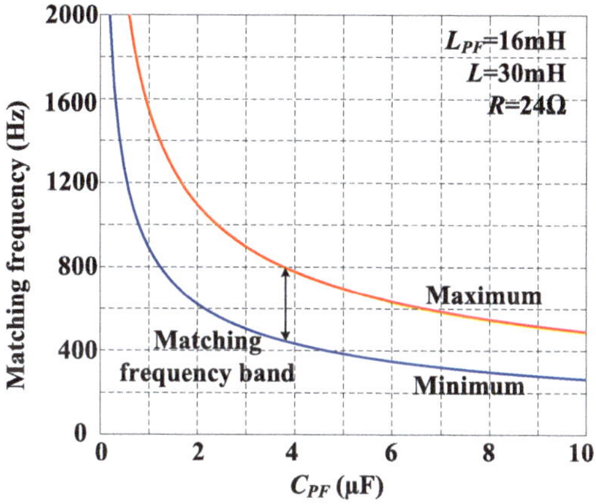

Fig. 6.3 Relationship between matching frequency and C_{PF}

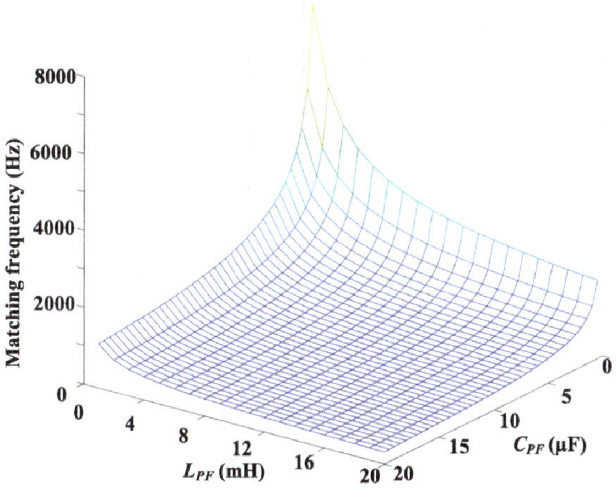

Fig. 6.4 Relationship among maximum matching frequency, L_{PF}, and C_{PF}

$$X_{TC}(\alpha)=\frac{\pi X_{L_{PF}} X_{C_{PF}}}{X_{C_{PF}}(2\alpha - \sin 2\alpha) - \pi X_{L_{PF}}} < 0 \tag{6.13}$$

$$(2\alpha - \sin 2\alpha) < 4\pi^3 f^2 \cdot L_{PF} C_{PF} \tag{6.14}$$

Based on (6.14), the relationship between X_{TC} and α in different frequencies and operating range α are drawn as Fig. 6.5. The negative and positive impedance express the capacitive and inductive X_{TC}, respectively. The positive X_{TC} should be avoided for the TCIMN firing angle selection. For example, when f is 400 Hz, α should be selected less than around 62°. Furthermore, from Fig. 6.5, the α controlling directions can also be concluded as (1) if f increases, α should also be increased; (2) if load impedance increases, α should also be increased.

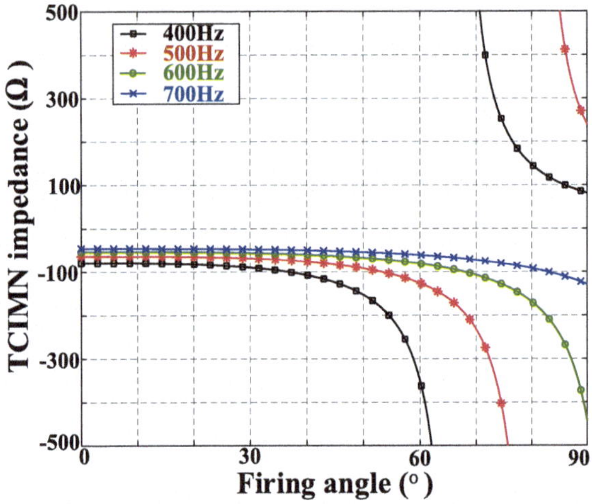

Fig. 6.5 Relationship between TCIMN impedance and α in different frequencies

E. *Harmonic Analysis*

The TCIMN simplified equivalent circuit is shown in Fig. 6.6. Harmonic current can only be brought by the switching of thyristor *S*. The function of i_L can be assumed as:

$$i_L(t) = i_{CPF}(t) + i_{LPF}(t) = \sum_{n=1}^{\infty} A_0 \cdot \cos(n\omega t) \tag{6.15}$$

where n is the harmonic order, and A_0 is the i_L current amplitude. Combing with (6.3), $i_{LPF}(t)$ and Fourier analysis are deduced as:

$$i_{LPF}(t) = A_1 \cdot \cos \omega t - A_1 \cdot \frac{\cos \alpha}{\cos k\alpha} \cdot \cos k\omega t \tag{6.16}$$

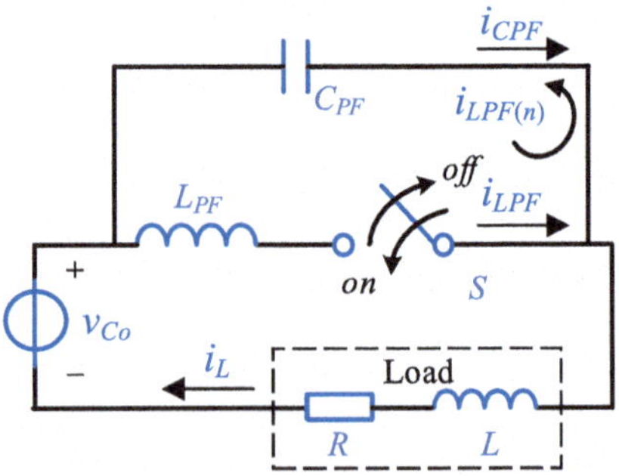

Fig. 6.6 Equivalent simplified circuit of proposed TCIMN

$$
\begin{aligned}
I_{LPF(n)} = \sqrt{2} \cdot \frac{A_1}{n}\left[\frac{\sin\alpha \cdot (n-1)}{n-1} + \frac{\sin\alpha \cdot (1+n)}{1+n}\right] - \\
\sqrt{2} \cdot \frac{A_1 \cos\alpha}{\pi \cos(k\alpha)}\left[\frac{\sin\alpha \cdot (k+n)}{k+n} + \frac{\sin\alpha \cdot (k-n)}{k-n}\right]
\end{aligned}
\tag{6.17}
$$

where $I_{LPF(n)}$ is the RMS value of $i_{LPF(n)}$, $A_1 = A_0\omega_0{}^2/(\omega_0{}^2 - \omega^2)$, $\omega = 2\pi f$, $k = \omega_0/\omega$, $\omega_0{}^2 = 1/L_{PF} - C_{PF}$, $n = 3, 5, 7, \ldots$From (6.17), with the increase of n, $I_{LPF(n)}$ is decreased largely, and the major harmonic current exists $n \leq 13$. According to the Kirchhoff current laws (KCL) theorem, the relationship between i_{LPF}, i_{CPF}, and i_L in this system can be assumed as follows:

$$
I_{CPF(n)} = K_1 I_{LPF(n)} \text{ and } I_{L(n)} = K_2 I_{LPF(n)} \tag{6.18}
$$

where $K_1 + K_2 = 1$. For n-order harmonics, the impedance of C_{PF} and load $X_{CPF(n)}$, and $X_{L(n)}$ can be calculated as:

$$
X_{CPF(n)} = 1/n\omega C_{PF} \text{ and } X_{L(n)} = n\omega L_{PF} \tag{6.19}
$$

Thus, the ratio between $I_{L(n)}$ and $I_{LPF(n)}$ is deduced as (6.20).

$$
I_{L(n)}/I_{CPF(n)} = K_2/K_1 = \left|X_{CPF}/\left(n^2 \cdot X_L\right)\right| \tag{6.20}
$$

Based on the above analysis and Fig. 6.5, $|X_{CPF}| \leq |X_{TC}| \leq |X_L|$, for n-order harmonic current $n = 3, 5, 7, \ldots$, the $I_{CPF(n)}$ can be considered as 9, 25, 49, ..., n^2 times of $I_{L(n)}$. Thus, the harmonic current loops into C_{PF} majorly and nearly not exists in i_L.

6.2.3 Control System

In this section, for the wide range matching purpose, the trigger signal calculation method is proposed to solve the long-transient problem of conventional PLL in part A. The proposed overall TCIMN control strategy, including dynamic reactive and impedance identification, is presented in part B.

A. *Proposed Trigger Signal Calculation Method*

Due to the long transient process of PID links in the PLL, the TCR part cannot produce the correct and accurate equivalent matching impedance. The comparisons between the conventional PLL method and the proposed method are shown in Fig. 6.7. Therefore, another trigger signal calculation method is proposed, and its calculation process is shown in Fig. 6.8. Each step can be realized by the digital signal process (DSP) in experiments. The first step is the v_{TC} sinusoidal wave sampling. Then, the step signals can be produced starting and ending at the v_{TC} zero-crossing points.

It should be noted that the amplitude of the step signals is 180°. The DSP can sample the time of zero-crossing points. The zero-crossing points going positive are successively assigned to t_k. On the contrary, the zero-crossing points going negative are successively assigned to t_{k+1}, where k is the count number of the zero-crossing points. At last, the triangle signals can be generated by multiplying the step signals by a slope A given as (6.21).

$$A = 180^{\circ}/(t_{k+1} - t_k) \tag{6.21}$$

Finally, the amplitude 180° triangle signals can be generated to compare with the TCR firing angle α and (180°−α).

B. *Proposed Overall TCIMN Control Strategy*

The overall control strategy of the proposed TCIMN can be summarized as (1) calculating dynamic load reactive power Q_L; (2) identifying the impedance X_L and X_{TC}; (3) calculating the firing angle α; (3) calculating the voltage phase φ with

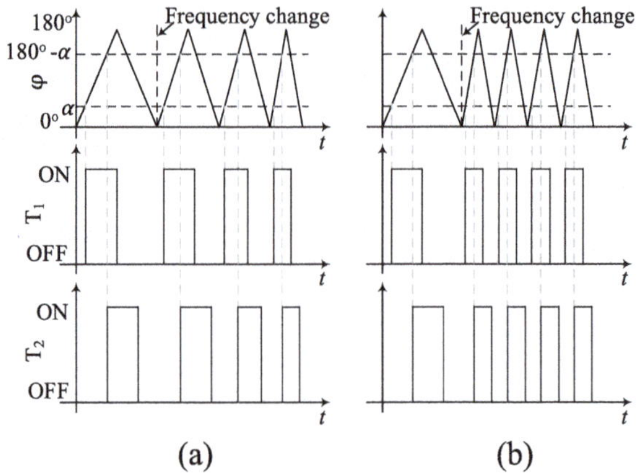

Fig. 6.7 Trigger signal calculation methods comparison during voltage frequency changing through **a** conventional PLL and **b** proposed method

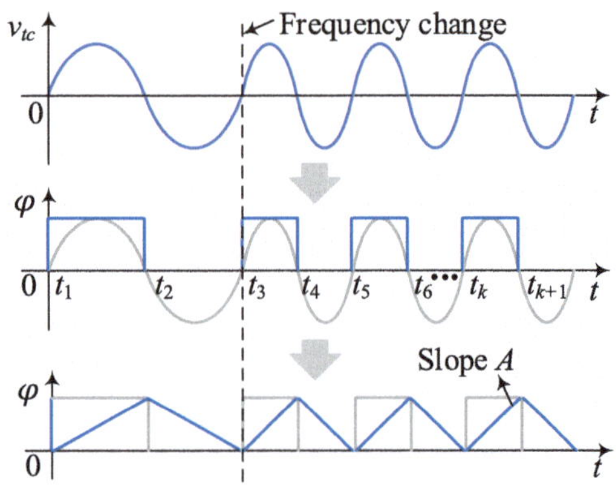

Fig. 6.8 Process of the proposed voltage phase calculation method

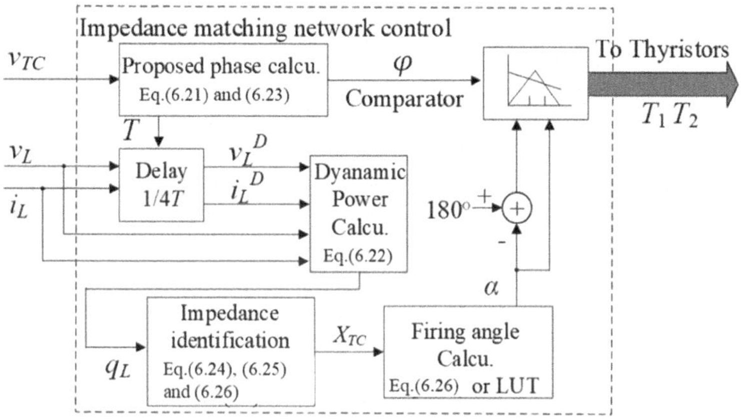

Fig. 6.9 Overall control strategy of proposed TCIMN and power amplifier

the proposed method; (4) comparing the α and the calculated triangle signals; (5) generating thyristors trigger signals. The overall control strategy can be drawn as Fig. 6.9. Corresponding functions are listed and explained as follows. The reactive power of the load can be calculated through single-phase instantaneous power theory, as shown in (6.22).

$$\begin{bmatrix} p_L \\ q_L \end{bmatrix} = \begin{bmatrix} v_L \cdot i_L + v_L^D \cdot i_L^D \\ -v_L^D \cdot i_L + v_L \cdot i_L^D \end{bmatrix} \tag{6.22}$$

where ${v_L}^D$ and ${i_L}^D$ are the delay signals of v_L and i_L, respectively. The delay time t_D is given in (6.23).

$$t_D = (1/4) \cdot T = (1/2) \cdot |t_{k+1} - t_k| \tag{6.23}$$

where T is the power amplifier output voltage period, the load root mean square (RMS) voltage V_L and reactive Q_L can be calculated through (6.24) and (6.25), respectively.

$$V_L = \frac{1}{2} \cdot \sqrt{v_L^2 + \left(v_L^D\right)^2} \tag{6.24}$$

$$Q_L = \frac{1}{2} \cdot \overline{q}_L \tag{6.25}$$

The required impedance of the impedance matching network X_{TC} can be determined by (6.4), and the α can be expressed as:

$$(2\alpha - \sin 2\alpha) = \frac{\pi X_{L_{PF}}}{X_{C_{PF}}} - \frac{\pi X_{L_{PF}} Q_L}{V_L^2} \tag{6.26}$$

Based on the discussion, X_{LPF} and X_{CPF} have already been determined. Thus, the α can be calculated by (6.26). For TCIMN, firing angles α relate to different thyristor turn-on periods. Thus, the firing angles influence the equivalent impedance of the impedance-matching network. Generally, in industry applications, based on the determined C_{PF} and L_{PF}, a look-up table (LUT) can be established and stored in DSP before application. The LUT presets a link between the α and X_{TC} for DSP quick calculation. The α is compared with the φ calculated by the proposed method. If φ is larger than α, the controller generates the thyristor T_1 turn-on signals. And if $\varphi > 180° - \alpha$, the controller generates the thyristor T_2 turn-on signals. Finally, the thyristors' control signals T1 and T_2 can be generated. Based on these control signals, TCIMN reactive power Q_{TC} can be generated to match the load reactive power Q_L. The overall control strategy is drawn in Fig. 6.9.

6.2.4 Simulation Results

In this section, the simulation case studies are provided to show the feasibility of the proposed TCIMN. The dynamic performances of the proposed TCIMN are illustrated under three different situations, namely: (1) power amplifier output voltage frequency changes, and (2) load condition changes. The final purpose is to prove the power utilization factor (PUF) is improved with the TCIMN, which can be calculated as follows:

$$\text{PUF} = \frac{P_s}{S_s} \times 100\% = \frac{P_s}{\sqrt{P_s^2 + Q_s^2}} \times 100\% \tag{6.27}$$

The parameters of the power amplifier, proposed TCIMN and loads in simulation and experiments are listed in Table 6.2.

From Figs. 6.10, 6.11, and Table 6.3, on the load side, it can be seen that v_L and i_L are not in phase. There is a large reactive power Q_L, nearly 6 kVar (varying with f), in the power amplifier system. Without impedance matching, the power amplifier is needed to transfer this part of reactive power to the loads. This part of reactive power

Table 6.2 Parameters of power amplifier and TCIMN in simulations and experiments

	Parameters	Physical values
Power amplifier	L_o, C_o	1 mH, 110 μF
	V_{Co}, f	220 V, 400 Hz–700 Hz
Load	L_1, L_2, R	30 mH, 20 mH, 24 Ω
Proposed TCIMN	L_{PF}, C_{PF}	16 mH, 5μF

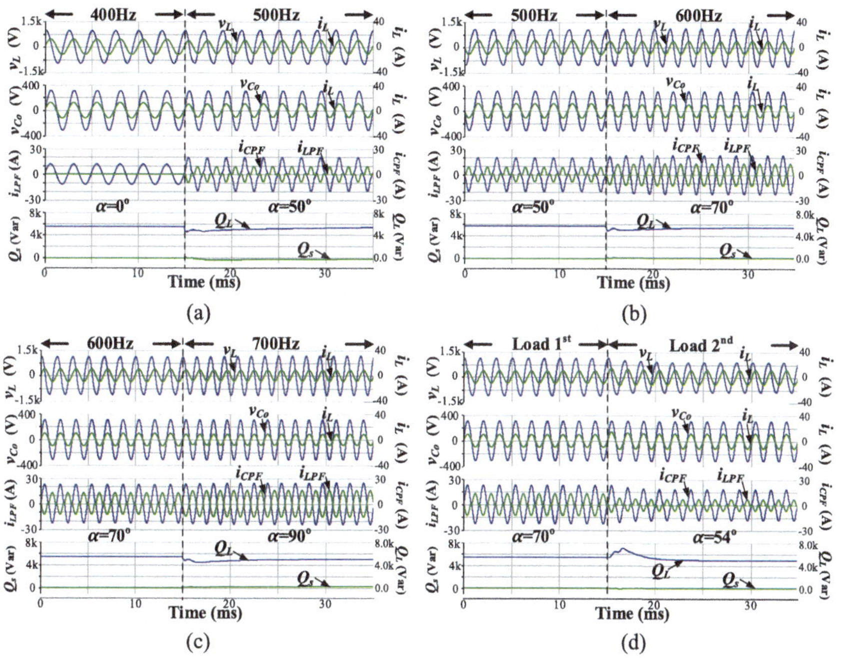

Fig. 6.10 Dynamic waveforms of v_L, i_L, v_{Co}, i_{CPF}, i_{PF}, Q_s, and Q_L in simulation with TCIMN impedance matched when power amplifier output voltage changes: **a** from 400 to 500 Hz; **b** from 500 to 600 Hz, **c** from 600 to 700 Hz, and **d** load changes from load 1st to load 2nd

also occupies the power amplifier capacity and reduces the power transfer PUF (PUF = 27.01% without TCIMN).

The TCIMN-matched results during frequency changing are shown in Figs. 6.10a, 6.11b, and Table 6.3. In steady states, no matter from 400 to 500 Hz in (a), from 500 to 600 Hz in (b), and from 600 to 700 Hz in (c) with the TCIMN impedance matching, v_{Co} and i_L are matched to be in phase. The reactive power in the power amplifier side Q_s is reduced to nearly 0. As for the dynamic transient processes, TCIMN only takes 5 ms to match the impedance with the firing angles changing. It should be noted that with the TCIMN matched, the V_{Co} can be largely reduced to 220 V, and V_L can still be maintained as 800 V due to the increasing PUF.

The TCIMN-matched results during load changing are as shown in Fig. 6.10d and Table 6.3. With the loads changing, the load Q_L is decreased from nearly 6 kVar to 5kVar. The α is required to reduce from 70° to 54°. With the 5 ms transient process, the Q_s can still keep as around 0.

The current harmonics spectrum of i_{LPF}, i_{CPF}, and i_L are shown in Fig. 6.10a–c, respectively. The harmonic components only exist in i_{LPF} and i_{CPF}. In detail, $I_{LPFn} = I_{CPFn}$ ($n = 3, 5, 7, 13$) proves that the harmonic current loops in the TCIMN and does not pollute the power amplifier. There is nearly no harmonic component in i_L (THDi_{LPF} = 51.9% and THD$_{iL}$ = 2.1%).

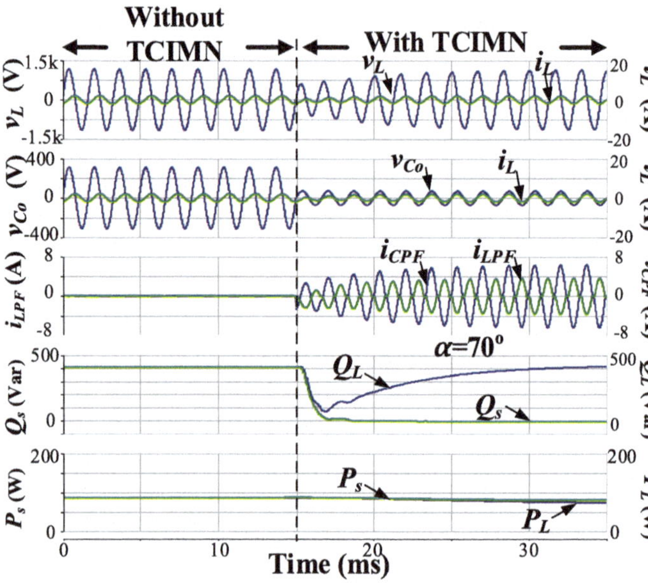

Fig. 6.11 Dynamic waveforms of v_L, i_L, v_{Co}, i_{CPF}, i_{PF}, Q_s, Q_L, P_s, and P_L in simulation without TCIMN matching and with TCIMN matching to keep load power unchanged

Table 6.3 Simulation results of TCIMN impedance matching in different situations

Match	Loads	f (Hz)	I_X (A)	V_{co} (V)	Q_L (Var)	P_L (W)	Q_s (Var)	P_s (W)	PUF (%)
Before	1st	600	6.9	800	5142	1150	5142	1150	27.01
	2nd	600	10.0	800	7703	2454	7703	2454	30.35
After	1st	400	8.2	220	5644	1855	47	1855	99.97
	1st	500	7.4	220	5440	1692	38	1692	99.97
	1st	600	6.9	220	5421	1524	40	1524	99.97
	1st	700	6.4	220	4951	1365	42	1365	99.95
	2nd	600	7.0	220	4890	1773	23	1773	99.98

Figure 6.11 shows the dynamic waveforms before and after TCIMN matching while keeping the load P_L and Q_L unchanged (i_L unchanged). The v_{Co} is largely reduced after the TCIMN connected with the PUF is increased from 22.95% (without TCIMN) to 99% (with TCIMN).

Through the impedance matching transient and steady states of the above conditions in Figs. 6.10, 6.11, and Table 6.3, the adaptive and broadband impedance matching characteristics of the proposed TCIMN can be verified. And after the TCIMN is matched, the PUF can be improved from nearly 27.01% (without TCIMN) to 99% (with TCIMN).

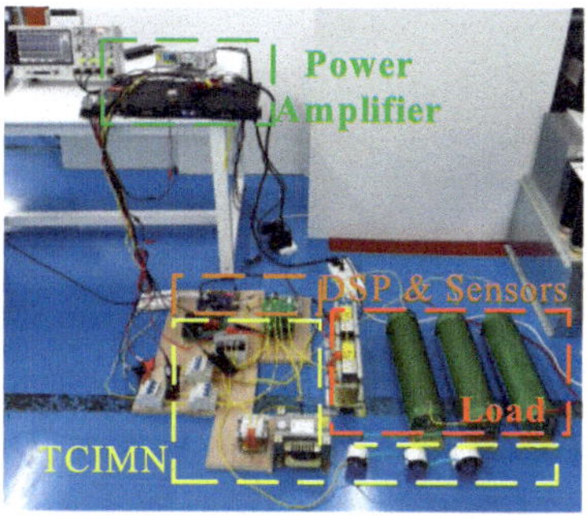

Fig. 6.12 Experimental setup of the TCIMN experimental prototype

6.2.5 *Experimental Results*

In this section, a scale-down experimental prototype of the proposed TCIMN with the power amplifier and loads is built in the laboratory, as shown in Fig. 6.12, to provide experimental results and to verify the above discussions. The digital control system of TCIMN is a digital signal processor (DSP) TMS320-F28335 controller. The DSP can process the signals as Fig. 6.9 and send trigger signals to the thyristor driver to control the fast-switching thyristor model SKKT 72/16E (operating frequency range 100 Hz–1 kHz). Among these experimental results, waveforms and harmonic spectrum are recorded by the Keysight DSO-X 3024 T oscilloscope.

In Fig. 6.13, the performances of the proposed TCIMN are illustrated under four representative dynamic situations. The corresponding dynamic waveforms are provided in Fig. 6.13. From Fig. 6.13, on the load side, v_L and i_L are not in phase. There is a large reactive Q_L, which can seriously affect power amplifier efficiency (PUF = 23%). It is shown in Fig. 6.13a–c, in the stable-states v_{Co} and i_L are in phase, which illustrates that load impedances are matched in these conditions (in 400 Hz, 500 Hz, 600, and 700 Hz). v_{Co} and i_L are in the same phase, although there is phase difference between v_L and i_L. The source-side reactive power Q_s is matched to nearly 0 (PUF = 99%). The V_{Co} can be reduced from 70 V (without TCIMN) to 30 V (with TCIMN). Therefore, the power amplifier efficiency has been significantly improved (up to PUF = 99%) with the proposed TCIMN.

6.2.6 *Section Summary*

A new broadband thyristor-controlled impedance match work (TCIMN) is proposed to be applied with a power amplifier for broadband impedance matching to improve power utilization. The TCIMN parameters design method and operating ranges are given to ensure that TCIMN can match load impedance in the required frequency range. With the mathematical proof, TCIMN can be guaranteed matching load impedance with less harmonic pollution. To realize the broadband adaptive matching,

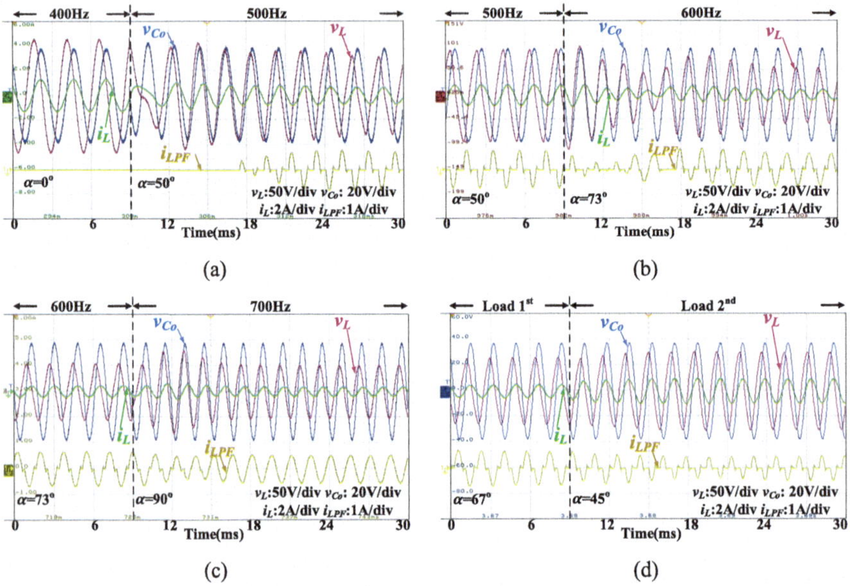

Fig. 6.13 Experimental dynamic v_L, i_L, v_{Co}, i_{CPF}, i_{PF} waveforms with impedance matching when power amplifier output frequency changes **a** from 400 to 500 Hz, **b** 500 Hz to 600 Hz, **c** from 600 to 700 Hz, and **d** from load 1st to load 2nd

a control method is proposed, including trigger signal calculation, dynamic reactive power calculation, and impedance identification. Finally, the proposed topology, control method, and all discussions are validated by dynamic simulation and experimental results, which clearly show that the proposed TCIMN can match load impedance in broadband range with superior dynamic and stable performance and increase the power amplifier PUF up to 99%.

6.3 A Shunt Adaptive Broadband Active Impedance Matching Network

6.3.1 Circuit Configurations and Impedance Functions of the Proposed SAIMN

In this section, the circuit configurations of the power amplifier, load network, and proposed adaptive broadband shunt active impedance matching network (SAIMN) are shown in Fig. 6.14. The load characteristics, matching principle, and range of the proposed SAIMN are analyzed in detail.

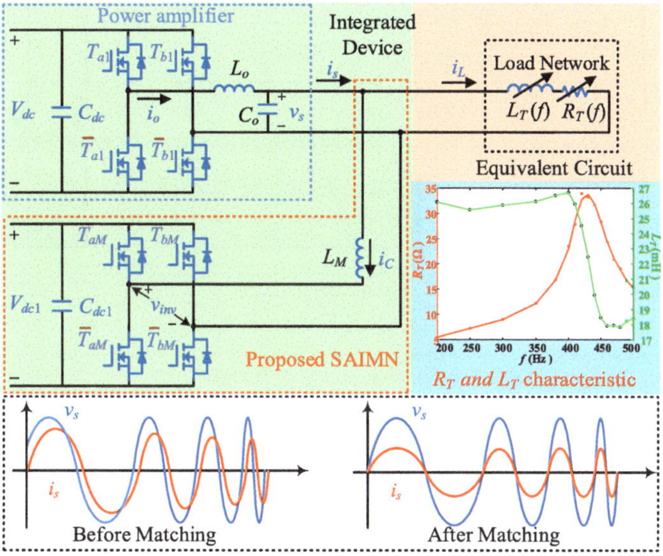

Fig. 6.14 Proposed SAIMN circuit configuration and matching performance

A. *Circuit Configuration of Power Amplifier and Proposed SAIMN*

As shown in Fig. 6.14, the power amplifier is a controllable voltage source with adjustable output voltage amplitude and frequency. The equivalent simplified circuit model of the system is shown in Fig. 6.15. The circuit model of the transducer is a series connection of a resistor and an inductor, which is unchanged in the operating process. The power amplifier consists of a dc voltage capacitor, a full-bridge inverter and an *LC* output filter. L_o and C_o are the output inductor and capacitor of the power amplifier. V_{dc} and v_s are the DC voltage source and the power amplifier output voltage. i_o, i_L and i_s are the output inductor current, load current, and power amplifier output current, respectively.

The proposed SAIMN consists of a full-bridge inverter with a DC capacitor $C_{\text{dc}1}$. The inverter in the proposed SAIMN is connected to the power amplifier's output via an output filtering inductor L_M. V_{inv} and V_{dc1} are the inverter output voltage at the fundamental frequency and the voltage of C_{dc1}, and i_C is the matching current. The traditional power amplifier and proposed SAIMN form an integrated system together. Therefore, the power amplifier is equipped with an impedance-matching function while supplying power to the load.

B. *Impedance Characteristics of Load Network*

The UAT is equivalent to a resistive–inductive linear load, and it can be expressed as follows:

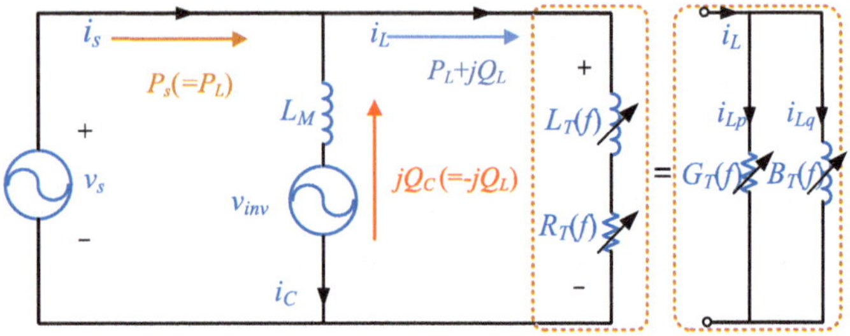

Fig. 6.15 Equivalent simplified circuit of proposed SAIMN

$$Z_L = R_T + jX_T = R_T + j(2\pi f \cdot L_T) \tag{6.28}$$

where R_T and L_T are the load equivalent resistance and inductance, X_T is the load equivalent reactance, f is the frequency of the power amplifier output voltage, and the angular frequency of the power amplifier output voltage is $\omega = 2\pi f$.

R_T and L_T vary with the change of working current, frequency, and conditions of the UAT. Among them, the working frequency has the most significant impact on the equivalent impedance of the load. Unlike the nonlinear loads, the load here only causes the variation of current amplitude but does not introduce harmonic currents into the system. Figure 6.16 shows the measured impedance-frequency characteristics of a new type of underwater acoustic low-frequency transducer under a certain operating condition.

To visualize the change of the load and analyze the matching process, the Smith chart of impedance is applied to describe the matching trajectories, as shown in Fig. 6.17. The arrow of the same color as the dashed cycle indicates the matching trajectories. Z_o represents the original load equivalent impedance. The arc length of the constant resistance between the point in the dashed box and the horizon axis represents the reactance value of the matching capacitor. The impedance matching here refers to the resonate matching or imaginary part of the load impedance matching. The point at the horizontal axis represents the equivalent impedance after matching.

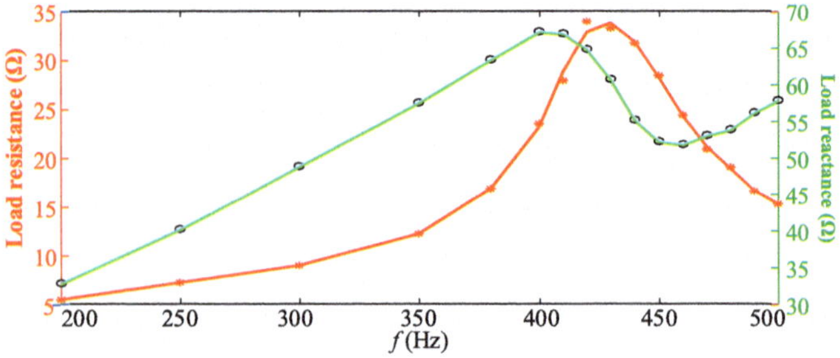

Fig. 6.16 Measured impedance-frequency characteristic of load

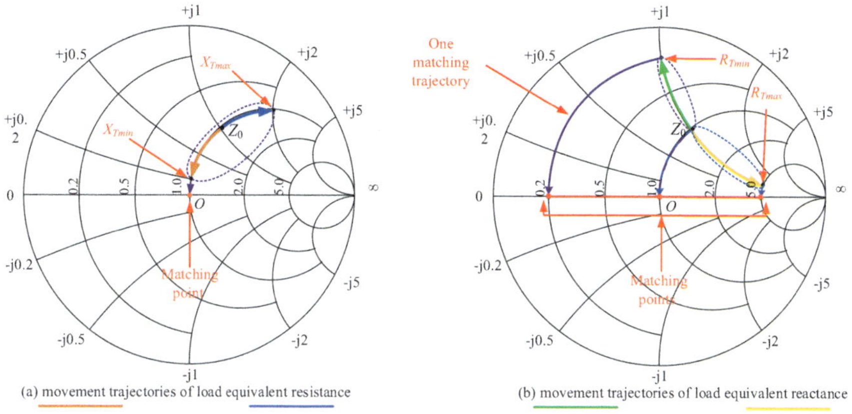

Fig. 6.17 Trajectories of load equivalent impedance move in the Smith chart considering potential parameter changes and matching methods

Figure 6.17a shows the change of load equivalent reactance due to the power amplifier frequency and load inductor change. When the load equivalent reactance changes, we can realize impedance matching by adding a series-connected (or shunt-connected) variable capacitor in the admittance circuit. The impedance can then be transformed counterclockwise along a constant resistance circle to the origin *O*. Figure 6.17 **b** shows the change of load equivalent resistance. If the equivalent reactance remains unchanged, we can add a series-connected or shunt-connected fixed capacitor to realize impedance.

C. *Matching Principle*

Since the UAT is a resistive–inductive load, the instantaneous current flowing through it can be decomposed into an active current component i_{Lp} and a reactive current component i_{Lq}, as shown in Fig. 6.15. The relationship between them can be expressed by (6.29):

$$i_L = i_{Lp} + i_{Lq} \tag{6.29}$$

The power amplifier output current is equal to the load current before the impedance matching. After the impedance matching, the impedance matching network injects a matching current i_C into the system. According to Kirchhoff's current law, the power amplifier output current, load current, and current matching are given as follows:

$$i_s = i_L + i_C \tag{6.30}$$

According to (6.29) and (6.30), if i_C equals -i_{Lq}, the power amplifier output current i_s is reduced to i_{Lp}. Actually, to stabilize the DC voltage V_{dc1}, it is necessary to inject a certain amount of active current into the matching network from the power amplifier.

Therefore, the matching current includes reactive and certain active components, which can be expressed as:

$$i_C = -i_{Lq} + k_{Lp} \cdot i_{Lp} \tag{6.31}$$

where $k_{Lp} < 1$ is a proportional constant, and the power amplifier output current i_s can be expressed as:

$$i_s = i_{Lp} + i_{Lq} + i_C = i_{Lp} + \mathrm{k}_{Lp} \cdot i_{Lp} \tag{6.32}$$

When the power loss of the inverter in the proposed SAIMN is not considered, the voltage, current, and power phasors before and after matching are shown in Figs. 6.18 and 6.19.

The apparent power S_s equals active power P_s after the impedance matching because the proposed SAIMN provides the reactive power required by the load. Therefore, in the case of a fixed power supply capacity, the power amplifier can

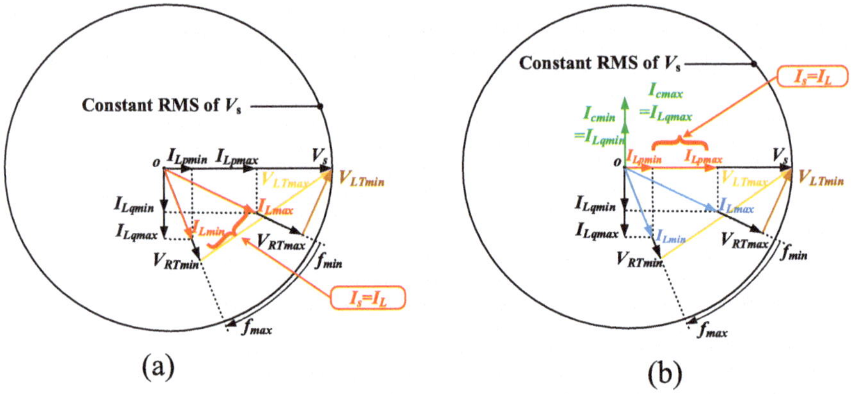

Fig. 6.18 Phasor representations of currents and voltages **a** before matching and **b** after matching

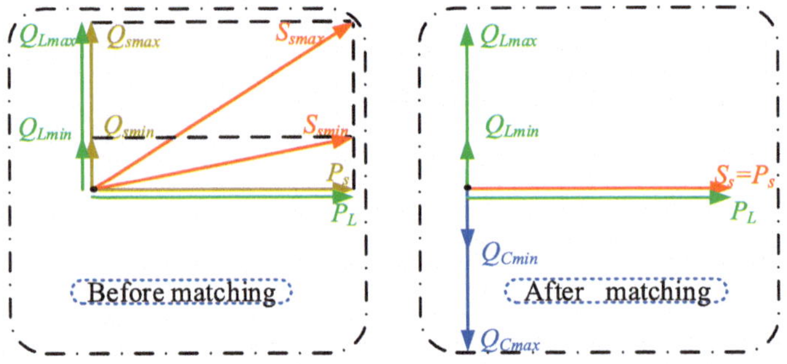

Fig. 6.19 Phasor representations of power

transfer more active power to the load. Compared with the condition without the impedance matching, the power amplifier current after impedance matching can be reduced.

D. *Matching Range Analysis*

The maximum compensated reactive power depends on the designed capacity of the proposed SAIMN hardware. This hardware capacity is designed based on the load impedance and operating frequency, especially in an underwater electroacoustic transduction system. To improve the system capacity utilization, the designed reactive power compensation capacity is based on the possible maximum load reactive power.

According to Fig. 6.6, the load reactive power Q_L and reactive current component are deduced as:

$$Q_L = \frac{V_s^2 \omega L_T}{R_T^2 + (\omega L_T)^2} \tag{6.33}$$

$$I_{Lq} = \frac{V_s \omega L_T}{R_T^2 + (\omega L_T)^2} \tag{6.34}$$

The relationship between the frequency and load reactive power, load reactive current component can be drawn in Fig. 6.20, where $R_T = 4\ \Omega$ and $L_T = 3$ mH. From Fig. 6.20, it is not that the higher the frequency, the greater the reactive power of the load.

Then, assuming R_T and L_T remain unchanged, the derivative of Q_L with respect to the angular frequency ω can be obtained as follows:

$$\dot{Q}_L = \frac{V_s^2 L_T \left(R_T^2 - (\omega L_T)^2\right)}{\left(R_T^2 + (\omega L_T)^2\right)^2} \tag{6.35}$$

Let (6.35) equal to 0, the relationship between the load impedance and angular frequency can be obtained as:

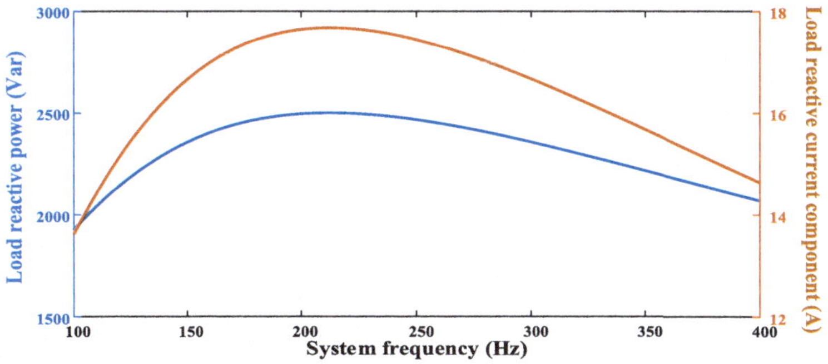

Fig. 6.20 The relationship between frequency and load reactive power, load reactive current component

$$R_T = \omega L_T \tag{6.36}$$

That is to say

$$f = \frac{R_T}{2\pi L_T} \tag{6.37}$$

Therefore, the maximum load reactive power and reactive current can be expressed as follows:

$$Q_{L(max)} = \frac{V_s^2}{2R_T} \tag{6.38}$$

$$I_{Lq(max)} = \frac{V_s}{2R_T} \tag{6.39}$$

According to Fig. 6.6, the matching current I_C and compensation reactive power Q_c can be deduced as:

$$I_C = \frac{V_{inv} - V_s}{\omega L_M} \tag{6.40}$$

$$Q_c = V_s I_C = V_s \frac{V_{inv} - V_s}{\omega L_M} \tag{6.41}$$

In theory, according to (6.41), when V_s remains constant, the maximum capacitive reactive power provided by the proposed SAIMN depends on V_{inv}. To improve the system capacity utilization, the designed reactive power compensation capacity is based on the possible maximum load reactive power or current. Therefore, to achieve impedance matching in the whole operating frequency range of the transducer, the relationship between the compensation capacity and the possible maximum load reactive power can be expressed as:

$$Q_{c(max)} \geq Q_{L(max)} = \frac{V_s^2}{2R_T} \tag{6.42}$$

Considering the 20% margin, the compensation capacity of the proposed SAIMN should satisfy:

$$Q_{c(max)} = 1.2 \cdot Q_{L(max)} = \frac{3V_s^2}{5R_T} \tag{6.43}$$

With the variable R_T and X_T considered, the relationship among the load reactive current, $R_{T,}$ and X_T can be drawn in Fig. 6.21 according to (6.29). Without the consideration of marginal, it is assumed that the impedance matching network has matching upper and lower limits (greater than 3 A and less than 9 A), as shown

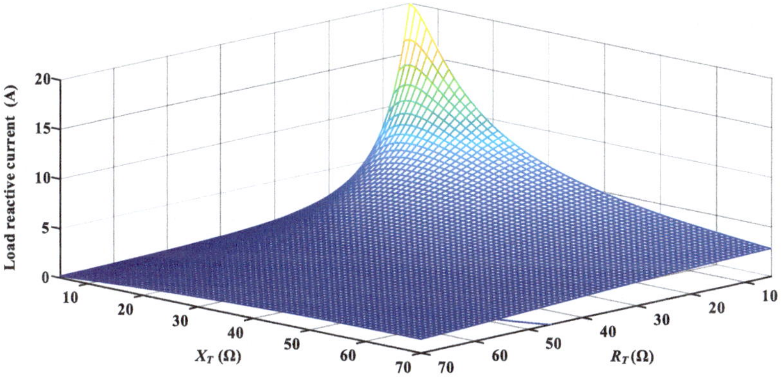

Fig. 6.21 Relationship among matching current, X_T, and R_T

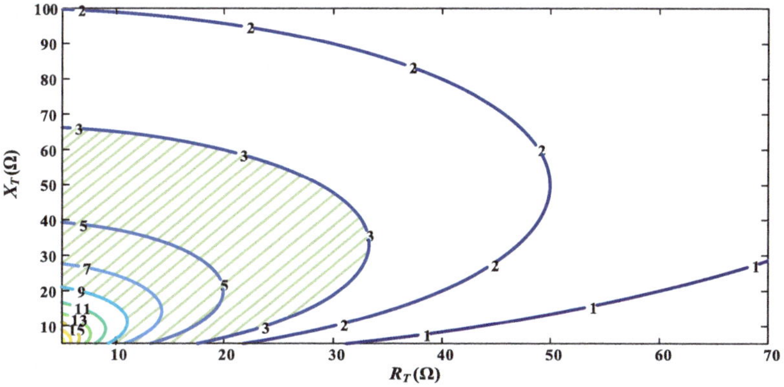

Fig. 6.22 The covered load impedance range in the different maximum matching currents of the proposed SAIMN

in Fig. 6.22, then the shaded part is the covered impedance range. Actually, if the load reactive current is less than the maximum matching current that the impedance network can output, the corresponding covered load impedance can be matched (the upper right area of the different maximum matching current lines) within the operating frequency range.

6.3.2 *Parameter Design of the Proposed SAIMN*

In this section, the parameter design strategy is discussed in two parts, namely, the design of V_{dc1} and the selection of L_M. The function of i_L can be assumed as:

$$i_L = \sqrt{2}I_L\sin[\omega t - \varphi(t)] = \sqrt{2}I_L\sin[2\pi ft - \varphi(t)] \tag{6.44}$$

where I_L is the RMS value of i_L, and $\varphi(t)$ is the impedance angle of the load. $\varphi(t)$ changes with time because of the change of system frequency. The function of $\varphi(t)$ can be expressed as:

$$\varphi(t) = \frac{\omega L_T}{R_T} = \frac{2\pi \mathrm{f}\, L_T}{R_T} \tag{6.45}$$

Expand (6.44) into two parts:

$$i_L = \underbrace{\sqrt{2}I_L \sin(\omega t) \cdot \cos\varphi(t)}_{i_{Lp}} + \underbrace{\left(-\sqrt{2}I_L \cos(\omega t) \cdot \sin\varphi(t)\right)}_{i_{Lq}} \tag{6.46}$$

Then, i_{Lp} and i_{Lq} can be expressed as:

$$i_{\mathrm{Lp}} = \sqrt{2}I_L \sin(\omega \mathrm{t}) \cdot cos\, \varphi(t) \tag{6.47}$$

$$i_{\mathrm{Lq}} = -\sqrt{2}I_L cos(\omega t) \cdot sin\, \varphi(t) \tag{6.48}$$

Therefore, the power amplifier output voltage can be deduced as:

$$\mathrm{v}_s = i_L \cdot Z_L = \sqrt{2}I_L \cdot \sqrt{R_T^2 + (\omega L_T)^2} \sin(2\pi \mathrm{ft}) \tag{6.49}$$

The current hysteresis PWM method is used to control the matching current i_C. Therefore, in the design of L_M, the hysteresis current band h should be taken into account. With the maximum system frequency, ripple current, hysteresis current band h, and the maximum value of switching frequency f_{swmax}, the value of L_M is designed by (6.50).

$$L_M = \frac{\mathrm{mV}_{smax}}{4hf_{swmax}} \tag{6.50}$$

where m is a proportional constant, V_{smax} is the maximum amplitude of v_s, $m = 1.6$, and the system frequency is 50 Hz. The value of m is selected according to the maximum value of the power amplifier frequency in this Section.

According to the KVL theorem, the relationship between v_s, v_{inv}, and i_C can be expressed as:

$$L_M \frac{\mathrm{d}i_C}{dt} + v_{inv} = v_s \tag{6.51}$$

The inverter output voltage can be expressed as:

$$v_{inv} = KV_{dc1} \tag{6.52}$$

where $K = 1$ or -1 depends on the switching of the inverter. Thus, combining (6.51) and (6.52), the required DC voltage V_{dc1} must satisfy:

$$V_{\mathrm{dc1}} \geq \left| v_s - L_M \frac{di_C}{dt} \right| \tag{6.53}$$

V_{dc1} is designed according to the maximum load reactive current as the frequency increases to the required maximum point to achieve impedance matching within the whole frequency range. Based on the discussion, $I_{Lq(max)}$ is not obtained at the maximum operating frequency. To simplify the analysis, it is assumed that only reactive current is included in the matching current. According to (6.48), the matching current i_C can be as large as $\sqrt{2}\, I_{Lq(max)}$ with $f = f_{max1}$:

$$\lim_{f \to f_{\max 1}} |i_C|_{\max} = \lim_{f \to f_{\max 1}} \left| -i_{Lq} \right|_{\max} = \sqrt{2} I_{Lq(max)} \tag{6.54}$$

where f_{max1} is the power amplifier output voltage frequency at which the load reactive current component is maximized. Therefore, the maximum voltage of inductor L_M is $\sqrt{2} I_{Lq(max)} \omega_{\max 1} L_M$, ω_{max1} is the angular frequency at which I_{Lq} reaches its maximum value. According to (6.49), the amplitude of the power amplifier voltage V_{sm} is $\sqrt{2} I_L \cdot \sqrt{R_T^2 + (\omega L_T)^2}$, , which is constant, and (6.53) can then be further expressed as:

$$V_{\mathrm{dc1}} \geq \left[\sqrt{2} I_L \cdot \sqrt{R_T^2 + (\omega L_T)^2} + \sqrt{2} I_{Lq(\max)} \omega_{\max 1} L_M \right] > \left| v_s - L_M \frac{di_C}{dt} \right| \tag{6.55}$$

To leave a certain safety margin, the V_{dc1} should satisfy (6.56):

$$\begin{aligned} V_{\mathrm{dc1}} &= 1.2 \cdot \left[\sqrt{2} I_L \cdot \sqrt{R_T^2 + (\omega L_T)^2} + \sqrt{2} I_{Lq(max)} \omega_{max1} L_M \right] \\ &= 1.2\left(V_{sm} + \sqrt{2} I_{Lq(max)} \omega_{max1} L_M \right) \end{aligned} \tag{6.56}$$

Therefore, the L_M and V_{dc1} are designed by (6.51) and (6.56).

6.3.3 Control of the SAIMN

In this section, the control strategy is discussed in two parts, namely, the control of power amplifier output voltage and impedance matching network current. The overall control block diagram is shown in Fig. 6.23.

The voltage and current double closed-loop control strategy is adopted to control the power amplifier output voltage. According to the matching principle in Sect. 6.3.1,

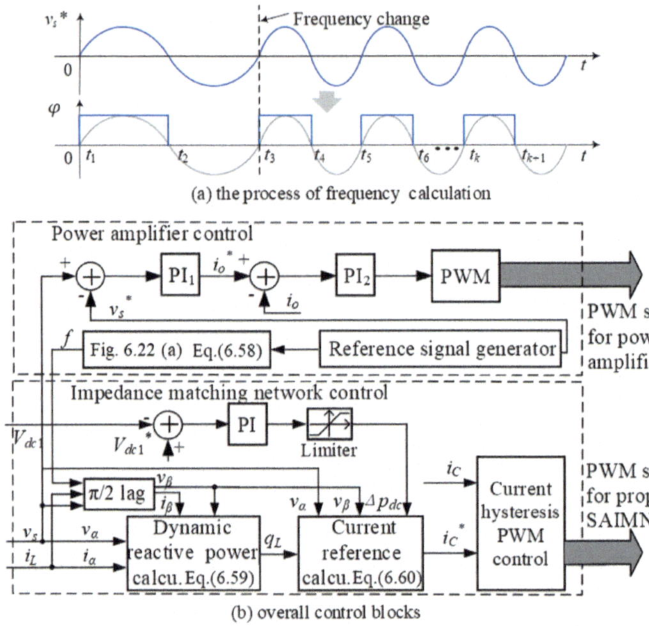

Fig. 6.23 Overall control strategy of the proposed SAIMN and power amplifier

the proposed SAIMN injects a matching current i_C to realize impedance matching. The reference matching current $i_C{}^*$ is calculated by the single-phase *p-q* theory.

The electroacoustic transduction system has high requirements for real-time performance. Therefore, to switch to different sound source levels, it is necessary to ensure that enough power is transferred to the UAT. Thus, the current hysteresis PWM method is applied to control the matching current i_C to its reference current $i_C{}^*$. It has the advantages of fast dynamic response, unconditional stability, robust current control performance, and less dependence on the control parameter.

(1) *Instantaneous reactive power compensation control*: The final goal of impedance matching is to improve the power factor by making the power amplifier only output active power. The load instantaneous reactive power q_L is calculated by:

$$q_L = v_\alpha \cdot i_\beta - v_\beta \cdot i_\alpha \tag{6.57}$$

where v_a and i_a are the original load voltage and current, respectively, v_β and i_β are the variables which are delayed $\pi/2$ by v_α and i_α, respectively. The system frequency is captioned by the zero-crossing signal calculation method, as shown in Fig. 6.22a:

$$f = \frac{1}{T} = \frac{1}{2|t_{k+1} - t_k|} \tag{6.58}$$

(2) *Instantaneous active power compensation control*: If V_{dc1} drops too much, the matching network will not achieve the matching effect. Therefore, it is important to keep V_{dc1} within a certain range. The PI control is applied to reduce the steady-state error caused by active power components. The active power Δp_{dc} controlling the DC voltage is expressed as follows:

$$\Delta p_{dc} = k_p \cdot (V_{dc1}^* - V_{dc1}) + k_I \cdot \int (V_{dc1}^* - V_{dc1}) \cdot dt \tag{6.59}$$

where k_p and k_I denote the proportional and integral gain of the PI controller, ${V_{dc1}}^*$ is the DC voltage reference value, then Δp_{dc} is used to maintain V_{dc1} as its reference.

(3) *Hysteresis current PWM control*: After the calculation of instantaneous reactive power and a coordinate transformation, the reference ${i_C}^*$ can be calculated by (6.60)):

$$\mathrm{i}_C^* = \frac{\mathrm{q}_L \cdot (-v_\beta) + \Delta \mathrm{p}_{dc} \cdot (-v_\alpha)}{\mathrm{v}_\alpha^2 + v_\beta^2} \tag{6.60}$$

Then, ${i_C}^*$ is compared with i_C to generate PWM signals by using hysteresis control, and the PWM signals are sent to the switching devices to ensure the matching performance.

6.3.4 Simulation Results

To verify the effectiveness of the proposed SAIMN, simulations are carried out using MATLAB. The dynamic performances of the proposed SAIMN are illustrated under two different situations: (1) load condition changes and (2) power amplifier output voltage frequency changes. The simulation parameters are listed in Table 6.4. The purpose is to prove that the power utilization factor (PUF) is improved with the SAIMN. PUF mainly focuses on the power factor. Therefore, the power factor (PF) is used to measure the performance of the proposed SAIMN. Moreover, to demonstrate that the power loss of the proposed SAIMN is small, the power conversion efficiency (PCE) is used to measure the performance.

A. *The change of loads*

The simulation results under different loads in 200 Hz with impedance matching are shown in Fig. 6.24 and Table 6.5. From Fig. 6.24a and Table 6.5, the load reactive power decreases gradually with the loads changing. Meanwhile, the reactive power Q_s can still keep around 0 after a short duration. From Fig. 6.24b, v_s and i_s are not in phase without SAIMN, and the power amplifier outputs a large amount of reactive power. With the adding of the SAIMN, v_s and i_s can be matched to be in phase after a very short transient process. Whether the load inductance change or resistance

Table 6.4 Parameters of power amplifier and SAIMN in simulations

	Parameters	Physical values
Power amplifier	L_o, C_o	8 mH, 110 uF
	f	100 Hz–400 Hz
	V_{sm}	200 V
Load	1st	$R_1 = 2\ \Omega$, $L_1 = 4$ mH,
	2nd	$R_2 = 5\ \Omega$, $L_2 = 4$ mH
	3rd	$R_3 = 2\ \Omega$, $L_3 = 8$ mH
	4th	$R_4 = 5\ \Omega$, $L_4 = 3$ mH
Proposed SAIMN	L_M	4 mH
	V_{dc1}	400 V

change, v_s and i_s can be matched to be in phase. Thus, the power factor of the power amplifier is almost always one during load changing.

B. *Power amplifier output voltage frequency change*

In this part, an additional set of load parameters ($R_T = 5\ \Omega$, $L_T = 3$ mH) are used in the simulation while keeping the amplitude of v_s unchanged. The simulation results of voltage frequency-changing are shown in Table 6.6 and Fig. 6.25. The load parameters in Table 6.6 are load 1st. After adding the SAIMN, v_s and i_s are matched to be in phase in steady states. Moreover, the power loss of the proposed SAIMN is the active power P_C, which is very small. Therefore, the efficiency of the system is high. The methods of calculating power loss and system efficiency (PCE) are listed in Table 6.5 in the appendix. The power amplifier output reactive power

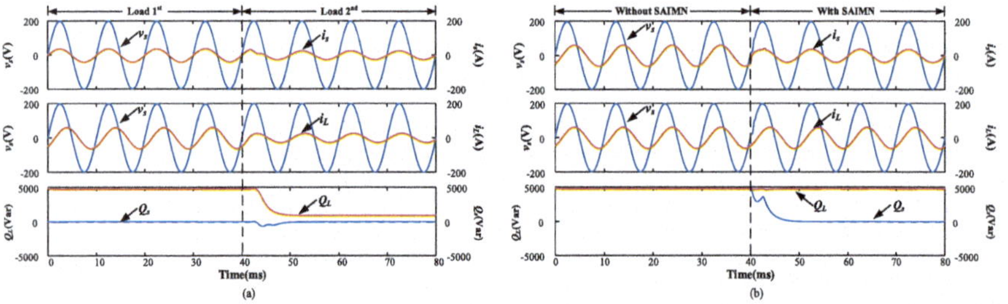

Fig. 6.24 Dynamic waveforms of v_s, i_s, i_L, Q_s, and Q_L in simulation: **a** with SAIMN impedance matched during loads changing, and **b** without SAIMN matching and with SAIMN matching

Table 6.5 SAIMN matching simulation results of different loads in steady states

Loads	$R_T(\Omega)$	L_T(mH)	Q_L (Var)	P_L (W)	Q_s (Var)	P_s (W)	PF
1st	2	4	3517	1408	2	1491	0.99
2nd	5	4	2029	2019	1	2036	0.99
3rd	2	8	1900	378	1	388	0.99

Q_s is reduced to nearly zero, and the PF is nearly unity, which greatly improves the capacity utilization of the power amplifier. With the SAIMN matching, the i_s can be largely reduced while keeping v_s unchanged. When transmitting the same active power to the load, a smaller capacity is required for the power amplifier equipped with an impedance-matching network than the power amplifier without impedance matching.

Table 6.6 Impedance matching performance comparison in steady states

Match	f (Hz)	I_L (A)	I_s (A)	Q_L (Var)	P_L (W)	Q_C (Var)	P_C (W)	Q_s (Var)	P_s (W)	PCE (%)	PF
Before	100	43.7	43.7	4792	3816	–	–	4792	3816	100	0.62
	200	26.5	26.5	3520	1407	–	–	3520	1407	100	0.37
	300	18.5	18.5	2575	683	–	–	2575	683	100	0.25
	400	13.9	13.9	1936	385	–	–	1936	385	100	0.20
After	100	43.7	27.2	4792	3816	4790	64	2	3880	98	0.99
	200	26.5	10.6	3517	1408	3515	83	2	1491	94	0.99
	300	18.5	4.9	2572	683	2751	13	1	696	98	0.99
	400	13.9	2.8	1901	379	1900	11	1	390	97	0.99

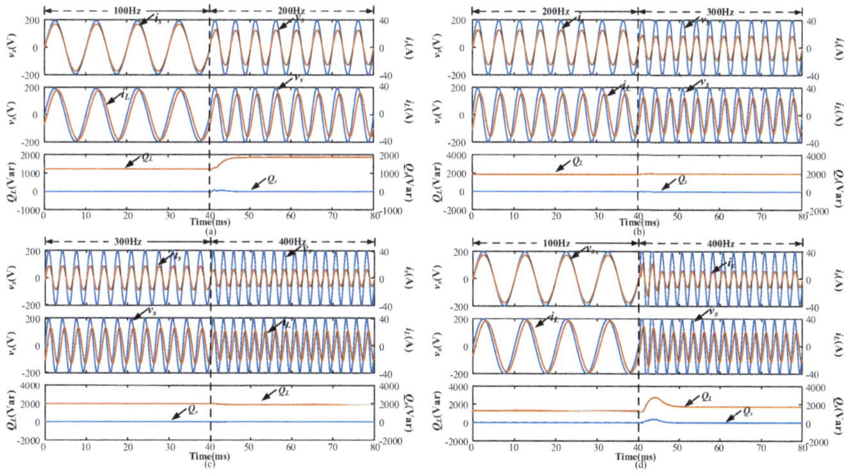

Fig. 6.25 Dynamic waveforms of v_s, i_s, i_L, Q_s, and Q_L in simulation with SAIMN when power amplifier output frequency changes **a** from 100 to 200 Hz, **b** 200 Hz to 300 Hz, **c** from 300 to 400 Hz, and **d** from 100 to 400 Hz

6.3.5 Experimental Results

In order to demonstrate the feasibility of the proposed method, experimental tests are carried out using a hardware-in-the-loop (HIL) platform consisting of a real-time model simulator and a real prototype controller, as shown in Fig. 6.26. The main circuit model is configured in the host computer software StarSim HIL and is performed on a Xilinx K7-325 T FPGA chip, which is embedded in the NI PXIe-7868R board of the real-time simulation system with a time step of 1 *us*. The system controller is designed in the host computer software StarSim RCP and is implemented in a rapid control prototyping (RCP) system based on the NI PXIe-8821 CPU with a sampling period of 100 *us*. The analog/digital signals interaction between the real-time simulation system and the RCP system is realized by an IO board, and the signals on the IO board are recorded by the Teltronix MDO3024 oscilloscope.

Figures 6.27 and 6.28 show the corresponding experimental results under some representative dynamic situations. The output filter inductor is designed to be 8 mH to reduce the ripple current in matching current because of the limit of system switch frequency. The other system parameters are the same as the simulations. In Fig. 6.27, the frequency of the power amplifier is 100 Hz, the 1st load is $R_1 = 5\ \Omega$ and $L_1 = 4$ mH, and the 2nd load is $R_2 = 5\ \Omega$ and $L_2 = 8$ mH, which shows the increase of load inductance. The 3rd load is $R_3 = 3\ \Omega$ and $L_3 = 8$ mH, and the 4th load is $R_4 = 6\ \Omega$ and $L_4 = 8$ mH, which shows the increase in load resistance.

From Fig. 6.27a, the phases of v_s and i_L are different before adding the SAIMN, and they become in phase in steady states after the SAIMN matching. With the SAIMN matching, the power amplifier output current reduces while keeping the

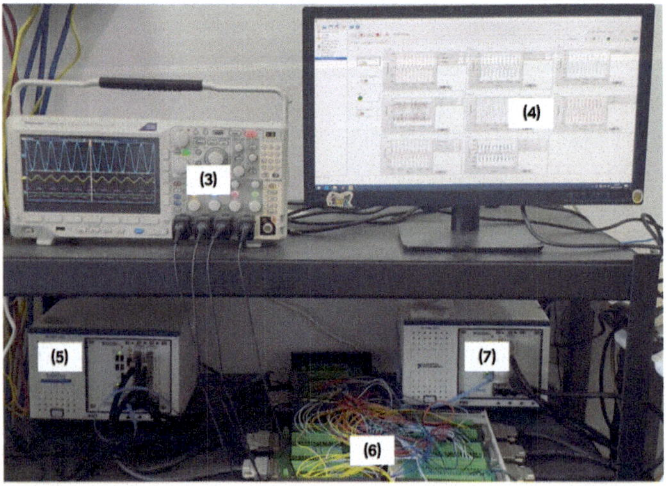

Fig. 6.26 StarSim Hardware-in-the-loop(HIL) experimental platform. ① Oscilloscope, ② Host PC, ③ NI-PXIe Target Rapid Control Prototype System, ④ I/O Board, ⑤ NI-PXIe Target Real-time Simulation System

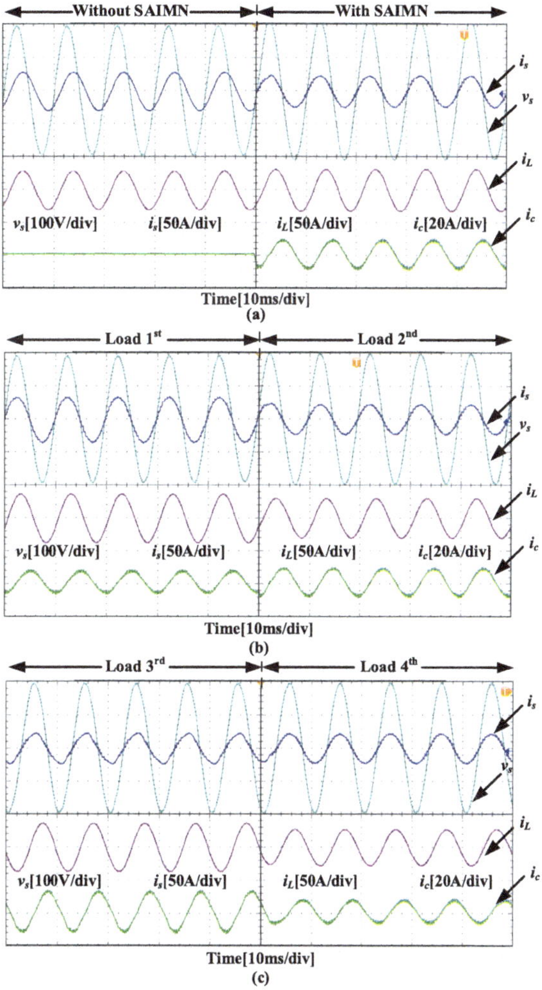

Fig. 6.27 Experimental dynamic v_s, i_s, i_L, i_c waveforms **a** without SAIMN matching and with SAIMN matching, **b** with SAIMN impedance matched when load inductance increases, and **c** with SAIMN impedance matched when load resistance increases

frequency and amplitude of the power amplifier voltage unchanged. Meanwhile, the design difficulty of the switching devices in the power amplifier can be reduced. From Figs. 6.27b, c, and 6.28, when the loads or power amplifier output frequency changes, v_s and i_s are matched to be in phase in steady states and the reactive power Qs is matched to nearly 0 (PF = 0.99). which verifies the effect of broadband impedance matching of the proposed SAIMN.

Moreover, the experimental results implemented by a converter prototype and real load are provided to verify the above discussions, as shown in Fig. 6.29. From Fig. 6.29a, there are small fluctuations in the voltage and current due to the fluctuation of the real load, but after a short transient process, vs and is are matched to be in phase.

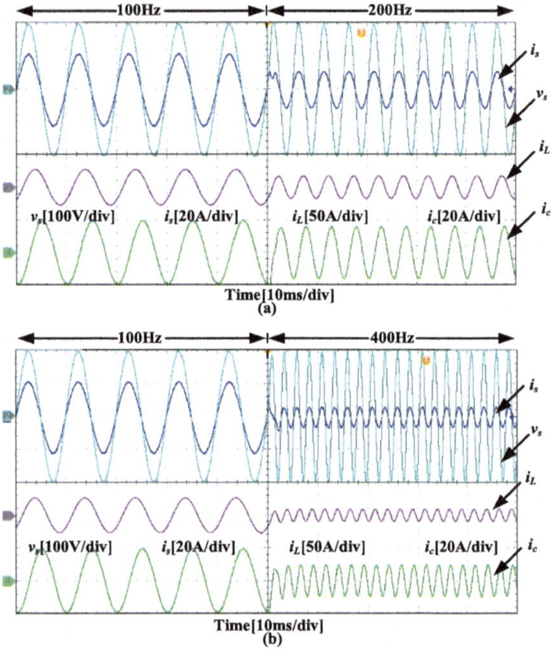

Fig. 6.28 Experimental dynamic v_s, i_s, i_L, i_c waveforms with impedance matching when power amplifier output frequency changes **a** from 100 to 200 Hz, and **b** 100 Hz to 400 Hz

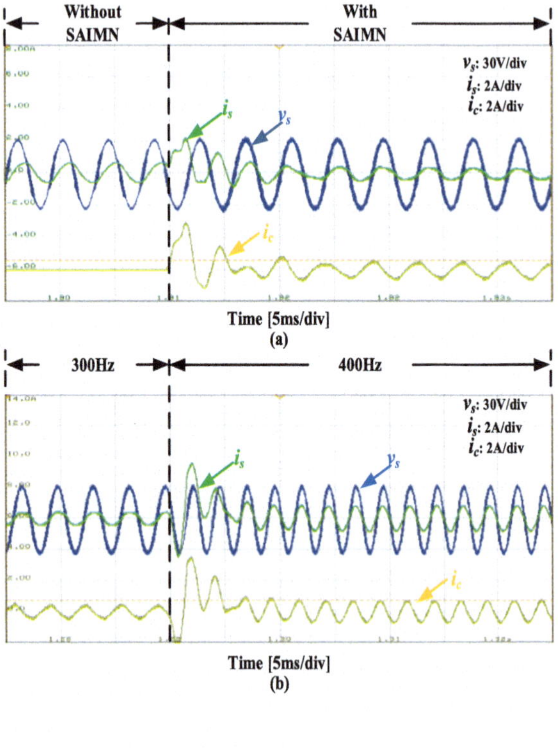

Fig. 6.29 Experimental dynamic v_s, i_s, i_c waveforms **a** without SAIMN matching and with SAIMN matching, and **b** when power amplifier output frequency changes from 300 to 400 Hz

The experimental results under the change of frequency are shown in Fig. 6.29b. From Fig. 6.29b, when frequency steps from 300 to 400 Hz, v_s and i_s are in phase after the 5 ms transient process. Therefore, the experimental results show that the proposed method can achieve a good impedance-matching performance in a wide frequency range.

6.3.6 Section Summary

A shunt active impedance matching network (SAIMN) was proposed to be applied with a power amplifier for the underwater electroacoustic transduction system to improve power utilization. The impedance characteristics of loads and the operation principle of the proposed SAIMN have been analyzed in detail. The SAIMN parameters design method and operating ranges have been given to ensure that SAIMN can match load impedance in the wide frequency range. A compound control method was presented to realize the adaptive broadband matching, including the dynamic reactive power calculation and injection of matching current. Finally, the simulations and experimental results have been presented to validate the proposed topology, control method, and all the detailed analyses, which clearly show the superior performance and desired effect of increasing the PF of the power amplifier up to 0.99.

6.4 A Series Adaptive Broadband Active Impedance Matching Network

6.4.1 Circuit Configurations of the Proposed FCIMN

In this section, the circuit configurations of the proposed FCIMN are given, as well as the operating principle.

A. *Proposed FCIMN Circuit Configuration*

As shown in Fig. 6.30, the proposed FCIMN consists of a full-bridge circuit. V_{dc} and C_{dc} are the DC power supply and the supporting capacitor, respectively, which are connected to the DC input side of the full bridge. L_M and C_M form a passive filter circuit connected to the output side of the full-bridge circuit. The capacitor C_M is connected in series between the power amplifier and load network. i_M and i_L are the currents of the inductor L_M and the load network, respectively. v_{CM} and v_L are the voltage of the FCIMN and the load network, respectively. The power amplifier provides power to the load and its matching network through C_o with the adjustable frequency voltage and amplitude.

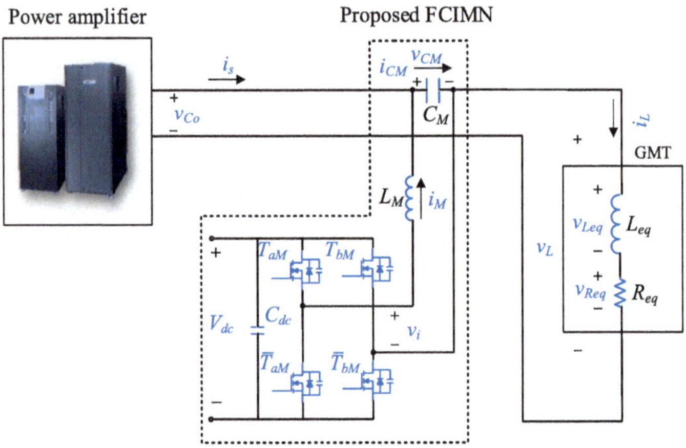

Fig. 6.30 Proposed FCIMN circuit configuration

B. *Operation Principles of Proposed FCIMN*

Normally, the loads connected with the power amplifier are all resistive-inductive. Thus, the matching network usually uses capacitors to match the inductive load. The proposed FCIMN performs impedance matching in an active way, that is, by generating a voltage that is opposite to the equivalent inductive load. Its matching principle is shown in Fig. 6.31. The load equivalent inductance voltage v_{Leq} is given as:

$$v_{Leq} = v_{Co} - v_{CM} - v_{Req} \tag{6.61}$$

In order to match the load equivalent inductance voltage v_{Leq}, the proposed FCIMN generates an opposite voltage, which can be expressed as:

$$v_{Leq} + v_{CM} = 0 \tag{6.62}$$

Combining (6.61) and (6.62) are deduced as:

$$v_{Co} = v_{Req} \tag{6.63}$$

Based on (6.63), the output voltage and current of the power amplifier are in the same phase, and the output voltage is all added to the equivalent resistance of the load.

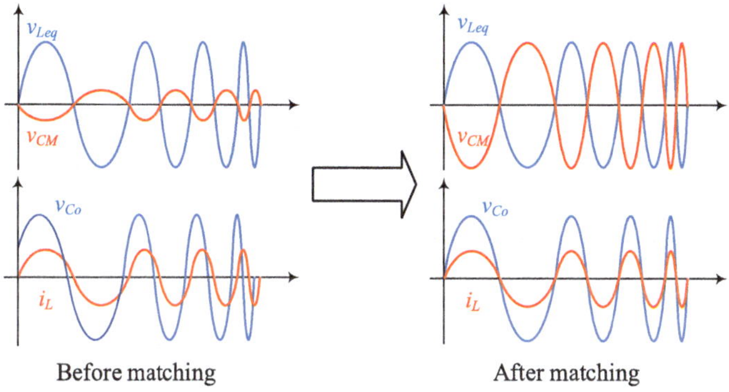

Fig. 6.31 Proposed FCIMN matching operation principles

6.4.2 *Control Method*

In this section, for the broadband matching purpose, the proposed overall FCIMN control strategy is presented in part A, including dynamic reactive power matching, active power calculation, and variable frequency calculation. Then, a segmented PR control is proposed in part B. In Part C, a combined feedback and feedforward voltage control is proposed to achieve a quick response of FCIMN.

A. *Proposed Overall FCIMN Control Strategy*

The overall control strategy of the proposed FCIMN can be summarized as (1) calculating the period T of power amplifier output voltage; (2) calculating dynamic load active power P_L and reactive power Q_L; (3) identifying the voltage $V_{CM}{}^*$; (4) doing voltage closed-loop control; (5) generating SiC MOSFETs driver signals. The overall control strategy can be drawn as Fig. 6.32. Corresponding functions are listed and explained as follows. The power amplifier output voltage period T is captured by the zero-crossing signal calculation method. It can be expressed as $2|t_{k+1}\text{-}t_k|$, where t_k and $t_{\mathrm{k}+1}$ are the time points of two adjacent zero-crossing points. The reactive power of the load can be calculated through single-phase instantaneous power theory (6.64).

$$\begin{bmatrix} p_L \\ q_L \end{bmatrix} = \begin{bmatrix} v_L \cdot i_L + v_L^D \cdot i_L^D \\ -v_L^D \cdot i_L + v_L \cdot i_L^D \end{bmatrix} \tag{6.64}$$

where $v_L{}^D$ and $i_L{}^D$ are the delay signals of v_L and i_L, respectively. The delay time t_{DI} is expressed as:

$$t_{D1} = (1/4) \cdot T = (1/2) \cdot |t_{k+1} - t_k| \tag{6.65}$$

The active power P_L and reactive Q_L can be calculated through (6.66) and (6.67), respectively.

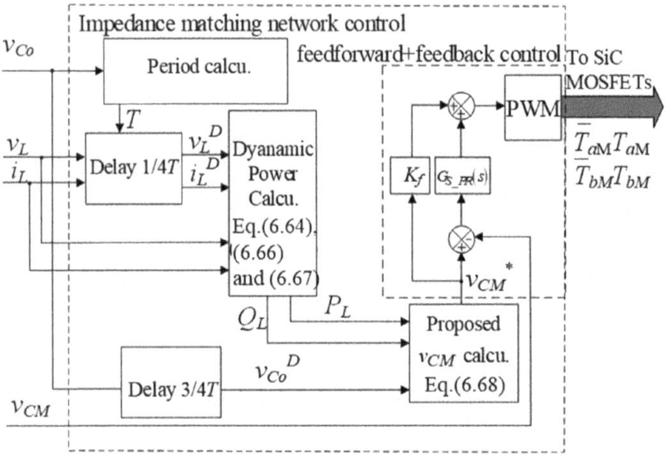

Fig. 6.32 Overall control strategy of proposed FCIMN

$$Q_L = \frac{1}{2} \cdot \overline{q}_L \tag{6.66}$$

$$P_L = \frac{1}{2} \cdot \overline{p}_L \tag{6.67}$$

Based on the analysis in Sect. 6.4.1, the final result of impedance matching is that the active power of the load is all provided by the power amplifier, and all the reactive power is provided by the matching network. That is, v_{Co} provides active power, v_{CM} provides reactive power, and v_{CM} is 90 degrees ahead of v_{Co}. Therefore. The required voltage $v_{\mathrm{CM}}{}^*$ of the impedance matching network can be deduced as:

$$v_{CM}^* = \frac{Q_L v_{Co}^D}{P_L} \tag{6.68}$$

where $v_{Co}{}^D$ is the delay signal of v_{Co}, which is in phase with v_{CM}. The delay time t_{D2} is given in (6.69).

$$t_{D2} = (3/4) \cdot T = (3/2) \cdot |t_{k+1} - t_k| \tag{6.69}$$

B. *Companions of Conventional Controllers and Proposed Segmented PR Controller*

The typical PR controller can be expressed in the s-domain as:

$$G_{PR}(s) = K_P + K_I \frac{s}{s^2 + \omega_0^2} = K_P + K_I R(s) \tag{6.70}$$

where K_p is the proportional gain, K_I is the integral coefficient, ω_0 is the fundamental angular frequency, and *R(s)* is a resonant controller. *R(s)* has infinite gain at the resonant frequency ($f = \omega_0/2\pi$), which assures perfect tracking for components that oscillate at f when implemented in a closed-loop. When $K_P = 10$, $K_I = 500$, $\omega_0 = 1000\pi$, the frequency characteristics of the PR controller are shown in Fig. 6.33a. In theory, the infinite gain of the PR controller occurs at the fundamental frequency of 500 Hz. However, the gain is finite in the real system, which is determined by K_P and is not affected by the resonant frequency. The output frequency of the power amplifier has fluctuated. Since the gain drops sharply when the output frequency of the power amplifier deviates from the resonance frequency, the output voltage error increases, and the transient response speed becomes slower. Therefore, the quasi-resonant (quasi-PR) controller was proposed. It can be expressed in the *s*-domain as:

$$G_{Q\_PR}(s)=K_P+K_I\frac{2\omega_c s}{s^2+\omega_0^2+2\omega_c s}=K_P+K_I R_Q(s) \tag{6.71}$$

where ω_c is the cutoff bandwidth around the resonant frequency.

Assuming that the output frequency variation margin of the power amplifier is ± 2%, then ω_c can be designed as ω_0*2%. The Bode plot of the quasi-PR controller is

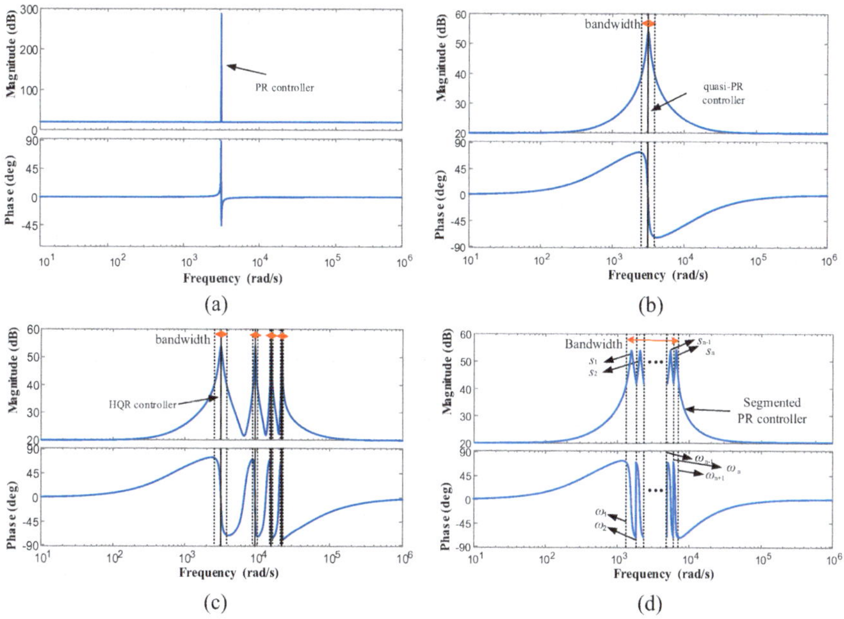

Fig. 6.33 Bode plot of different controllers, **a** PR controller, **b** quasi-PR controller, **c** HPR controller, and **d** proposed segmented PR controller

shown in Fig. 6.33b. Compared with the conventional PR in Fig. 6.33a, the quasi-PR in Fig. 6.33b is less sensitive to the fluctuation in the frequency of the power amplifier and can avoid the stability problems attributed to infinite gain. However, this controller only works near the fundamental frequency. To compensate the 3rd, 5th, 7th… order harmonics, the multiple harmonic quasi-resonant (HQR) controllers was proposed, which can be expressed in the *s*-domain as (6.72). Its frequency response is shown in Fig. 6.33c.

$$G_{M\_PR}(s) = K_P + \sum_{n=1,3,5,7} \frac{2\omega_c K_{I_n} s}{s^2 + n^2\omega_0^2 + 2\omega_c s} = K_P + K_{I_n R_M(s)} \tag{6.72}$$

However, none of the above three controllers can achieve broadband controller. Therefore, a segmented PR controller is proposed in the *s*-domain as (6.73), which divides the entire system into S_1, S_2, … and S_n. The working bandwidth of the system is set to ω_1–ω_{n+1}. That is, a high gain must be maintained in this frequency band to achieve tracking without static error. Divide ω_1–ω_{n+1} into n parts, namely ω_1–ω_2, ω_2–ω_3, … and ω_n–ω_{n+1}. When ω_0 is less than or equal to ω_2, it works in S_1, when ω_0 is greater than ω_2 and less than ω_3, it works in S_2, …, when ω_0 is greater than ω_{n-1} and less than ω_n, it works in S_{n-1}, and when ω_0 is greater than ω_n, it works in S_n. S_1, S_2, … and S_n are quasi-PR controllers with resonance points of $\omega_1 + \omega/2$, $\omega_2 + \omega/2$, … and $\omega_n + \omega/2$, respectively, where ω is the bandwidth of each small segment.

Table 6.7 shows the data related to (6.73). After designing the controller parameters for each segment, the frequency response is shown in Fig. 6.33d. It can be seen that the proposed segmented PR control can achieve high gains in the required operating frequency bands ω_1–ω_{n+1}.

$$G_{S\_PR}(s) = \begin{cases} K_{P_1} + K_{I_1}\frac{2\omega_c s}{s^2\omega_0^2+2\omega_c s}(\omega_0 \le \omega_2), & S_1 \\ K_{P_2} + K_{I_2}\frac{2\omega_c s}{s^2\omega_0^2+2\omega_c s}(\omega_2 < \omega_0 \le \omega_3), & S2 \\ \cdots & \cdots \\ K_{P_{n-1}} + K_{I_{n-1}}\frac{2\omega_c s}{s^2\omega_0^2+2\omega_c s}(\omega_{n-1} < \omega_0 \le \omega_n), & S_{n-1} \\ K_{P_n} + K_{I_n}\frac{2\omega_c s}{s^2\omega_0^2+2\omega_c s}(\omega_n < \omega_0), & S_n \end{cases} \tag{6.73}$$

C. *Combined Feedback and Feedforward Voltage Control*

In order to verify the effectiveness of the proposed topology and control strategy, a combined feedback and feedforward voltage control (see Fig. 6.32) is applied. This control method can save costs (on current sensors) while still maintaining good transient response and steady-state performances. The feedforward control strategy uses a proportional controller, and the feedback control strategy uses a segmented PR controller. K_f is the feedforward coefficient. It is usually small and is used to improve system responsiveness.

Table 6.7 The LUT of (6.73)

ω_0	Resonance points	Segmented
$< \omega_1$	$\omega_1 + \omega/2$	S_1
$\omega_1 - \omega_2$		
$\omega_2 - \omega_3$	$\omega_2 + \omega/2$	S_2
...	...	...
$\omega_{n-1} - \omega_n$	$\omega_{n-1} + \omega/2$	S_{n-1}
$\omega_n - \omega_{n+1}$	$\omega_n + \omega/2$	S_n
$> \omega_{n+1}$		

6.4.3 Parameter Design

In this section, the parameter design strategy is discussed and explained in two parts. In part A, the parameter design of the filter is proposed. In part B, the design of the controller parameter is given.

A. *Design of Filter*

The proposed active impedance matching network is equivalent to a voltage source inverter, which works at a high switching frequency. A low-pass LC filter is designed to filter out the harmonics of the switching frequency and its nearby frequency bands. The cutoff frequency of the LC filter determines the control bandwidth of the impedance matching network and the attenuation against the switching harmonics from matching inverters.

For high-frequency PWM converters, the carrier frequency f_s is much greater than ten times the fundamental frequency f_o (200–500 Hz), the resonant frequency f_n is greater than ten times the f_o, and is 1/10–1/5 of the f_s. Therefore, the f_n can be expressed as:

$$10f_o < f_n\left(=\frac{1}{2\pi\sqrt{L_M L_C}}\right) < f_s/5 \tag{6.74}$$

The maximum ripple current Δi_{s_max} of the filter inductor L_M, which can be calculated from the output current value. Therefore, the filter inductor L_M can be expressed as:

$$L_M \geq \frac{V_{dc}}{8\Delta i_{s\_max} f_s} \tag{6.75}$$

Table 6.8 Parameters of power amplifier, load, SVGIMN and FCIMN in simulations and experiments

	Parameters	Physical values
Power amplifier	L_o	1 mH
	C_o	110 μF
	V_{Co}	220 V
	f	200 Hz–500 Hz
Load	L_{eq}	20 mH
	R_{eq}	24 Ω
SVGIMN	V_{DC}	800 V
	f_s	40 kHz
	L_M	2 mH
Proposed FCIMN	V_{DC}	800 V
	f_s	40 kHz
	L_M	1 mH
	C_M	2 μF

Based on (6.74) and (6.75), the filter parameters can be calculated. However, there are infinite combinations of L_M and C_M for a given filter resonance frequency. When the L_M increases, the transient current of the matched inverter may decrease. In addition, a large filter inductance increases the cost and the weight of the output filter, as well as the voltage stress of the inverter. Moreover, the larger the C_M, the lower the output voltage distortion rate. Therefore, this article chooses the f_s to be 40 kHz, Δi_{s_max} is 30% of the RMS value of the output-rated current, and the f_n is 3.5 kHz. The selected filter parameters are shown in Table 6.8.

B. *Controller Parameter Design*

The source impedances have little effect on the response of the system at the LC resonant point. Therefore, this article regards the power amplifier as an ideal voltage source and does not consider its impact on the system response. The single-loop voltage closed-loop control block diagram of the proposed FCIMN with an LC-filter is shown in Fig. 6.34 a, and its equivalent form is shown in Fig. 6.34b, where $G_{S\_PR}(s)$ is the segmented PR voltage controller. $G_D(s)$ is the system delay consisting of one sample delay and an equivalent half sample delay of zero-order hold (ZOH) delay. $G_{vc}(s)$ is the transfer function from converter side voltage $v_i(s)$ to output capacitor voltage $v_i(s)$. $G_{vi}(s)$ is the transfer function from converter side voltage $v_i(s)$ to output capacitor voltage $v_i(s)$. K_f is the feedforward coefficient. An embedded two-step Posicast controller is applied to dampen the LC resonance. The transfer function of the two-step Posicast controller is $G_{tsp}(s)$. Therefore, the open-loop transfer function $T_o(s)$ can be expressed as:

$$T_o(s)= (K_f + G_{S\_PR}(s))G_D(s)G_{tsp}(s)G_{vc}(s) \tag{6.76}$$

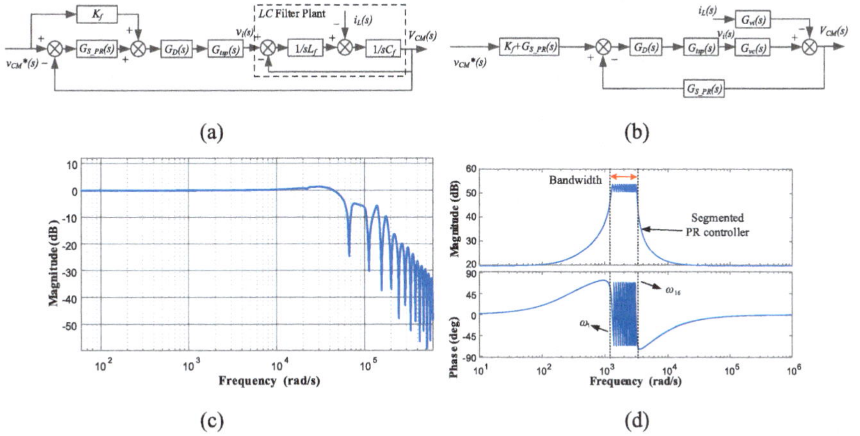

Fig. 6.34 **a** the combined feedback and feedforward control block diagram, **b** equivalent control diagram, **c** bode plots of the combined feedback and feedforward voltage control with Posicast controller, and **d** bode plot after designing the parameters of segmented PR controller

where

$$\begin{cases} G_{vc}(s) = \frac{v_{CM}(s)}{v_i(s)} = \frac{\omega_r^2}{s^2+\omega_r^2} & ,\omega_r = \sqrt{\frac{1}{L_M C_M}} \\ G_{vi}(s) = \frac{v_{CM}(s)}{i_L(s)} = \frac{sL_f\omega_r^2}{s^2+\omega_r^2} & ,\omega_r = \sqrt{\frac{1}{L_M C_M}} \\ G_{tsp}(s) = \frac{1}{1+\delta} + \frac{\delta}{1+\delta}e^{-sT_d} \\ G_D(s) = e^{-1.5T_s s} \end{cases} \tag{6.77}$$

where T_d is half of the resonance period ($T_r = 2\pi/\omega_r$) and δ is the step response overshoot of the lightly damped system. For digital implementation, the discrete expression of the transfer function of the two-step Posicast controller is:

$$G_{Ztsp}(z) = \frac{1}{1+\delta} + \frac{\delta}{1+\delta}z^{-\text{int}(T_d/T_s)} \tag{6.78}$$

The performance of the combined feedback and feedforward voltage control with the posicast controller is shown in Fig. 6.34c. Clearly, the control strategy can achieve good tracking of sinusoidal reference. It has good attenuation characteristics in the high-frequency range.

The working frequency range of GMT studied in this paper is 200–500 Hz, and the simulated equivalent load will be given in the next Section. We set $\omega = 40\pi$. That is, the bandwidth is 20 Hz. The 200–500 Hz frequency bandwidth is 300 Hz; that is, 15 segment PR controllers need to be configured. The transfer function of this segmented PR control in the s-domain can be expressed as:

$$G_{S\_PR}(s) = \begin{cases} K_{P_1} + K_{I_1}\frac{2\omega_c s}{s^2\omega_0^2+2\omega_c s}(\omega_0 \le \omega_2), & S_1 \\ K_{P_2} + K_{I_2}\frac{2\omega_c s}{s^2\omega_0^2+2\omega_c s}(\omega_2 < \omega_0 \le \omega_3), & S2 \\ \cdots & \cdots \\ K_{P_{14}} + K_{I_{14}}\frac{2\omega_c s}{s^2\omega_0^2+2\omega_c s}(\omega_{14} < \omega_0 \le \omega_{15}), & S_{14} \\ K_{P_{15}} + K_{I_{15}}\frac{2\omega_c s}{s^2\omega_0^2+2\omega_c s}(\omega_{15} < \omega_0), & S_{15} \end{cases} \tag{6.79}$$

Based on the above analysis, the working frequency band of the system is ω_1–ω_{16} (200–500 Hz). Among them, ω_1 is 400π (200 Hz), ω_2 is 440π (220 Hz), … and ω_{16} is 1000π (500 Hz). The resonance frequency points of S_1, S_2, … and S_{15} segmented PR controllers are 210 Hz, 230 Hz, … and 490 Hz, respectively. The bode plot after designing the parameters is shown in Fig. 6.34d. It shows that the proposed segmented PR controller can have a high amplitude margin and phase angle margin in the entire frequency band.

6.4.4 Simulation Results

In order to verify the feasibility of the proposed FCIMN, simulation case studies verification are carried out. The main parameters of simulation are shown in Table 6.8.

The dynamic performances of the FCIMN are illustrated under power amplifier output voltage frequency changes. The final purpose is to certify the power utilization factor (PUF) are improved with the FCIMN, which can be expressed as:

$$\text{PUF} = \frac{P_{PA}}{S_{PA}} \times 100\% = \frac{P_{PA}}{\sqrt{P_{PA}^2 + Q_{PA}^2}} \times 100\% \tag{6.80}$$

PUF mainly focuses on the power factor, and the goal is to reduce the reactive power of the system.

Firstly, we compared the parallel and series-connected impedance matching networks, namely the proposed adaptive broadband full-bridge converter impedance matching network (FCIMN) and SVG impedance matching network (SVGIMN). The matched results during the power amplifier output voltage frequency-changing are shown in Table 6.9. We can conclude that both FCIMN and SVGIMN can achieve impedance matching within the working frequency band. FCIMN can make the power amplifier output a larger current than SVGIMN when the output voltage and frequency of the power amplifier are the same. Therefore, GMT can obtain larger power, and the series matching method can improve the utilization rate of the power amplifier.

The FCIMN matched results during the power amplifier output voltage frequency change are shown in Fig. 6.35, Fig. 6.36, and Table 6.9. In steady states, no matter from 200 to 220 Hz, from 220 to 240 Hz, from 240 to 260 Hz, from 260 to 280 Hz, from 200 to 300 Hz, from 300 to 400 Hz, from 400 to 500 Hz, and from 200 to

Table 6.9 Simulation results of FCIMN and SVGIMN impedance matching in different situations

Match	f_O (Hz)	V_{co} (V)	I_L (A)	Q_L (Var)	P_L (W)	Q_{PA} (Var)	P_{PA} (W)	Q_{IM} (Var)	P_{IM} (W)	PUF (%)
Before	200	220	6.3	1007	962	1007	962	–	–	69.08
	300	220	4.9	913	582	913	582	–	–	53.75
	400	220	3.9	784	374	784	374	–	–	43.06
	500	220	3.3	672	257	672	257	–	–	35.72
SVGIMN	200	220	4.4	1007	961	0.4	972	1006	10	99.99
	300	220	2.7	913	581	0.4	592	913	10	99.99
	400	220	1.7	784	374	0.1	385	784	11	99.99
	500	220	1.2	672	257	0.6	267	672	10	99.99
Proposed FCIMN	200	220	9.1	2068	1975	28	1995	2096	19	99.99
	300	220	9.2	3156	2010	23	2012	3179	2	99.99
	400	220	9.2	4248	2030	15	2022	4263	8	99.99
	500	220	9.2	5328	2036	8	2025	5335	11	99.99

Note The unsatisfying characteristics are shaded

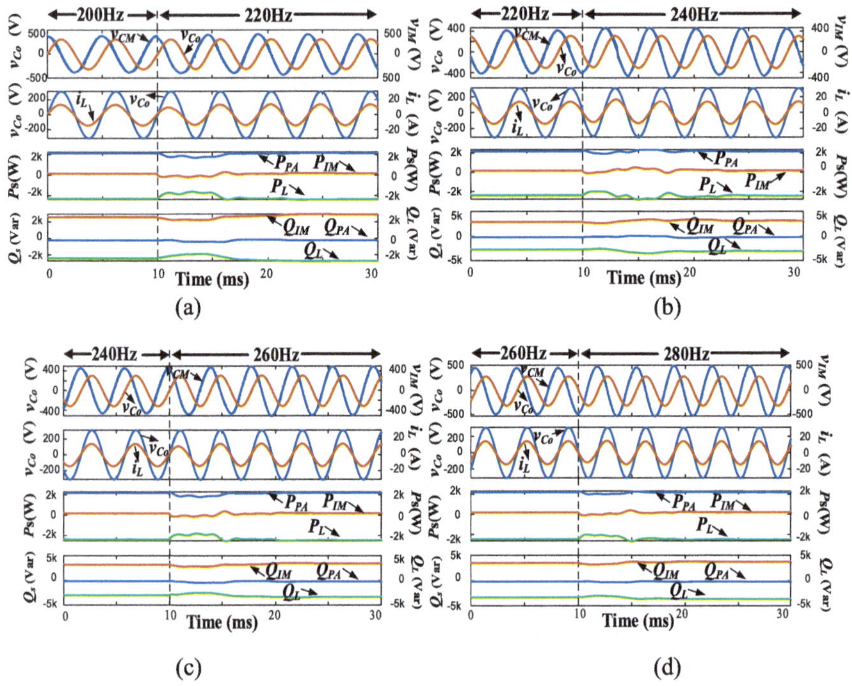

Fig. 6.35 Dynamic waveforms of v_{CM}, i_L, v_{Co}, P_{PA}, P_{IM}, Q_L, and Q_{PA} in simulation with FCIMN impedance matched when power amplifier output voltage frequency changes: **a** from 200 to 220 Hz; **b** from 220 to 240 Hz, **c** from 240 to 260 Hz, and **d** from 260 to 280 Hz

500 Hz in Fig. 6.35 and Fig. 6.36 with the FCIMN impedance matching, i_L and v_{Co} are matched to be in phase. FCIMN only takes less than 1.5 fundamental periods to match the impedance with the matching network voltage v_{CM} changing for the dynamic transient processes. The Q_{PA} is reduced to nearly 0, the P_{PA} is nearly equal to P_L, and the Q_{IM} is nearly equal to Q_L.

A group of comparative simulations, including a fixed capacitor, switch-controlled capacitors, TCIMN, and proposed FCIMN, are summarized in Table 6.10. The fixed capacitor is designed for 200 Hz. The fixed capacitor cannot match the impedance at both 200 Hz and 500 Hz. The TCIMN can achieve matching within the working frequency band. However, this method only adjusts once in a cycle, so the transient time is relatively long. However, the proposed FCIMN can quickly and accurately match the impedance of 200 Hz and 500 Hz.

The above simulation results verify that the adaptive and broadband impedance matches the characteristics of the FCIMN. After the FCIMN matches the load impedance, the reactive power Q_{PA} in the power amplifier system is significantly reduced, and the PUF can be improved to 99%.

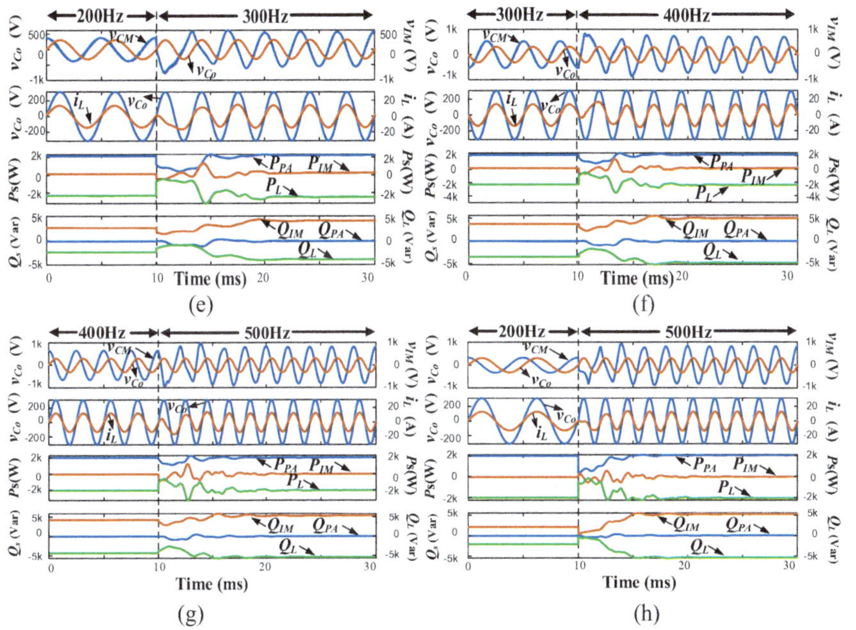

Fig. 6.36 Dynamic waveforms of v_{CM}, i_L, v_{Co}, P_{PA}, P_{IM}, Q_L, and Q_{PA} in simulation with FCIMN impedance matched when power amplifier output voltage frequency changes: **a** from 200 to 300 Hz; **b** from 300 to 400 Hz, **c** from 400 to 500 Hz, and **d** from 200 to 500 Hz

Table 6.10 Impedance matching performance comparison

Match		f (Hz)	V_{co} (V)	I_L (A)	Q_L (Var)	P_L (W)	Q_{PA} (Var)	P_{PA} (W)	PUF (%)
Before		200	220	6.3	1007	962	1007	962	69.08
		500	220	3.3	672	257	672	257	35.72
After	Fixed Cap	200	220	9.1	2103	2008	117	2008	99.83
	Switch-controlled capacitors		220	9.1	2102	2007	117	2008	99.83
	TCIMN		220	9.1	2108	2013	45	2013	99.97
	Proposed FCIMN		220	9.1	2068	1975	28	1995	99.99
	Fixed Cap	500	220	3.8	916	349	762	351	41.83
	Switch-controlled capacitors		220	8.7	4758	1817	598	1818	94.99
	TCIMN		220	9.1	5267	2012	70	2012	99.94
	Proposed FCIMN		220	9.2	5328	2036	8	2025	99.99

Note The unsatisfying characteristics are shaded

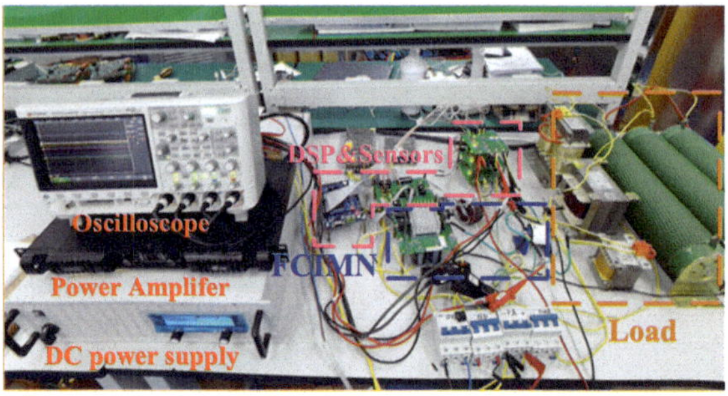

Fig. 6.37 Experimental setup of the proposed FCIMN

6.4.5 Experimental Results

In order to verify the feasibility of the proposed FCIMN, experimental case study verification is carried out. The main parameters of the experimental prototype are shown in Table 6.8. The scale-down experimental step has been built, as shown in Fig. 6.37.

The control unit of FCIMN used is a digital signal processor (DSP) TMS320-F28335, and the sampling frequency of the control system is 40 kHz. For the FCIMN part, the full-bridge power components use CREE C2M0040120, and the capacitors use thin-film capacitors. For the source part, the DC power supply is EA-PS 9750–20 of Elektro-Automatik, and the power amplifier is HD-2500 of AOSHENG. The proposed FCIMN is dedicated to GMT, which can be considered as pure linear loads under the required operating frequency range. The DSP can process the signals and send driver signals to the gate driver to control the SiC. Among these experimental results are recorded by the Keysight DSO-X 3024 T oscilloscope.

Figure 6.38 illustrates the dynamic matching waveforms of power amplifier output voltage v_{co}, matching voltage v_{CM}, and load current i_L by applying FCIMN under four representative dynamic situations. This shows that the closed-loop controller designed in this article has a good dynamic performance. In the stable-states, v_{co} and i_L are in phase, which illustrates that load impedances are matched in these conditions (in 200 Hz, 300 Hz, 400 Hz, and 500 Hz). The power amplifier reactive power Q_{PA} is matched to nearly 0. Therefore, the power amplifier efficiency has been significantly improved (up to PUF = 99%) with the proposed FCIMN.

6.4.6 Section Summary

In this Section, a new FCIMN was proposed between a power amplifier and the GMT for broadband impedance matching to improve PUF. A control method was proposed to achieve adaptive broadband impedance matching, including dynamic active power calculation, reactive power calculation, segmented PR controller, and combined feedback and feedforward voltage control. In addition, The FCIMN parameters design

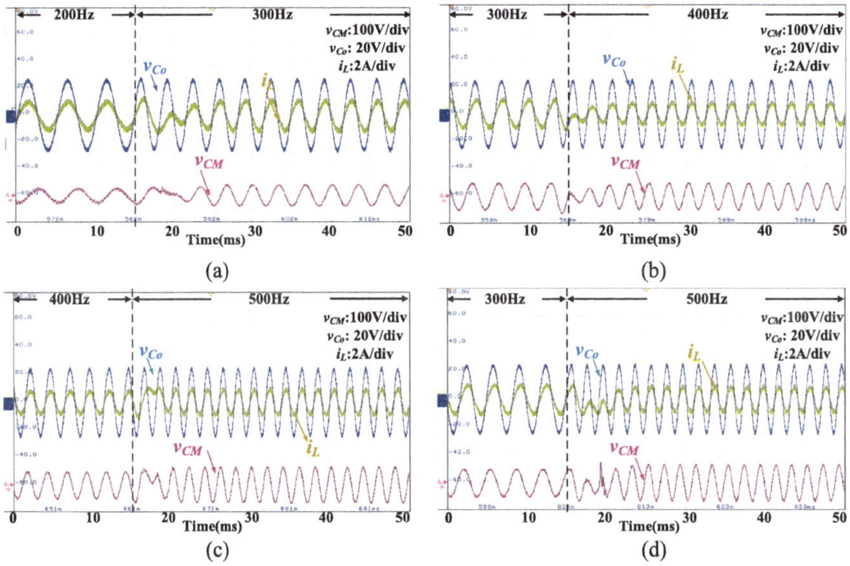

Fig. 6.38 Experimental dynamic v_{CM}, i_L, v_{Co}, i_M waveforms with impedance matching when power amplifier output frequency changes **a** from 200 to 300 Hz, **b** 300 Hz to 400 Hz, **c** from 400 to 500 Hz, and **d** from 300 to 500 Hz

method was given to ensure that FCIMN can match load impedance within the required operating frequency range of GMT. Finally, representative simulation and experimental results were provided to show that the proposed FCIMN can match load impedance in a broadband range with excellent dynamic and stable performance and increase the PUF of the power amplifier to 99%.

6.5 A Hybrid Adaptive Broadband Active Impedance Matching Network

6.5.1 Circuit Configuration of the Proposed HIMN

This section presents the circuit configuration of the power amplifier, HIMN, and load network. The equivalent model of the UAT and the operating principle of the HIMN are provided. In addition, matching models and matching functions are introduced.

A. *Circuit Configuration of Power Amplifier and Proposed HIMN*

As shown in Fig. 6.39, the power amplifier is a controllable voltage source that can vary the output voltage amplitude and frequency. It consists of n cascaded MOSFETs, an H bridge, and an output filter circuit. v_{Co} is the output voltage, and V_{Co} is the amplitude of v_{Co}. The proposed HIMN consists of a passive impedance matching network

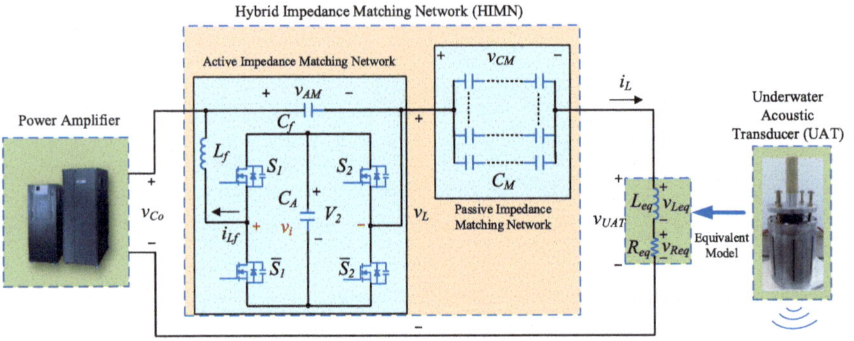

Fig. 6.39 Main circuit structure of the active sonar acoustic wave emitting part

and an active impedance matching network. The passive impedance matching network consists of series and parallel capacitors. The equivalent capacitance value and voltage of the capacitor bank are C_M and v_{CM}, respectively. The active impedance matching network is a full-bridge converter, which consists of a MOSFETs H-bridge and an LC filter circuit. v_i, v_{AM}, V_2, v_{UAT}, v_{Leq}, and v_{Req} are the voltages of H-bridge, C_f, C_A, UAT, L_{eq}, and R_{eq}, respectively. v_{HIMN} is the sum of v_{AM} and v_{CM}. The magnitudes of v_{AM}, $v_{CM,}$ v_{UAT}, v_{HIMN}, v_{Leq}, and v_{Req} are V_{AM}, V_{CM}, V_{UAT}, V_{HIMN}, V_{Leq}, and V_{Req}, respectively.

Figure 6.40 shows the simplified equivalent circuit model of Fig. 6.39. The power amplifier is represented by a voltage source, and the output active and reactive power are represented by P_{PA} and Q_{PA}, respectively. The capacitive reactance of the matching capacitor of the proposed HIMN is expressed as X_{CM} $(-1/\omega C_M)$. The reactive powers of active impedance matching and passive impedance matching are represented by Q_{AM} and Q_{CM}, respectively. The reactive power and active power of HIMN are represented by Q_{HIMN} and P_{HIMN}, respectively. The equivalent inductive reactance, reactive power and active power of UAT are represented by X_{eq} (ωL_{eq}), $Q_{UAT,}$ and P_{UAT}, respectively.

B. *UAT Equivalent Model*

Low-frequency and high power UAT has been widely used in marine engineering. Giant magnetostrictive transducers have attracted the research of many scholars because of their advantages of large strain output and high energy density. The

Fig. 6.40 Simplified main circuit model

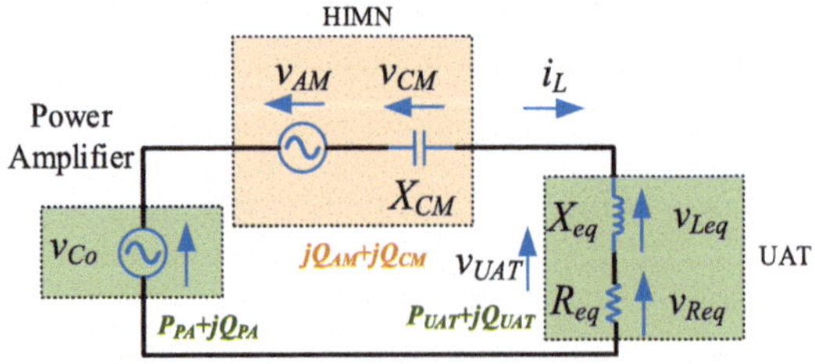

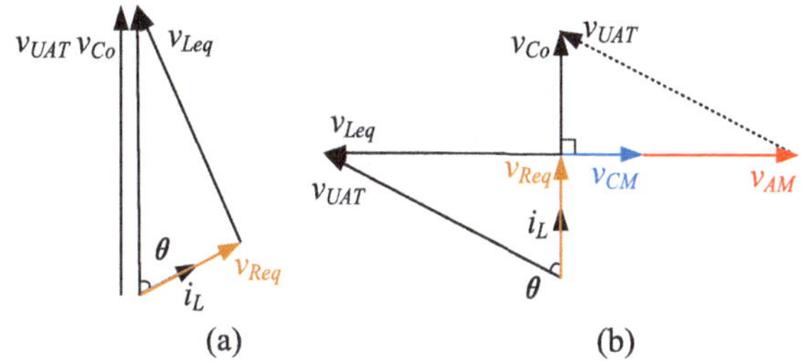

Fig. 6.41 Phasor diagrams of before and after impedance matching: **a** before matching, **b** after impedance matching

acoustic radiation is obtained by adding an AC source at both ends of the coil to make the giant magnetostrictive rod drive the front mass to vibrate.

The electromechanical coupling relationship of giant magnetostrictive UAT is shown in Fig. 6.41. U is the input voltage, i is the input current, and Z_e is the electrical impedance. T_{em} and T_{me} represent the conversion coefficients of the electro-mechanical coupling between the electrical and mechanical sides, respectively. Z_m is the mechanical impedance, F is the mechanical end output, and v is the speed of mechanical vibration. The electromechanical coupling relationship can be represented by (6.81) and (6.82).

$$U = Z_e i + T_{em} v \tag{6.81}$$

$$F = Z_m v + T_{me} i \tag{6.82}$$

After replacing the equivalent mechanical circuit in Fig. 6.42 with the electrical equivalent circuit, the electrical equivalent circuit of the UAT is shown in Fig. 6.43. Z_{m2} is the motional impedance. The total impedance at the UAT electrical terminal can be expressed as:

$$\begin{aligned} Z &= Z_e + Z_{m2} \\ &= \underbrace{\left(R_e + \frac{R_m \omega^2 L_m^2}{\left(R_m - R_m \omega^2 L_m C_m\right)^2 + \omega^2 L_m^2}\right)}_{R_{eq}} \\ &+ j \underbrace{\left(\omega L_e + \frac{R_m \omega L_m \left(R_m - R_m \omega^2 L_m C_m\right)}{\left(R_m - R_m \omega^2 L_m C_m\right)^2 + \omega^2 L_m^2}\right)}_{X_{eq}} \end{aligned} \tag{6.83}$$

where $R_m = \frac{A^2}{r}$, $L_m = \frac{A^2}{k}$, $C_m = \frac{m}{A^2}$, m is the equivalent mass, k is the stiffness, r is the equivalent mechanical resistance, and A is the mechanical factor.

The first item and the second item of (6.83) represent R_{eq} and $X_{eq,}$ respectively. To realize the impedance matching of UAT, this paper matches a certain type of

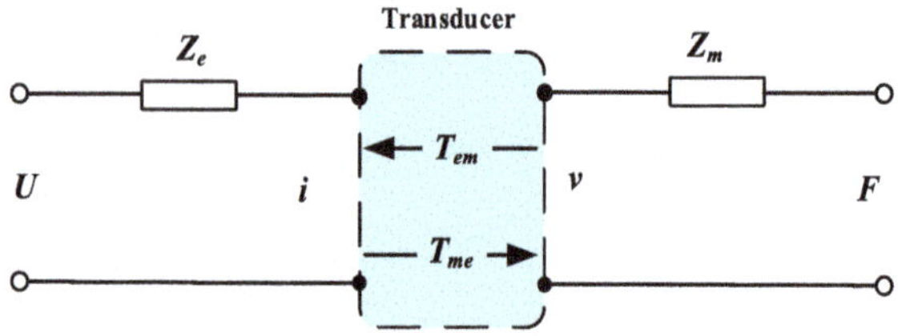

Fig. 6.42 Transduction mesh with electrical and mechanical sides

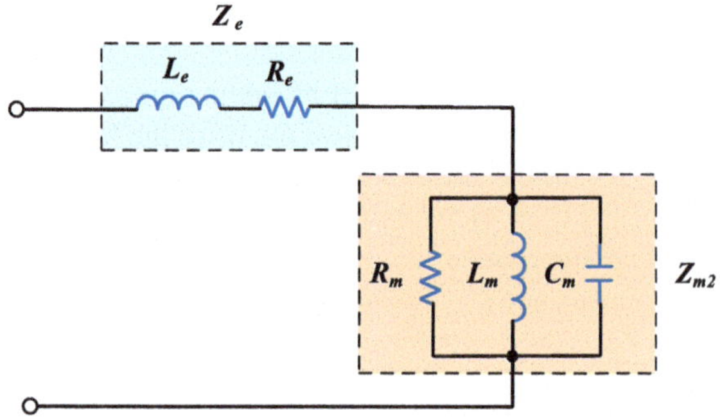

Fig. 6.43 Electrical equivalent circuit of UAT

real UAT, and its measured impedance characteristics are shown in Fig. 6.44. In this measurement band, the UAT impedance tends to increase with frequency. The resonant frequency of the UAT is actually defined by the mechanical resonant frequency, which refers to the resonant frequency when the imaginary part of the UAT mechanical end impedance is zero. In the actual measured UAT impedance characteristic diagram, the resonance point is when R_{eq} has a maximum value. The resonance point in Fig. 6.44 is 370 Hz. Because the UAT has a higher output gain than the adjacent frequency at this resonance point, the UAT generally works near the resonance point. According to (6.83) and the measured data, the magnitude of electrical impedance and mechanical impedance can be determined.

C. *Modeling of Active Matching Network*

In this part, the active matching network applies a single-phase full-bridge inverter with an LC filter. The output voltage v_i of the converter is determined by the drive signals S_1 and S_2 as follows.

$$v_i = \begin{cases} V_2, \text{ if } S_1 \text{ on and } S_2 \text{ off }, S(1) \\ 0, \text{ if } S_1 \text{ on and } S_2 \text{ on }, S(2) \\ 0, \text{ if } S_1 \text{ off and } S_2 \text{ off}, S(3) \\ -V_2, \text{ if } S_1 \text{ off and } S_2 \text{ on}, S(4) \end{cases} \tag{6.84}$$

where $S(x)$ is used to represent four switch combinations, and x is the independent variable.

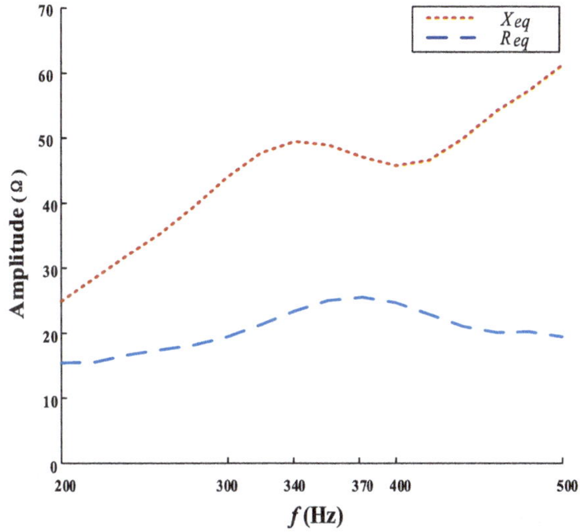

Fig. 6.44 Impedance characteristics of a UAT actually measured

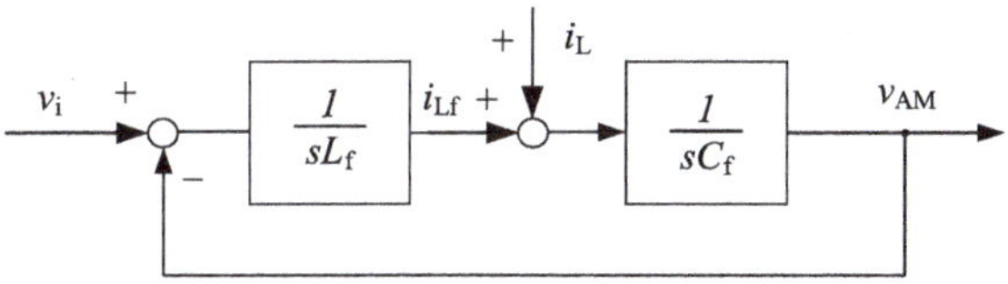

Fig. 6.45 S-domain LC filter model

The model of the LC filter in the *s*-domain is shown in Fig. 6.45. Based on Fig. 6.45, the continuous-time state-space equation can be expressed as

$$\frac{d}{dt}\begin{bmatrix} i_{Lf} \\ v_{AM} \end{bmatrix} = \begin{bmatrix} 0 & -\frac{1}{L_f} \\ \frac{1}{C_f} & 0 \end{bmatrix}\begin{bmatrix} i_{Lf} \\ v_{AM} \end{bmatrix} + \begin{bmatrix} \frac{1}{L_f} & 0 \\ 0 & \frac{1}{C_f} \end{bmatrix}\begin{bmatrix} v_i \\ i_L \end{bmatrix} \tag{6.85}$$

D. *Operation Principles of HIMN*

Based on the analysis in the previous section, the UAT load is a resistive and inductive load. Therefore, the inductive load needs to be matched. The HIMN proposed in this paper is matched by a combination of passive and active. Passive matching uses fixed capacitors for impedance matching, where the matching range is not adjustable. Active matching uses active converters for matching, where the matching range can be adjusted. According to the KVL theorem, the voltage relation of the system can be expressed as

$$-v_{Co} + v_{AM} + v_{CM} + v_{Leq} + v_{Req} = 0 \tag{6.86}$$

The reactive power of the UAT is all provided by the impedance matching network. Therefore, the relationship among v_{Leq}, v_{AM}, and v_{CM} can be deduced as

$$v_{Leq}+v_{AM}+v_{CM}=0 \tag{6.87}$$

Combining (6.86) and (6.87), it can be deduced as

$$v_{Co}=v_{Req} \tag{6.88}$$

Based on (6.88), the output voltage of the power amplifier is directly supplied to the UAT equivalent resistance. The matching voltages v_{CM} and v_{AM} in (6.87) are generated by matching capacitors and active converters, respectively. Therefore, the output reactive power Q_{PA} of the power amplifier is 0, and the active power can be expressed as

$$P_{PA}=V_{Co}I_L/\sqrt{2} \tag{6.89}$$

The active and reactive power of the UAT can be expressed as

$$Q_{UAT}=I_L^2X_{eq} \tag{6.90}$$

$$P_{UAT}=I_L^2R_{eq} \tag{6.91}$$

The active and reactive power of the HIMN can be expressed as

$$Q_{HIMN}=Q_{CM}+Q_{AM} \tag{6.92}$$

where $Q_{CM}=I_L^2X_{CM}$, $Q_{AM}=V_{AM}I_L/\sqrt{2}$.

Combining (6.88), (6.89), and (6.91), P_{PA} can be deduced as

$$P_{PA}=P_{UAT} \tag{6.93}$$

Combining (6.87), (6.90) and (6.92), Q_{HIMN} can be deduced as

$$Q_{HIMN}=-Q_{UAT} \tag{6.94}$$

The phasor diagrams before and after impedance matching are shown in Fig. 6.41.

E. *Impedance Functions*

This Section uses capacitors for initial matching. After matching, the system still works in the inductive region. Because the capacitive reactance decreases with the increase of the frequency, the UAT inductance reactance generally increases with the increase of the frequency, and the matching capacitor matches the UAT inductance at the lowest frequency. The relationship between the capacitive reactance of the matching capacitor and the inductive reactance of the load can be expressed as

$$(X_{CM})_{Max}=\left(X_{eq}\right)_{Min} \tag{6.95}$$

The secondary matching is achieved by the active converter to match the remaining inductive reactance. The remaining inductive reactance voltage and the UAT equivalent resistance voltage are provided by the active converter and the power amplifier, respectively. It is exactly matched when the ratio of the remaining inductance to the UAT equivalent resistance is equal to the ratio of the active converter output voltage to the power amplifier output voltage. In order to achieve full-band matching, (6.96) is to be fulfilled.

$$\left(\frac{X_{eq} - X_{CM}}{R_{eq}}\right)_{Max} \leq \frac{(V_{AM})_{Max}}{(V_{Co})_{Max}} \tag{6.96}$$

6.5.2 Parametric Design Methods

This section mainly includes three parts. Part A analyzes the matching range of HIMN. Part B introduces how to choose a C_M. Part C introduces the parameter design of the active matching part.

A. *Matching Range Analysis*

The matching range of the HIMN is the sum of the passive and the active matching range. The reactive voltage that needs to be matched is V_{Leq} when I_L is constant. This matching voltage is the sum of V_{CM} and V_{AM}. V_{Leq} and impedance matching range are shown in Fig. 6.46. The red dotted line is the V_{Leq}. The minimum frequency minimum matching voltage and the maximum frequency maximum matching voltage of HIMN match the minimum value $(V_{Leq})_{min}$ and the maximum value $(V_{Leq})_{max}$ of V_{Leq}, respectively. The green dotted line represents the voltage value of passive matching. As the frequency increases, the voltage of passive matching becomes smaller and smaller, where the matching value cannot be adjusted. The area filled by the blue vertical line is the matching voltage value that the active matching can provide, where the matching value can be adjusted. The combination of passive matching and active matching constitutes the range of HIMN matching, as shown in the area filled by an orange slash. In this figure, the main circuit current I_L is kept constant, and the reactive capacity of the active matching part of the FCIMN and HIMN is of the same size.

B. *Selection of* C_M

The matching capacitor C_M serves to provide capacitive reactance to match part of the UAT inductive reactance. In other words, matching capacitors provide reactive power Q_{CM}. It is used to match part of the reactive power of the UAT. Their relationship can be expressed as

$$X_{CM} = \frac{\left(V_{CM}/\sqrt{2}\right)^2}{Q_{CM}} = -KX_{eq} = -K\frac{\left(V_{Leq}/\sqrt{2}\right)^2}{Q_{UAT}} \tag{6.97}$$

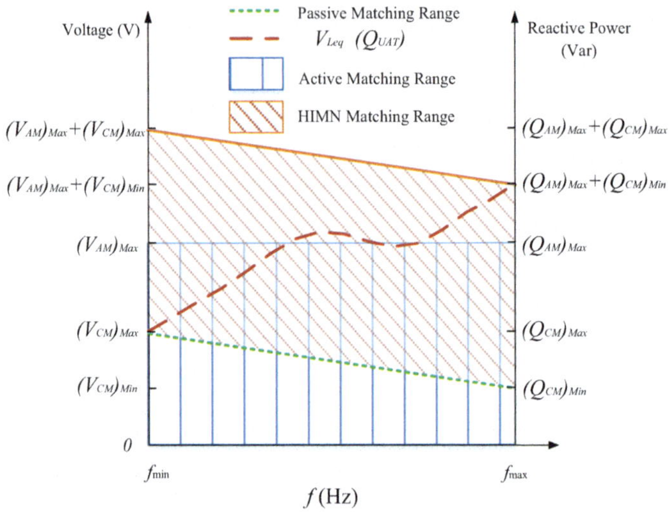

Fig. 6.46 V_{Leq} and impedance matching range

where K is the capacitance matching coefficient.

UAT inductive reactance is the smallest at the minimum frequency f_{min}, so take K as one here. Then the expression of C_M can be deduced as

$$C_M = \frac{1}{\omega(X_{eq})_{Min}} \tag{6.98}$$

C. *Active Converter Parameter Design*

Based on Fig. 6.46, it can be seen that the matching voltage of passive matching is the smallest, and the active matching voltage is the largest at the maximum frequency of the system. Therefore, the maximum active matching voltage $(V_{AM})_{max}$ is required at the maximum frequency of the system. It can be expressed as

$$(V_{AM})_{Max} = (V_{eq})_{Max} - (V_{CM})_{Min} \tag{6.99}$$

6.5.3 Control Method

The overall control block diagram is shown in Fig. 6.47. The proposed control method based on reference signal synchronization mainly includes two parts. Part A proposes a reference voltage calculation method based on the ratio of X_{eq} to R_{eq} of the load to achieve a fast calculation of the reference value when the frequency jumps. Part B

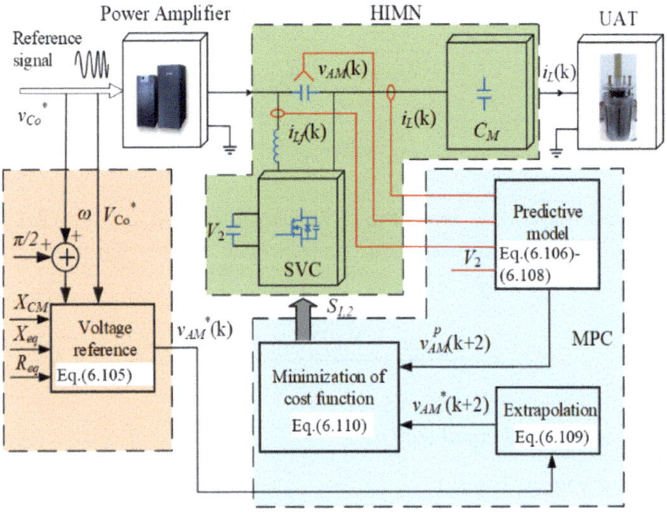

Fig. 6.47 Overall control strategy of proposed HIMN

proposes an MPC voltage control strategy to achieve fast-tracking of the reference voltage.

A. *Reference Voltage Calculation Method*

This traditional reference compensation voltage calculation method needs to calculate the load reactive power. However, when the frequency jumps, because the reactive power needs to be processed by a low-pass filter, the reactive power calculation process has a transient process of several cycles. To reduce the transient process of reference compensation voltage calculation, this section proposes a calculation method based on the ratio of X_{eq} to R_{eq} of the load. This power amplifier reference voltage is expressed as

$$v_{Co}^* = V_{Co}^* \cos \omega t \tag{6.100}$$

where $v_{Co}{}^*$ is the reference voltage of the power amplifier, V_{Co} is the amplitude, ω is the angular frequency, and t is the time. After impedance matching, the power amplifier only provides active power to the UAT, and the HIMN only provides reactive power to the UAT. Therefore, the load and the power supply have the following relationship

$$\frac{V_{AM}}{V_{Co}} = \frac{X_{CM} + X_{eq}}{R_{eq}} \tag{6.101}$$

$$\frac{V_{AM}}{V_{HIMN}} = \frac{X_{CM} + X_{eq}}{X_{eq}} \tag{6.102}$$

$$\frac{V_{CM}}{V_{HIMN}} = \left| \frac{X_{CM}}{X_{eq}} \right| \tag{6.103}$$

$$\frac{V_{CM}}{V_{Co}} = \left| \frac{X_{CM}}{R_{eq}} \right| \tag{6.104}$$

The compensation voltage leads the power amplifier voltage by π/2. Combining (6.100) and (6.101), the expression of the reference compensation voltage can be derived as:

$$v_{AM}^{*} = -\frac{\left(X_{CM} + X_{eq}\right) V_{Co}^{*} \cos(\omega + \pi/2)t}{R_{eq}} \tag{6.105}$$

where V_{Co}* is considered equal to V_{Co}.

B. *MPC-Based Voltage Control*

An MPC voltage control strategy is proposed to achieve broadband voltage tracking. In order to obtain the discrete-time model of the filter output voltage, the forward Euler method is used to discretize (6.82), and the predictive mode is formulated as:

$$\begin{aligned} \begin{bmatrix} i_{Lf}^{p}(k+1) \\ v_{AM}^{p}(k+1) \end{bmatrix} &= \begin{bmatrix} 0 & -\frac{T_s}{L_f} \\ \frac{T_s}{C_f} & 0 \end{bmatrix} \begin{bmatrix} i_{Lf}(k) \\ v_{AM}(k) \end{bmatrix} \\ &+ \begin{bmatrix} \frac{T_s}{L_f} & 0 \\ 0 & \frac{T_s}{C_f} \end{bmatrix} \begin{bmatrix} v_i(k) \\ i_L(k) \end{bmatrix} + \begin{bmatrix} i_{Lf}(k) \\ v_{AM}(k) \end{bmatrix} \end{aligned} \tag{6.106}$$

where T_s is the sampling period, i_{Lf}(k), v_{AM}(k), $i_{Lf}^{p}(k+1)$ and $v_{AM}^{p}(k+1)$ represent the filter inductor current in k, the filter output voltage in k, the one-step prediction of the filter inductor current, and the one-step prediction of filter output voltage, respectively.

In the implementation of digital control, the algorithm execution takes some time. To compensate for the delay time, a two-step predicted MPC is used. This method uses (6.106) to predict the k + 2 instant value of the filter output voltage.

$$\begin{aligned} v_{AM}^{p}(k+2) =& \frac{T_s^2}{L_f C_f} v_i(k) + \left(1 - \frac{T_s^2}{L_f C_f}\right) v_{AM}(k) + \frac{2T_s}{C_f} i_{Lf}(k) \\ & - \frac{T_s}{C_f} i_L(k) + \frac{T_s}{C_f} i_L^{*}(k+1) \end{aligned} \tag{6.107}$$

where $i_L^{*}(k+1)$ is the future filter output current value, which can be derived using the Lagrange extrapolation formula as:

$$i_L^{*}(k+1) = 4i_L(k) - 6i_L(k-1) + 4i_L(k-2) - i_L(k-3) \tag{6.108}$$

The two-step prediction reference filter output voltage value can be estimated by the fourth-order Lagrange extrapolation method as:

$$\begin{aligned} v_{AM}^{*}(k+2) = {} & 10v_{AM}^{*}(k) - 20v_{AM}^{*}(k-1) + 15v_{AM}^{*}(k-2) \\ & -4v_{AM}^{*}(k-3) \end{aligned} \tag{6.109}$$

In order to achieve the tracking of the reference voltage, the voltage cost function is established as follows.

$$g_i = \left| v_{AM}^{*}(k+2) - v_{AM}(k+2) \right| \tag{6.110}$$

Find the optimal switch combination to minimize g_i. Figure 6.48 shows the flowchart of the MPC algorithm. For a single-phase H-bridge, there are four switch combinations. Therefore, the cost function g_i needs to be calculated four times in one control cycle, and then the switch combination with the smallest cost function g_i is selected to drive the H-bridge. After the next sampling time arrives, the previous steps are repeated.

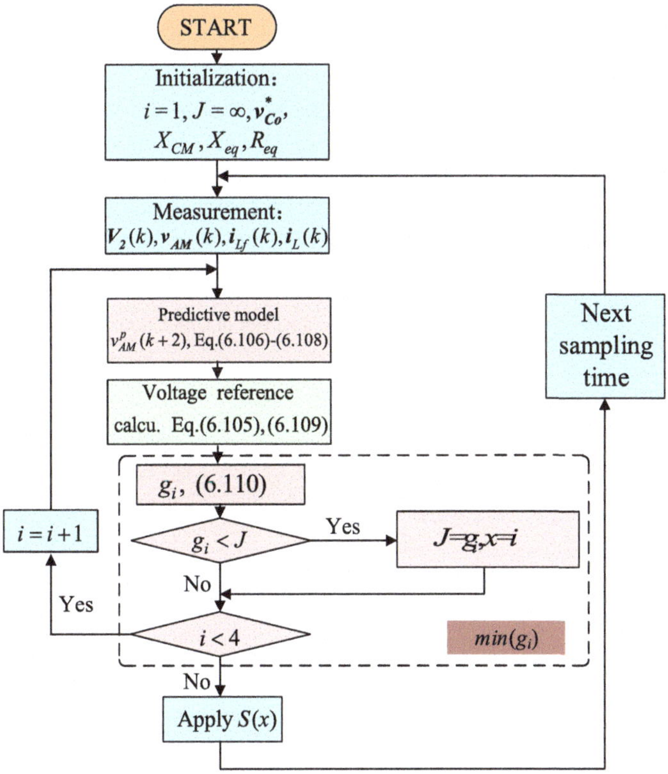

Fig. 6.48 Flowchart of the MPC algorithm

C. *Comparison with Prior Control Scheme*

The proposed MPC control strategy is compared with FCIMN's control strategy. The equation for calculating the reference voltage $v_{AM}{}^{*}$ for FCIMN control is:

$$v_{AM}^{*} = \frac{Q_{UAT} v_{Co}^{D}}{P_{UAT}} \tag{6.111}$$

where $v_{Co}{}^{D}$ is obtained by the power amplifier output voltage v_{Co} with a delay of 3/4 period, and there is also a delay of 1/4 period in the calculation of instantaneous reactive power. In addition, the segmented PR control proposed in FCIMN relies on the system period calculated by the zero-crossing comparison method, which has a delay of 1/2 period.

The reference voltage calculation method of the control strategy proposed in this section has no delay, as shown in (6.105). The MPC control method directly controls the converter switching devices according to the optimization of the cost function g_i, and it does not need to rely on the system period calculated in real-time.

6.5.4 Simulation Results

To verify the effectiveness of the proposed HIMN and control strategy, the simulation is designed. The main parameters of the simulation are shown in Table 6.11.

Simulation verification is based on the actual measured UAT impedance values, as Fig. 6.44. According to the UAT impedance characteristic curve, its inductive reactance is rising in the whole working frequency band, which is the smallest at

Table 6.11 Parameters of power amplifier, load, and HIMN in simulations AND experiments

	Parameters	Physical values
Power amplifier	V_{Co}	311 V
	f	200 Hz–500 Hz
UAT Load characteristics	X_{eq}, R_{eq} (200 Hz)	25 Ω, 15 Ω
	X_{eq}, R_{eq} (340 Hz)	50 Ω, 23.5 Ω
	X_{eq}, R_{eq} (400 Hz)	46 Ω, 25 Ω
	X_{eq}, R_{eq} (500 Hz)	63 Ω, 20.5 Ω
Proposed HIMN	V_2	800 V
	T_s	20 μs
	L_f	5 mH
	C_f	20 μF
	C_M	32 μF
	f_s	4.9 kHz

the minimum frequency and the largest at the maximum frequency. It reaches the extreme points at 340 Hz and 400 Hz. Therefore, 200 Hz, 340 Hz, 400 Hz, and 500 Hz points are selected for simulation and experiment for verification. The UAT equivalent resistances are respectively 15 Ω, 23.5 Ω, 25 Ω and 20.5 Ω at 200 Hz, 340 Hz, 400 Hz and 500 Hz, where the UAT equivalent inductive reactance are 25.12 Ω, 50 Ω, 46 Ω and 62.8 Ω, respectively. Plug the UAT load characteristics of Table 6.11 into the (6.98), and the passive matching capacitance is calculated as 32 μF. Further, based on (6.101)–(6.104), the percentage of fixed capacitance and active converter of HIMN can be calculated. The calculation results are shown in Table 6.12.

In order to verify that the HIMN proposed in this section has a wider matching range and dynamic performance than the original method, three sets of comparative simulations are conducted. The first group is HIMN, which uses the MPC control strategy proposed in this paper. The second group is the active matching part of the HIMN (the passive part has been removed) using the MPC control strategy proposed in this paper. The active matching is the same as the FCIMN topology. The third group is HIMN, which uses the segmented PR control.

The first set of simulation results are shown in Fig. 6.49, Table 6.13, and Table 6.14. In steady-state, HIMN can achieve impedance matching from 200 to 340 Hz, from 200 to 400 Hz, from 340 to 400 Hz, and from 400 to 500 Hz. v_{Co} and i_L are matched to be in phase. v_{AM} is close to 0 at 200 Hz, so only the passive matching part performs impedance matching at this time. For the impedance matching transient process, the maximum transient time of HIMN is one fundamental cycle when the frequency is from 200 to 400 Hz. Q_{PA} and P_{HIMN} are close to 0. The value of Q_{HIMN} is approximately equal to Q_{UAT}. The value of P_{PA} is approximately equal to P_{UAT}.

The second set of simulation results are shown in Fig. 6.50 and Table 6.14. In steady-state, it cannot achieve impedance matching from 200 to 340 Hz and from 400 to 500 Hz.

The third set of simulation results are shown in Fig. 6.51. In steady-state, it can achieve impedance matching from 200 to 340 Hz and from 400 to 500 Hz. v_{Co} and i_L are matched to be in phase. For the impedance matching transient process, the

Table 6.12 The ratio of active matching and passive matching to HIMN

	200 Hz	340 Hz	400 Hz	500 Hz
X_{eq}	25 Ω	50 Ω	46 Ω	63 Ω
R_{eq}	15 Ω	23.5 Ω	25 Ω	20.5 Ω
X_{CM}	−24.88 Ω	−14.63 Ω	−12.44 Ω	−9.95 Ω
V_{AM}/V_{Co}	0	1.51	1.34	2.58
V_{AM}/V_{HIMN}	0	0.71	0.73	0.84
V_{CM}/V_{HIMN}	1	0.29	0.27	0.16
V_{CM}/V_{Co}	1.66	0.62	0.50	0.48

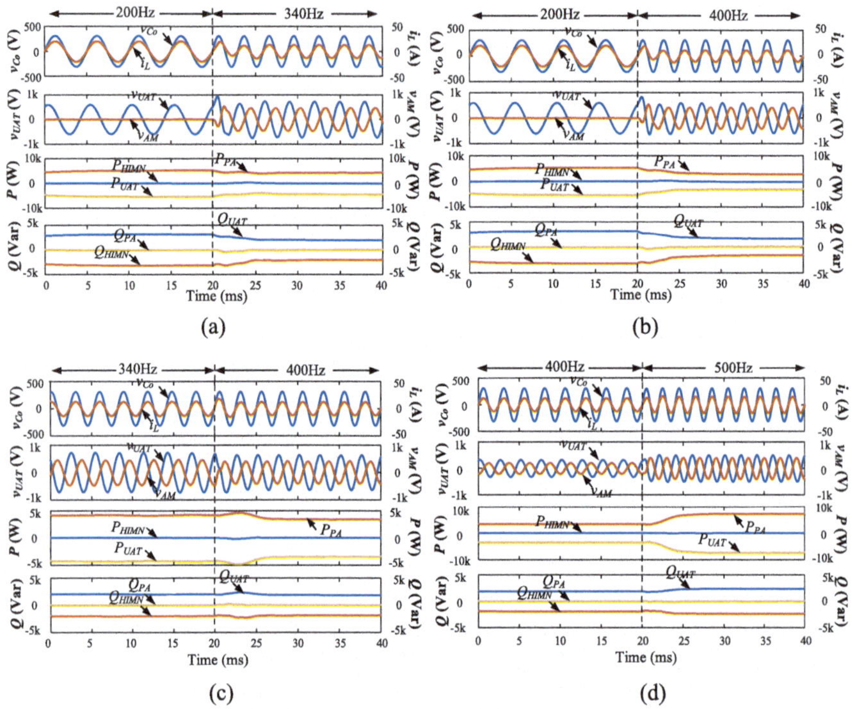

Fig. 6.49 Dynamic waveforms of v_{Co}, i_L, v_{UAT}, v_{AM}, P_{PA}, P_{HIMN}, P_{UAT}, Q_{UAT}, Q_{PA}, and Q_{HIMN} in simulation with HIMN when power amplifier output voltage frequency changes: **a** from 200 to 340 Hz, **b** from 200 to 400 Hz, **c** from 340 to 400 Hz, and **d** from 400 to 500 Hz

maximum transient time of segmented PR control is two fundamental cycles when the frequency is from 200 to 340 Hz.

The results in Table 6.14 show that the HIMN has a higher output-matching voltage than the FCIMN. Therefore, the HIMN proposed in this paper has a higher matching range than the FCIMN. Based on the simulation results in Fig. 6.49 and Fig. 6.51, the MPC control strategy proposed in this paper has better dynamic performance than the segmented PR control.

6.5.5 Experimental Results

To verify the effectiveness of the proposed HIMN and control strategy, the experiments are designed. The main parameters of the experiment are shown in Table 6.11. The experimental platform is shown in Fig. 6.52. To further verify the effectiveness of the proposed HIMN, control strategy, and simulation, this hardware-in-the-loop (HIL) experimental platform is built based on TMS320F2812 and RT-LAB (OP5600, Opal-RT). As shown in Fig. 6.52, the HIMN model is carried out in the RT-LAB

Table 6.13 Simulation results of HIMN impedance matching in different situations

Approach	f (Hz)	V_{Co} (V)	Q_{UAT} (Var)	P_{UAT} (W)	Q_{PA} (Var)	P_{PA} (W)	Q_{HIMN} (Var)	P_{HIMN} (W)	PF
Before	200	311	1418	847	1418	847	–	–	0.51
	340	311	792	372	792	372	–	–	0.43
	400	311	810	439	810	439	–	–	0.48
	500	311	697	222	697	222	–	–	0.30
Proposed HIMN	200	311	5212	3113	35	3163	−5176	50	1
	340	311	4270	2028	18	2035	−4270	8	1
	400	311	3448	1883	23	1902	−3426	19	1
	500	311	7240	2354	23	2357	−7217	3	1

Table 6.14 Comparison between proposed HIMN and prior FCIMN

Approach	f (Hz)	V_{co} (V)	V_{AM} (V)	V_{UAT} (V)
Before	200	311	–	311
	340	311	–	311
	400	311	–	311
	500	311	–	311
FCIMN	200	311	0	310
	340	311	463	563
	400	311	413	523
	500	311	792	870
Proposed HIMN	200	311	0	595
	340	311	463	725
	400	311	413	645
	500	311	792	1004

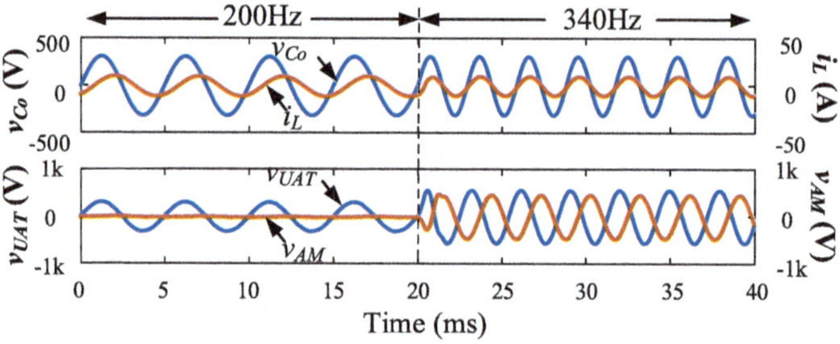

(a)

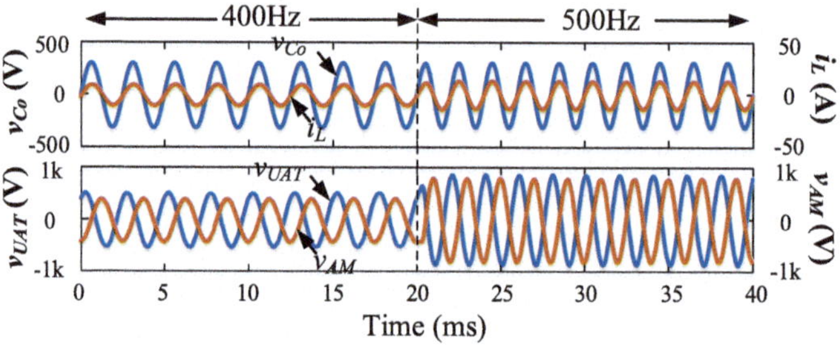

(b)

Fig. 6.50 Dynamic waveforms of v_{Co}, i_L, v_{UAT}, and v_{AM} in simulation with only the active matching part of the HIMN (the passive part has been removed) when power amplifier output voltage frequency changes: **a** from 200 to 340 Hz, **b** from 400 to 500 Hz

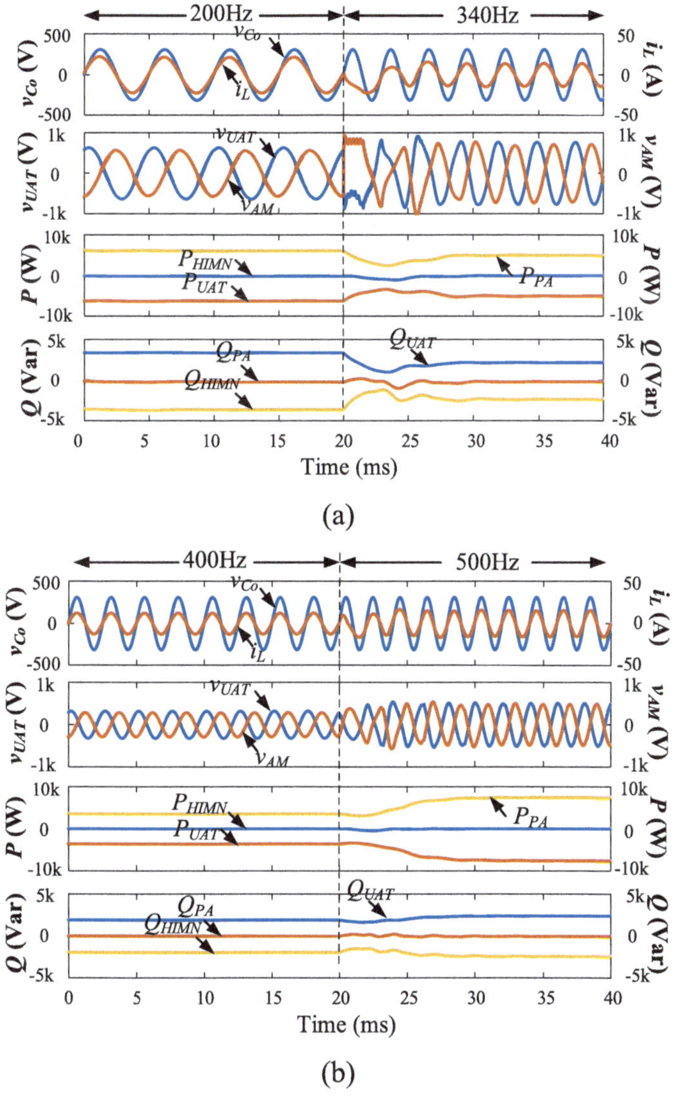

Fig. 6.51 Dynamic waveforms of v_{Co}, i_L, v_{UAT}, v_{AM}, P_{PA}, P_{HIMN}, P_{UAT}, Q_{UAT}, Q_{PA}, and Q_{HIMN} in simulation with FCIMN control scheme when power amplifier output voltage frequency changes: **a** from 200 to 340 Hz, **b** from 400 to 500 Hz

with a step of 20 μs, and all signals (i.e., the current and voltage) are measured with 25 kHz sampling frequency and outputted through the interface card to the I/O port of an external DSP controller. After the HIMN control algorithm is programmed and implemented in the DSP, the PWM signals are received by the RT-LAB via the input interface card.

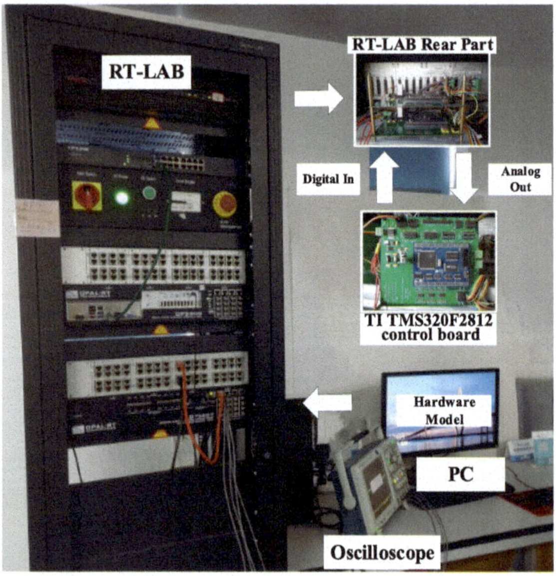

Fig. 6.52 Experimental platform of proposed HIMN

Figure 6.53 illustrates the dynamic matching waveforms of v_{co}, i_L v_{AM}, and v_{UAT} when only the active matching part of the HIMN remains (the passive part has been removed) using the MPC control strategy proposed in this paper under two representative dynamic situations. In steady-state, it cannot achieve impedance matching from 200 to 340 Hz and from 400 to 500 Hz.

Figure 6.54 illustrates the dynamic matching waveforms of v_{co}, i_L v_{AM}, and v_{UAT} by applying HIMN under four representative dynamic situations. The experimental results show that the proposed HIMN can achieve wide-band matching, and the control strategy has good dynamic performance. v_{co} and i_L are in phase in the stable states, proving that in these conditions (in 200 Hz, 340 Hz, 400 Hz, and 500 Hz), load impedances are matched.

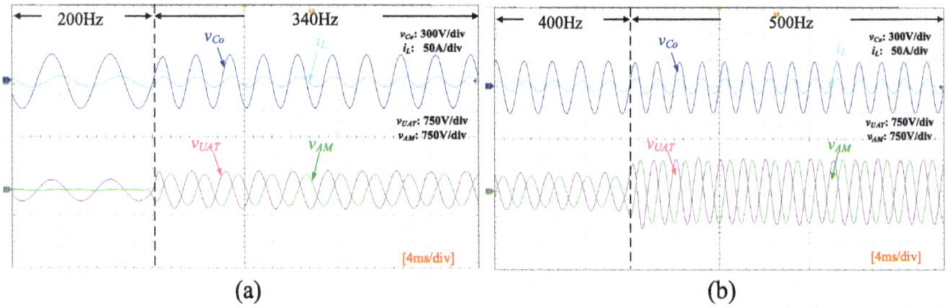

Fig. 6.53 Dynamic waveforms of v_{Co}, i_L, v_{UAT}, v_{AM} in experimental with only the active matching part of the HIMN (the passive part has been removed) when power amplifier output frequency changes: **a** from 200 to 340 Hz, **b** from 400 to 500 Hz

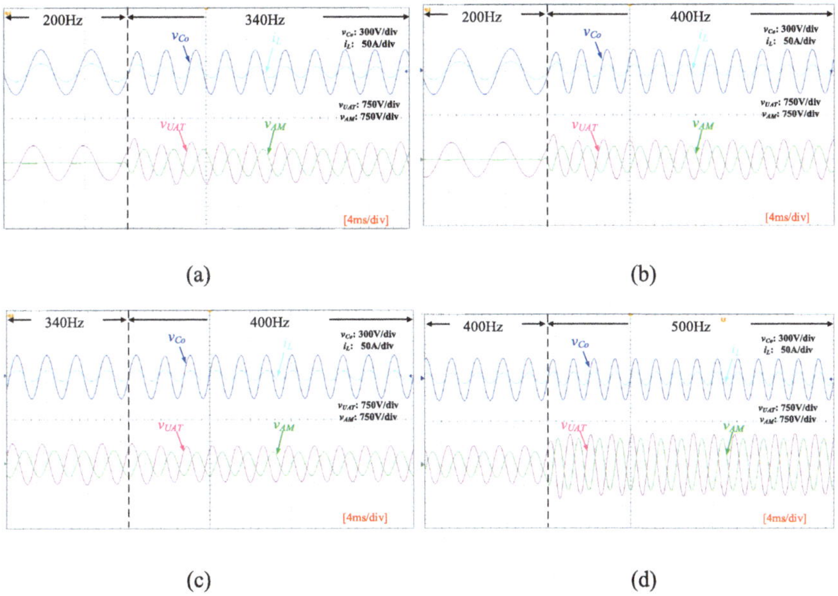

Fig. 6.54 Dynamic waveforms v_{Co}, i_L, v_{UAT}, v_{AM} in experimental with HIMN when power amplifier output frequency changes: **a** from 200 to 340 Hz, **b** from 200 to 400 Hz, **c** from 340 to 400 Hz, and **d** from 400 to 500 Hz

6.5.6 Section Summary

This section presents a new HIMN between the power amplifier and the UAT for broadband impedance matching to improve the power factor of the system. The proposed HIMN topology consists of ctive and passive matching to improve the matching range. Control methods based on reference signal synchronization, including reference compensation voltage calculation method and two-step prediction MPC, have been used to reduce the system's matching delay. The representative simulation and experimental results have been presented. The results show that the proposed HIMN can achieve broadband impedance matching and improve the power factor of the system to approximately. Compared with the conventional method, the proposed HIMN has a wider matching range and better dynamic and steady-state performance.

6.6 Summary

This chapter summarized the applications of adaptive impedance networks, including passive-controlled adaptive impedance and active adaptive impedance networks for power amplifiers. The TCIMN is firstr to realize the adaptive broadband impedance matching for a wide range of load impedance and improve the power factor with superior dynamic and stable performance. To achieve better matching performance, a SAIMN is proposed by injecting a capacitive reactive current into the system, which can reduce the current stress of the power amplifier switching devices while supplying the same active power to the load. Then, a series of active adaptive broadband FCIMN and its segmented PR controller are proposed for the underwater electroacoustic transduction system to improve power utilization. Furthermore, a HIMN and its two-step MPC control are proposed for broadband impedance matching and power factor improvement with a wider matching range and better dynamic and steady-state performance.

References

1. G. Gong, H. Ertl, J.W. Kolar, Novel tracking power supply for linear power amplifiers. IEEE Trans. Ind. Electron. **55**(2), 684–698 (2008)
2. D. Guanghai Gong, Hassler, J.W. Kolar, A comparative study of multicell amplifiers for AC-power-source applications. IEEE Trans. Power Electron. **26**(1), 149–164 (2011)
3. K. Upamanyu, G. Narayanan, Improved accuracy, modeling, and stability analysis of power-hardware-in-loop simulation with open-loop inverter as power amplifier. IEEE Trans. Ind. Electron. **67**(1), 369–378 (2020)
4. R. Han et al., Thermal stress balancing oriented model predictive control of modular multilevel switching power amplifier. IEEE Trans. Ind. Electron. **67**(11), 9028–9038 (2020)
5. K. Song, F. Zhang, S. Hu, Y. Fan, Ku-band 200-W pulsed power amplifier based on waveguide spatially power-combining technique for industrial applications. IEEE Trans. Ind. Electron. **61**(8), 4274–4280 (2014)
6. F. Shi, Z. Chen, X. Cheng, Behavior modeling and individual recognition of sonar transmitter for secure communication in UASNs. IEEE Access **8**, 2447–2454 (2020)
7. Y. Zhang, H. Wang, Z. Wang, Y. Yang, F. Blaabjerg, Simplified thermal modeling for IGBT modules with periodic power loss profiles in modular multilevel converters. IEEE Trans. Ind. Electron. **66**(3), 2323–2332 (2019)
8. L. Wang, C.-S. Lam, M.-C. Wong, Modeling and parameter design of thyristor-controlled LC-coupled hybrid active power filter (TCLC-HAPF) for unbalanced compensation. IEEE Trans. Ind. Electron. **64**(3), 1827–1840 (2017)
9. A. Al Bastami, A. Jurkov, D. Otten, D. T. Nguyen, A. Radomski, D.J. Perreault, A 1.5 kW radio-frequency tunable matching network based on phase-switched impedance modulation. IEEE Open J. Power Electron. **1**, 124–138 (2020)
10. M. Added, N. Boulejfen, Variable impedance matching network based on varactor diodes, in Proceedings 2015 IEEE MMS, pp. 1–4 (2015)
11. E.H. Newman, Real frequency wide-band impedance matching with nonminimum reactance equalizers. IEEE Trans. Antennas Propagat. Mag. **53**(11), 3597–3603 (2005)
12. J. Kim, J. Jeong, Range-adaptive wireless power transfer using multiloop and tunable matching techniques. IEEE Trans. Ind. Electron. **62**(10), 6233–6241 (2015)
13. N.J. Smith, C.-C. Chen, J.L. Volakis, An improved topology for adaptive agile impedance tuners. IEEE Antennas Wireless Propagat. Lett. **12**, 92–95 (2013)
14. X. Liu, C. Sun, Y. Yang, J. Zhuo, Low sidelobe range profile synthesis for sonar imaging using stepped-frequency pulses. IEEE Geosci. Remote Sens. Lett. **14**(2), 218–221 (2017)
15. N.H. Klausner, M.R. Azimi-Sadjadi, Manifold-based classification of underwater unexploded ordnance in low-frequency Sonar. IEEE J. Oceanic Eng. **45**(3), 1034–1044 (2020)

16. D. Chatterjee, A. Ghosh, TCSC control design for transient stability improvement of a multi-machine power system using trajectory sensitivity. Elect. Power Syst. Res. **77**(5–6), 470–483 (2007)
17. H. Liu, X. Xie, C. Zhang, Y. Li, H. Liu, Y. Hu, Quantitative SSR analysis of series-compensated DFIG-based wind farms using aggregated RLC circuit model. IEEE Trans. Power Syst. **32**(1), 474–483 (2017)
18. P.P. Kulkarni, N.D. Ghawghawe, Hybrid DA-PSO approach for optimal location and parameter settings for TCSC controller in power system. IET Gener. Transmiss. Distrib. **14**(7), 1222–1232 (2020)
19. X. Xu, J. Zhao, Z. Xu, S. Chai, J. Li, Y. Yu, Stochastic optimal TCSC placement in power system considering high wind power penetration. IET Gener. Transmiss. Distrib. **12**(12), 3052–3060 (2018)
20. J. Shen et al., A high-frequency high-power test bench for 11 MW/595 Hz drives With 1.25 MW grid capability. IEEE Trans. Ind. Applicat. **53**(5), 4744–4756 (2017)

Chapter 7
Conclusions and Prospective Further Work

7.1 Chapter Summaries

This book presents and discusses the adaptive impedance converters for different applications, including power quality compensation in distribution systems, grid-connected systems for renewable energy integration, interlink converters in AC and DC microgrids, co-phase railway traction supply systems, and impedance matching networks for power amplifiers. According to different applications, adaptive impedance converters have different topologies, mathematical models, and control systems. The key contributions and applications of each chapter are summarized in the following Table 7.1.

7.2 Achievements

This book presents and discusses the adaptive impedance converters for different applications. The achievements of this book are summarized as follows:

- The centralized TCLC-DSTATCOM PTT integrates the DSTATCOM with PTT by TCLC filters, which distinguishes it from existing topologies in two ways: One is the coordination of TCLC filter and DSTATCOM for superior voltage-reactive compensation performances. The other is the PTT integration method for equilibrium and comprehensive performances. By adjusting the tap ratio, the compensation current, device overheating, and power losses can be largely reduced.
- The delta-connected AI-HAPF with its parameter design is proposed to have the desirable characteristics of a wide compensation range (from inductive to capacitive) with low dc-link operating voltage.

L. Wang et al., *Design and Applications of Adaptive Impedance Converters*,
https://doi.org/10.1007/978-981-97-5319-2_7

Table 7.1 Summary of adaptive impedance converters proposed in each Chapter

	Application	Proposed topologies	Contributions/main work
Chapter 1	–	–	Present an overview of different coupling impedance converters and give the potential advantages of adaptive impedance converters
Chapter 2	Power quality compensation in distribution systems	TCLC-DSTATCOM PTT; delta-connected AI-HAPF; SVC// HAPF	The adaptive passive coupling impedance is used to effectively provide dynamic reactive power and reduce the DC side voltage and power loss of the converter with effectively compensating reactive current, harmonic current, and unbalanced current
Chapter 3	Grid-connected system for renewable energy integration and power quality	HGCI; SVC//CGCI	In addition to injecting the active power and providing reactive power, unbalanced power, and harmonic power compensation, the HGCI can enlarge the operational range of and lower the DC operating voltage of the inverter, while the SVC// CGCI has great promise for wide operation range with both low voltage and current rating of active inverter
Chapter 4	Interlink converter for AC/DC microgrids	HCIC; three-port power hub	The HCIC with droop control has the ability to have good power flow control and good power quality compensation. Moreover, the selective droop control can effectively avoid the over-capacity of HCIC. The three-port power hub not only achieves the bidirectional active power flow control between the three-phase AC grid and DC grid but also enables the reactive, unbalanced, and harmonic power compensation for the utility grid
Chapter 5	Co-phase railway traction supply	AI-HPQC	The AI-HPQC has the characteristics of a wide compensation range and low DC-link voltage. Also, the control strategy for the AI-HPQC was proposed for the coordination control of the adaptive impedance part and the active inverter part

(continued)

Table 7.1 (continued)

	Application	Proposed topologies	Contributions/main work
Chapter 6	Impedance matching networks	TCIMN, SAIMN, FCIMN, HIMN	The passively controlled TCIMN can realize adaptive broadband impedance matching for a wide range of load impedance and improve the power factor with superior dynamic and stable performance The shunt active SAIMN is proposed by injecting a capacitive reactive current into the system, which can reduce the current stress of the power amplifier switching devices while supplying the same active power to the load The series active FCIMN is proposed for the underwater electroacoustic transduction system to improve power utilization The HIMN and its two-step MPC control are proposed for broadband impedance matching and power factor improvement with a wider matching range, better dynamic, and steady-state performance
In this Chapter	–	–	Summary of the adaptive impedance converters proposed in this thesis

- The modeling, operation principle, and parameter design of SVC//HAPF as a cost-effective compensator are analyzed to compensate the reactive power dynamically while providing harmonic compensation.
- The structure, parameter design, and control method of HGCI for PV power generation with power quality conditioning are proposed to inject the active power and provide reactive power, unbalanced power, and harmonic power compensation at the same time.
- The SVC//CGCI and it control system are proposed to dynamically compensate the reactive power and balance the source active power, which has the great promise for wide operation range with both low voltage and current rating of the active inverter.
- A new droop control method has been developed for the HCIC system, and its fundamental active and reactive power flow in microgrids has been controlled by the proposed P-δ and Q-P-V *droop* controls in the converter part of HCIC, whereas P-Q-α droop has been proposed for the SVC part of HCIC. It demonstrates that the proposed control of HCIC in the hybrid AC/LVDC microgrid has the ability to provide good power flow control and good power quality compensation.

- To avoid the over-capacity situation, a selective power droop control of HCIC is proposed to regulate active/reactive powers as well as improve power quality based on the tradeoff among active, reactive, and harmonic power.
- A three-port power hub, which consists of a V/V transformer, a back-back inverter, and a static VAR compensator with adaptive impedance, and its control algorithm for power regulation is proposed for remote area residential power supply and generalized to smart city application.
- The effect of different coupled impedances in conditioners is discussed. Moreover, an adaptive impedance-coupled hybrid power quality conditioner is proposed for the coordination control of the adaptive impedance part and the active inverter part in the co-phase railway supply.
- The SAIMN parameters design method and operating ranges have been analyzed to ensure that SAIMN can match load impedance in the wide frequency range. A compound control method was presented to realize the adaptive broadband matching.
- A control method of FCIMN is proposed to achieve adaptive broadband impedance matching, including dynamic, active power calculation, reactive power calculation, segmented PR controller, and combined feedback and feedforward voltage control.
- A control method of HIMN based on reference signal synchronization, including reference compensation voltage calculation method and two-step prediction MPC, is proposed to reduce the system's matching delay.

7.3 Prospective Further Work

Based on the research works done in this book, there are numerous issues meriting for further exploration in future research:

1. The necessary protection is very important for high voltage application. Otherwise, the system components can be easily damaged due to the suddenly voltage and current changes. The different protection techniques and protection devices are necessary to be included in the future.
2. More PWM control strategies can be further investigated for adaptive impedance converters.
3. The stability of proposed system is an important point. The detailed closed loop control chart and stability analysis is required to be studied in the future.
4. The industrial control controller/system is required to further develop to improve the control efficiency.

SPRINGER NATURE

GPSR Compliance

The European Union's (EU) General Product Safety Regulation (GPSR) is a set of rules that requires consumer products to be safe and our obligations to ensure this.

If you have any concerns about our products, you can contact us on ProductSafety@springernature.com

In case Publisher is established outside the EU, the EU authorized representative is:

Springer Nature Customer Service Center GmbH
Europaplatz 3
69115 Heidelberg, Germany

The manufacturer's authorised representative in the EU is Springer
Nature Customer Service Centre GmbH, Europaplatz 3, 69115 Heidelberg,
Germany. If you have any concerns regarding our products, please
contact ProductSafety@springernature.com

Printed and bound by CPI Group (UK) Ltd, Croydon, CR0 4YY
01/10/2026
02225926-0006